THE MAKING OF MODERN SCIENCE

THE MAKING OF MODERN SCIENCE

SCIENCE, TECHNOLOGY, MEDICINE AND MODERNITY: 1789–1914

DAVID KNIGHT

polity

The right of David Knight to be identified as Author of this Work has been asserted in accordance with the UK Copyright, Designs and Patents Act 1988.

First published in 2009 by Polity Press

Polity Press
65 Bridge Street
Cambridge CB2 1UR, UK

Polity Press
350 Main Street
Malden, MA 02148, USA

ISBN-13: 978-0-7456-3675-7
ISBN-13: 978-0-7456-3676-4 (pb)

A catalogue record for this book is available from the British Library.

Typeset in 10.5 on 14 pt Adobe Janson
by Servis Filmsetting Ltd, Stockport, Cheshire
Printed and bound by MPG Books Group, UK

The publisher has used its best endeavours to ensure that the URLs for external websites referred to in this book are correct and active at the time of going to press. However, the publisher has no responsibility for the websites and can make no guarantee that a site will remain live or that the content is or will remain appropriate.

Every effort has been made to trace all copyright holders, but if any have been inadvertently overlooked the publisher will be pleased to include any necessary credits in any subsequent reprint or edition.

For further information on Polity, visit our website: www.politybooks.com

CONTENTS

List of Illustrations vi
Preface: The Age of Science viii
Acknowledgements xiii

Introduction: Approaching the Past 1
1 Science in and after 1789 12
2 Science and its Languages 33
3 Applied Science 56
4 Intellectual Excitement 82
5 Healthy Lives 105
6 Laboratories 129
7 Bodies, Minds and Spirits 151
8 The Time of Triumph 172
9 Science and National Identities 195
10 Method and Heresy 217
11 Cultural Leadership 238
12 Into the New Century 264

Timeline 283
Notes and References 289
Index 353

Illustrations

1.   Dens for the wild beasts at the Paris zoo   17
2.   Molecular models: A.W. Hofmann   46
3.   Blue and yellow macaw   49
4.   Pattern plate, annotated lithograph   50
5.   Mr Walker's improved steam engine, 1802   58
6.   Sale catalogue of 2,665 of James Watt's books (1849)   73
7.   An intricately designed fossil creature – exploded drawing of a
     fossil 'stone lily'   94
8.   Buckland lecturing in Oxford, 1822   96
9.   Dinosaur footprint   98
10. Photographic illustration: C. Darwin, *The Expression of the
     Emotions in Man and Animals*   101
11. Notes taken by Robert Pughe, a medical student, beside text   108
12. Portable laboratory, by Faraday   133
13. Teaching laboratory at University College, London, 1846   141
14. The latest thing – a section through the new Berlin chemical
     laboratory, 1866   145
15. Phrenological heads   156
16. Our cousin the gorilla being fitted for a suit   161
17. Skeletons of the primate family   164
18. The Queen and Prince Albert inspecting machinery   174
19. Warning signals for impending bad weather   182
20. Lord Rosse's six-foot telescope   203

21. Spiral nebula   204
22. A naval survey voyage – burying a shipmate beneath the Arctic
     ice, under a portentous 'mock Moon'   214
23. Lecture theatre of the London Institution, 1828   241
24. A field naturalist at work, pond-dipping   247
25. Lecture prospectus, Royal Institution, 1825   250
26. Prospectus for the new weekly journal, *Nature*, 1869   252
27. Popular science and its nightmarish possibilities   256
28. Facing up to the astonishing prospect of women graduates   260
29. The prestige of science equivocally indicated in an advertisement
     for cocoa   261
30. Cathode rays   270

PREFACE: THE AGE OF SCIENCE

Of all the inventions of the nineteenth century, the label 'scientist' was one of the most striking. In 1789 there had been natural philosophers and natural historians, and at a slightly less gentlemanly level, chemists, anatomists and instrument makers. Their childish curiosity continued into adulthood, when solving problems and finding explanations could be a leisure activity, maybe sociable. But science was, as Humphry Davy (1778–1829) put it to the young Michael Faraday (1791–1867) in 1812, a harsh mistress from the pecuniary point of view. Outside medicine, where scientific knowledge might help as a back-up to the all-important bedside manner and clinical experience, science was a vocation or hobby rather than a job or profession. The word 'scientist' was invented in 1833: the thing came later.

Science required skills: mechanical, mathematical, manual, logical, observational and organizational ability were in various degrees necessary. Medicine and mathematics were taught in universities, but in 1789 there were no degrees in science (or indeed in modern history, or modern languages and literature): Greek and Latin classics formed a liberal and traditional education for mandarins, and for the scholarship boys who would become clergy, tutors, judges and schoolmasters. Engineers and the majority of medical men learned on the job, as apprentices; and science similarly might best be learned, as by Faraday with Davy, in a kind of informal apprenticeship. There were in France, Prussia and Russia a few salaried posts to compete for in Academies of Sciences, but generally patronage was the key – in 1789 usually from the nobility and gentry, and by 1914 from established scientists. Those who scorned patronage, or could not find it, might live by their pens, as Grub Street hacks, or by giving public lectures;[1] but for most, unless they were independently well-off like Charles Darwin (1809–82), some kind of profession was necessary to keep the wolf from the door, and science had to be a spare-time activity.

The science student was invented in the 1790s with the founding of the École Polytechnique in revolutionary France, entered through competitive examination, with formal courses in science and engineering taught by academicians active in research. But the graduate student came from Germany, where Justus von Liebig (1803–73) at the small University of Giessen opened a teaching laboratory in the 1820s and began PhD programmes in chemistry – the first notable research school.[2] The German model of the research university soon spread over Europe, and in the second half of the century reached Britain and the USA.[3] Graduates found jobs in teaching and also in industry and commerce. The number of established posts in universities and technical institutions (including government laboratories) increased rapidly, especially after the Franco-Prussian War of 1870 when the better-educated nation won.[4] From this point 'applied science', long promised, actually began to deliver the goods. With qualifications came group solidarity,[5] career structures, professions and social recognition. The scientist was no longer an absent-minded virtuoso or dilettante but a valuable citizen transforming economies and world-views. This consolidation happened at the same time as nationalities in Europe were cohering, and Australians began to stop thinking of themselves as Britons who happened to be in the southern hemisphere.[6]

Science is thus a practical, intellectual and social activity that became momentous during the nineteenth century. We must therefore be aware not just of geniuses and their discoveries but of scientific societies: open, learned, professional, formal or informal (maybe just networks), and in all these categories cutting across lines of class, region and nation, and increasingly specialized. Science was both public knowledge and meritocratic, a route to social mobility and a vehicle of modernity. Scientific education might be critical, with more or less conscious consideration of 'scientific method', but often it was and is dogmatic, perhaps necessarily so, to prevent today's students falling into yesterday's errors.[7] There are right answers in elementary, established science: the teaching may be hands-on but cannot be open-ended, and there is perforce much to be learned more or less by rote. Notoriously, simplified explanations must be 'unlearned' as students progress. Overall, a way of thinking is promoted – cautious and sceptical, involving ideal situations, controlled conditions as in a laboratory, models and general laws. Scientists wrestle with questions difficult to answer, and very occasionally with questions difficult to ask, but they hope for answers that are definite or have definite probabilities. This seems very different from ethical or political questions. The supposed 'two cultures' and 'science wars' that set science against other ways of thinking, and the scientism that supposes scientific methods are the only valid ones in any field, emphasize this difference.

In the nineteenth century science made a huge impact on the public, or publics, as it impinged on food production, transport, communications, fashion, art, literature, politics, religion, and even on national prestige and confidence. Although many were religious believers, scientists became part of a clerisy, a secular intellectual elite taking over the roles of prophet and instructor previously played by the clergy.[8] Not everybody was happy about this. Grim industrial cities, mines and factories appalled observers, but it was through improved technology that they were bettered, as governments gradually dropped laissez-faire attitudes and occupied themselves with clean water, sewers, factory inspection, legislative control of the purity of food and drugs, and the safety of railways and shipping. Mary Shelley's *Frankenstein* (1818), taking issue with the confidence of the poet of Enlightenment, Erasmus Darwin (1731–1802), and of Davy, struck contemporaries as just another gothic-horror novel; but its vision of the scientist as a sorcerer's apprentice, playing God, has proved to be astonishingly resonant.[9] It provided a counterweight to the mass of writings, lectures, great exhibitions, botanic gardens and illustrations emphasizing the progressive and beneficent nature of science.[10] Both popularization of science and strong reactions against it in occultism, magic and spiritualism, as well as opposition to vaccination and vivisection, were prominent features of the nineteenth century. Alfred Russel Wallace (1823–1913), co-discoverer of natural selection, was unusual among scientists in managing to embrace all these things, along with socialism.[11]

Popular science was not always what academicians would have wanted: phrenology, mesmerism and animal magnetism were more popular than the austere and exact sciences (and indeed the last of them remains popular, though the theory was exploded by Benjamin Franklin, 1706–90). We must also not forget, in our enthusiasm for thoughts never thought before and sights never seen, those at the receiving end of science, who paid the price for its progress. Traditional values and ways of life were lost. New explosives such as gun-cotton blew up and killed people making it, chemical works poisoned people and animals living nearby, officials classified people (perhaps as atavistic or degenerate),[12] and many found their skills becoming useless in the face of mechanization. Triumphs of metropolitan science like synthesizing indigo, and growing rubber and quinine at Kew so that seedlings could be sent to plantations in the colonies, damaged the economies of India, Brazil and Peru where the plants originated. Those far from centres of science in Europe and North America were expected to provide, as it were, raw materials upon which those at 'home' could reason, and to whom credit would be given. Theorizing, and even naming, was reserved for those in metropolitan centres, to whom provincials and colonials should defer. Only at the end of this period were Asians and Africans able to play the full part in science that their ancestors once had. Similarly, women were widely excluded from the active end of science during this period of separate spheres (She knows but matters of the house, / And he, he knows a thousand things[13]); but one reason for the word 'scientist' displacing 'man of science' was that by 1900 women were at last able to break in and be recognized.

The nineteenth century was an age of technological progress, of empire-building, of new nations, new specialisms and professions, new possibilities, and new understanding of the natural world. In a brief telling of this age of science, much must be left out. Science has had its context as well as its content; I have tried here to preserve a balance. In looking at science, I believe that we can easily tip too much towards the content, the theories, experiments and creative individuals, and away from the social and historical context – the ethical and religious beliefs, the educational systems, the societies and clubs, the publishers, the popularizers and the publics. Accordingly, while I have tried to do justice to the intellectual aspects of science, some would have wanted more about evolution and conservation of energy, those great generalizations that brought a new coherence to science in the 1850s. Evolution and energy do indeed permeate the story, but I have tried to place them in contexts of progress, religious doubt, ethical debates, scientists' claims for status and respect, and improvements to dyes, lighting, transport, explosives and battleships. In the early part of the nineteenth century, science lagged behind technology, and men of science sought to explain how inventions worked. But science then got ahead, and applied science became the key to new industries. In the later years of the century, 'pure' science was distinguished from applied: but really it is hopeless to try to separate science, technology and medicine – all aspects of one intellectual, practical and social activity that we are concerned with here. Interpreting nature involved thinking about what to do with that knowledge. This is well exemplified in chemistry, the science most studied in the nineteenth century, vital in medicine and in industry, the source of wealth and power, which gave us atomic theory, molecular models, an understanding of matter and its relations to electricity, and photography. I have tried to give chemistry its due in the story.

Science, as an element of Western culture, transformed the world in the nineteenth century, but we cannot focus upon its context everywhere. Western Europe has to be the centre of our story: and while I have tried to do justice to France, the centre of things in 1789, and to Germany, the most important scientific nation in 1914, I have used examples from Britain out of greater familiarity with that context, and more than their cosmic importance warrants. A different perspective would no doubt give a somewhat different picture. The intellectual and practical achievements associated with science, and the interpretation of nature, will naturally fill our story, but the coming of the scientist will be its core. And we shall be looking at how science and scientists were received and perceived, as well as how they were made.

ACKNOWLEDGEMENTS

I would like to thank my colleagues in Durham who have invited me to lecture and to give seminar papers, and have listened patiently and suggested new thoughts to me; and to those in the Society for the History of Alchemy and Chemistry, the Historical Section of the Royal Society of Chemistry, the British Society for the History of Science and the Geological Society, who have also invited me to hold forth, and given me valuable feedback. I have also benefited much from being invited to lecture to local groups and societies, not specialists in the history of science: such occasions have not only been a great pleasure, but the audiences' warm welcome, wide and curious knowledge, and demand for a broad view have been a delight.

Many thanks to Leigh Priest, who prepared the very full and efficient index.

This book has had a long gestation, and my family has been extremely patient throughout the pregnancy: I am very grateful.

MAN THE INTERPRETER OF NATURE[14]

SAY! When the world was new and fresh from the hand of its
  Maker,
Ere the first modelled frame thrilled with the tremors of life,
Glowed not primeval suns as bright in yon canopied azure,
Day succeeding to day in the same rhythmical march;
Roseate morn, and the fervid noon, and the purple of evening –
Night with her starry robe solemnly sweeping the sky?
Heaved not ocean, as now, to the moon's mysterious impulse?
Lashed by the tempest's scourge, rose not its billows in wrath?
Sighed not the breeze through balmy groves, or o'er carpeted
  verdure
Gorgeous with myriad flowers, lingered and paused in its flight?
Yet what availed, alas! These glorious forms of Creation,
Forms of transcendent might – Beauty with Majesty joined,
None to behold, and none to enjoy, and none to interpret?
Say! Was the WORK wrought out! Say, was the GLORY
  complete?
What could reflect, though dimly and faint, the INEFFABLE
  PURPOSE
Which from chaotic powers, Order and Harmony drew?
What but the reasoning spirit, the thought and the faith and the
  feeling?
What, but the grateful sense, conscious of love and design?
Man sprang forth at the final behest. His intelligent worship
Filled up the void that was left. Nature at length had a soul.

                                        Sir John Herschel

'Homo, naturæ minister et interpres.' – *Bacon* – Herschel's note

INTRODUCTION: APPROACHING THE PAST

Ours is an epic story. Most people in 1789 saw science as a hobby: but by 1914 they could not doubt that it was a crucial part of 'Western civilization', empirical knowledge that not only lay behind material progress, prosperity and power, but had also transformed world-views, fine art and literature, and the way humans thought about themselves and each other – how science was perceived is as important as how it was achieved. This book is the story of how that intellectual and social revolution happened, through a series of themes that follow each other with some chronological rationale. It is not a critical bibliographical essay, but the references will guide anyone who wants to pursue things further and disagree creatively with what they find here. Nevertheless, it is right to begin with an assessment of how we historians have got to where we are, and how we work. Science began with curiosity, through experience in homes, workshops and libraries.[1] Taken more seriously, it involves observing, recording, testing, tinkering, pondering, arguing and generalizing, so as to interpret and (within limits) to control nature. It isn't obvious that anyone should do or take it very seriously. Most people never have. But given that since 1800 more and more people (first in the West, now everywhere) have done so, made careers out of it and transformed the way we live, its history is important for us all. Much history of science was and is written for and by these curious folk, the scientists. Textbooks, especially in chemistry,[2] used to contain potted histories. Scientific journals publish reviews of 'literature', with bibliographies, sometimes going back many years, on particular topics of current interest to their readers. They also publish elegies or obituaries, to be taken with due doses of caution; and sometimes brief biographies of their readers' great predecessors. In journals, glossy house magazines and websites, scientific societies often include historical articles, as well as the reminiscences of distinguished elderly scientists, who may also write their memoirs or reprint their essays and addresses. These are augmented by published interviews, sometimes surprisingly candid, with practitioners reflecting upon lives in science.[3] Such writings may be inaccessible to outsiders, especially when algebraic or chemical equations feature largely; but if they can forget their training in using the abstract noun and the passive verb, scientists write well – better than many historians of science. Theirs is not the pretentious, turgid academic prose of those trained in the social sciences: like Joseph Priestley (1733–1804)[4] and the popularizer William Paley (1743–1805),[5] scientists writing for the public generally try never to write an obscure sentence, though they don't always succeed.

Until the middle of the twentieth century, therefore, history of science and medicine was mostly written by scientists, insiders describing their world and its past. Like lawyers, clergy and other professionals, they dwelt chiefly upon precedents, case studies and examples useful to them that were instructive to their readers. Alchemy, the earth-centred cosmos, René Descartes' (1596–1650) planetary vortices, and phlogiston might be discussed as warnings of pitfalls to be avoided, wanderings from the royal road to truth. This was applied history, pragmatic, didactic, placing the student (like the author) in a long and great tradition, and indicating a glorious future. Seeing how we got where we are, ruefully noting defeats but celebrating victories rather than dwelling upon ways not taken or the losers, is reputable if incomplete. And when well done – sometimes by serious historians of science – this approach pays off, yielding a good story, perhaps an epic.[6] Humphry Davy, as President of the Royal Society in 1820 soon after the defeat of Napoleon, saw himself as a general in the army of science.[7] Military metaphors remain popular: science is a war, against hunger, disease, ignorance and superstition (maybe religion), and intellectual torpor.[8] Wars against abstractions can never be won: the show must go on, and on.

WHIG HISTORY

Such history is now, like imperialism, out of academic fashion. It can be uncritical, anecdotal and accessible only to insiders: but the main criticism was that it is 'Whig' history. The Whigs were the party that welcomed

William of Orange and Mary Stuart in place of King James II in the 'Glorious Revolution' of 1688. Very suspicious of royal prerogatives, sympathetic to the Americans' fight for independence in 1776, a later generation (fearing and detesting inglorious and bloody revolution) pushed through the Reform Bill of 1832, giving votes to the middle classes in the new industrial cities of Britain. The party leader Charles James Fox (1749–1806), in his *History of James the Second* (a tyrant and arch-villain in Whig eyes),[9] and the former Cabinet minister Thomas Babington Macaulay (1800–59) in his best-selling and beautifully written *History of England* (1848–61), saw history as a great unfolding political drama, gradually leading to representative constitutional government, and foresaw further progress. This is the sort of history that governments would like schoolchildren to learn: things are getting better and better, and the past is viewed through the eyes of the present, with opponents of progress denounced in the terms of a later political correctness. Whig history was itself denounced by Herbert Butterfield in 1931 as contrary to seeing the past as it was, the ideal of historians since Johann Gottfried Herder (1744–1803),[10] and thereafter academic historians have been conscious of losers and suspicious of fine writing.[11] Historiographic traditions matter to us because they affect how scientists see each other, and are perceived.

Butterfield also wrote about the origins of modern science, an intellectual revolution that he perceived as perhaps more important than the Renaissance and Reformation so beloved of historians in their categorizations.[12] It is curious that this book is just as Whiggish in its celebration of the Scientific Revolution as Fox and Macaulay were in their delight in the Glorious Revolution of 1688. This Scientific Revolution of the sixteenth and seventeenth centuries, the time of the new astronomy of Galileo (1564–1642) and Isaac Newton (1642–1727), became a great object of study.[13] It was also the time of Francis Bacon and René Descartes, with the beginnings of modern philosophy. And historians of science, just emerging as a professional group, found it easy, especially in France, to approach history of science as a branch of intellectual history.[14] Alexandre Koyré (1892–1964) was the great exemplar,[15] in a tradition that went back to William Whewell (1794–1866), an omniscient pundit in Victorian Cambridge. Whewell's hefty *History of the Inductive Sciences* (1837) was written to show how method – getting the right end of the stick and the appropriate idea – was crucial in the development of science.[16] 'Inductive epochs', when a new idea brought disparate observations together, punctuated the development of science, and could (and should) be weighed up by the historian. Whewell thus (rather idiosyncratically) judged the work of Faraday as on balance more important in chemistry than that of Antoine Lavoisier (1743–94), usually taken as its founding father. It seemed as if the history and philosophy of science formed one subject, a part of what was more loosely called History of Ideas. Historians neglected traditions messier and more empirical than astronomy, such as medicine, chemistry and natural history – and natural magic, too.[17] Revisionists have revealed how very important these things were in early modern science, in an age of geographical discovery, religious and political ferment, and also of witch-hunting in Europe and the USA.[18]

FLIRTING WITH PHILOSOPHERS

Thomas Kuhn (1922–96) gave a new twist to this philosophical tradition. He perceived the history of science as an alternation of periods of quiet normal science developing within a framework established by an exemplary figure, a Newton or a Lavoisier, that he called a 'paradigm', and occasional revolutionary upsets when, after anomalies had accumulated, someone saw that with different assumptions far more could be explained.[19] If such people succeeded in persuading their sceptical and conservative contemporaries and the next generation that they were right, then their ideas became the new paradigm, and the science set off on a new tack. Theirs, then, are the names we remember in the history of science. There was not just one Scientific Revolution, but many, some upsetting more apple-carts than others. This big picture implies a focus upon great men and women, geniuses, and upon brief and dramatic episodes.[20] But in fact Kuhn's emphasis upon the scientific community and the need to convince it helped reinforce the idea of science as public knowledge. Historians, whose special concern is with the particular but who need to lift their eyes from it, were stimulated to look afresh at institutions, their members (in what was called prosopography, or collective biography), rules, activities and publications: in fact, not only at Kuhnian revolutions but at normal science.[21]

Scientists mostly disliked Kuhn's view of their activities – normal science could seem a bit like careerism and painting by numbers – and became on the whole less interested in history as syllabuses grew fuller. Then, with the 1960s, we entered the heady days of the DNA spiral, the space race and the promise of nuclear power in the glowing future. They found more attractive the view of the philosopher Karl Popper, in which science was a matter of conjectures and refutations.[22] Hypotheses are dreamed up somehow, but are scientific insofar as they are testable. Science is provisional, consisting of generalizations that have survived all attempts so far to refute them. The task of scientists is to test their hypotheses and those of others in the cool light of reason and experiment: they must be sceptical. It may seem psychologically implausible to see scientists actively seeking evidence against their ideas. Indeed, the pattern does not fit actual humans in the past, who defended their views forcefully in sometimes furious debate.[23] But it might be a basis for revisionism, always important in history, making heroes of Pope Urban VIII and Bishop Samuel Wilberforce because of their searching questioning of the sun-centred cosmos and of evolution. That would be ironic because Popper intended to distinguish science from other activities, like psychoanalysis or politics (or indeed natural theology), which are not directly falsifiable: his method was a filter to exclude 'pseudo-science'. Such categorizing may be misleading for the historian, but after all science does have features that make it distinct from other activities, and that is why its history and philosophy are special and worth studying.

Philosophers, not too worried about how actual scientists had behaved, were prepared to 'rationally reconstruct' the past,[24] and then moved for a time into increasing abstraction, wondering if the existence of a white handkerchief might confirm 'all ravens are black', and what might happen if blue and green turned into grue and bleen. Historians of science, who had already found that academic historians were not (despite Butterfield) very open to their advances, now broke off their affair with philosophers also and began a flirtation with sociologists. Auguste Comte (1798–1857) and Herbert Spencer (1820–1903), founders of this discipline, had both been much involved in science, and hoped that its route to positive knowledge could be followed in human sciences.[25] Comte had seen three stages of thought – religious, metaphysical and then positive – in the development of individuals and of civilization, and this positivism was very attractive in the nineteenth and twentieth centuries. Then, in the same context, Marxist ideas – introduced to Anglophone historians of science by the Russian delegation at the Second International Congress for the History of Science in 1931[26] – proved powerful as well as fashionable in the 1960s and 1970s. Bacon in the seventeenth century had written that printing, gunpowder and the magnetic compass had made the modern world. Later scholars concerned with big scientific ideas had nevertheless neglected technology, seeing it as inseparable from science but depending on it. Marxists reversed that.

INTERNAL OR EXTERNAL HISTORY?

Historians of science thus found themselves debating how far they should be 'internal' or 'external' in their emphases. Were scientific advances brought about by people solving problems inherited from their predecessors, dwarfs on the shoulders of giants maybe but able to see a bit further? Or were they consequences of economic and social situations, of a zeitgeist where the individuals concerned were of minor importance and there were practical problems to be solved? Interest in the history of specific sciences over long periods, a feature of both scientists' history and international congresses, had promoted internalism. The fact of simultaneous discovery in science when the times were ripe, not uncommon even when discoverers are not conscious of being in a race and do not know of each other, favoured externalism. This encouraged focus upon social factors and on the constraints under which scientists operated in different societies, and led, for example, to an illuminating study of the Royal Institution in which Davy and Faraday were portrayed as tools of the ruling class.[27] But those of us attracted to history of science through the excitement of ideas, and interested in kinds of minds and what makes individuals tick, could not accept that this was the whole story. The sources (letters, wills, notebooks, sketches, reports, papers, books, specimens, apparatus, buildings, mountains) showed that the development, the content, of science cannot be reduced to, but must not be divorced from, its wider context.[28] And sources are for historians what observations, experiments and calculations are for scientists: there, connections are perceived, bright ideas arise and hypotheses are tested.

Partly in reaction to arid debate about internal and external factors, biography with its focus upon individuals entered upon a new lease of life. Not confined to world-historical individuals, Hegelian heroes who defied the ordinary rules, but looking hard also at representative lives, biography in the hands of serious historians of science was much more fun to read than treatises. It used to be the case that writers might separate the science from the life of their subject, in distinct chapters; or write one without the other. Sometimes, as with Priestley's very long entry in the Victorian *Dictionary of National Biography*, where different authors wrote on the chemist and the cleric, it seems as though they were describing two people of the same name who lived at the same time. Nevertheless, getting the balance right between the science that made a person famous, and the ordinary things that made him or her typical, setting the scene without overloading it, is not straightforward.[29] Here also there is a problem with sales. Lives of Galileo and Darwin are legion, and yet publishers cry out for new ones. What's new about a new biography is a new biographer, and that interaction is important, but sometimes that's all that can be said about it. Lives of scientists less prominent in the public mind do not sell – unless, as with the surprising case of John Harrison (1693–1776) in *Longitude*,[30] a story of pertinacity and persistence in the face of inertia in high places can be made into a drama. Nevertheless, recently we have had important biographies of Darwin's ally Wallace, and of his even-less-known opponent John Phillips (1800–74);[31] and because science and technology have been routes to social mobility, biographers since Samuel Smiles (1812–1904) have made them accessible through the genre of prospering self-help. The converse, unfair neglect, has also promoted interest in Gregor Mendel (1822–84),[32] the pioneer of genetics, whose sad story is a nightmare for scientists who feel disregarded, and an encouragement to cranks who hope for ultimate recognition (though posterity never did anything for anyone).

Historians feel queasy about biography in reaction to hero-worshipping Victorians like Thomas Carlyle (1795–1881), but do write them, to our great profit because they bring out the full context and force us to appreciate that people in the past lived under different circumstances with different expectations – the past is indeed a foreign country, and 'science' has by no means always been the same thing. Biographies of Thomas Henry Huxley (1825–95) – making his way in the world, involved far more than Darwin in public science and taking on the religious establishment – have wonderfully illuminated the world of Victorian science and intellectual life.[33] Biographies are best written after the death of their subject, but it is possible to write a biography of a science as of a living person.[34] Histories of institutions are not unlike biographies, and will contain biographical sketches and insights. As one historian put it, 'natural history has always been a sociable business . . . Books dealing with individual naturalists rarely give a sense of these entangled webs, the contacts that have always made the world of natural history work.'[35] The *Darwin Correspondence* wonderfully illuminates all this.[36]

Universities, companies, publishers, professional institutes and scientific societies love to publish histories of themselves,[37] which may be bland, anecdotal, glossy and self-congratulatory, but when written by those with serious historical concerns are extremely valuable, reminding us that science is a social activity (even if carried on by an absent-minded professor in an ivory tower). Here, sources new to the historian of science are available: minutes of meetings, treasurers' and bursars' papers, contracts, drafts of books and papers, referees' and readers' reports in the process of peer review, correspondence, papers received and filed but never published, memoranda, laboratory and field notebooks, and diaries may all cry out to be used. When they are, a new perspective is opened up for us: we meet clashes, ambition and personalities, but also institutional loyalties and visions of what science is and what it's for. This may sometimes look like history of science with the science left out, but that view is too narrow: scientific institutions have their special characteristics as well as their resemblance to churches, Masonic lodges, industries and the civil service.

HISTORY FROM BELOW

As writing history from below became an objective for many twentieth-century historians, those concerned with science began to look seriously at popularization. A major project to sample science in nineteenth-century periodicals in English led to the publication of three books, and a much increased understanding of what most people read about science. For, after all, those who study it seriously, attend learned lectures and meetings and read scientific journals, were and are a very small minority.[38] The publics – from the powerful and well-educated to the barely literate – who support science and technology through taxes and purchases, need to know about it and ought to have some say in how this expensive activity is carried on. We usually think about those doing science, but there are others – all of us much of the time – who have science done to them, who are at the receiving end as medieval congregations were for theology. We may be subjected to tests, find our empirical knowledge despised and disregarded by so-called experts, our crops being patented by someone else, or our livelihood taken away by some innovation that makes our skills obsolete. The science that people want to know, or that those in the media think they want to know, is rather different from that which academicians and professors feel they ought to know: public understanding of science is a fascinating and confused topic.[39] 'Breakthroughs' and 'eureka moments' of discovery are announced, then mostly come to nothing. The history of science, like that of relationships, is full of broken promises. Health is and was always a preoccupation, and so is the feeling that science threatens not just jobs but also values. Science is and was often approached through the personal: rows between celebrities are good fun. Ballyhoo, in the form of great international exhibitions and hands-on museums, resulting in the setting-up of educational quarters such as London's 'Albertopolis' in South Kensington (following Berlin's Museum Island), must not be disregarded.[40]

As well as those general periodicals in which science featured but was not prominent, there were scientific journals, and from the late eighteenth century some were private ventures, separate from societies and academies. They might bring those engaged in science in different ways into a self-conscious community, coming to think of themselves as scientists. Thus in Germany chemists engaged in crafts, industry and medical activity were brought together by a journal, while in France Lavoisier and his disciples launched a journal to promote their vision and version of chemistry, and to publish faster than the Academy did.[41] In Britain, private journals began a little later, and were at first general: one, *Philosophical Magazine*, still continues though it has mutated into a physics publication.[42] It was rapidly joined by *Nicholson's Journal*, then by *Annals of Philosophy*, but it absorbed these and other competitors, so that as science became more professional these informal outlets were increasingly replaced by more specialized journals, often published by newly specific societies (for geology, astronomy, chemistry, zoology and so on) and increasingly inaccessible to outsiders.[43] In contrast to those in laboratory and mathematical sciences, publications in the widely popular science of natural history remained readable.[44] Meanwhile, there had been a revolution in publishing, greatly reducing the costs so that books ceased to be luxuries in a world of increasing literacy.[45] Study of books and publishing is crucial for understanding how science was and is carried on.[46]

WHAT THEN IS HISTORY OF SCIENCE?

General history used to be a matter of kings and queens, of popes and prime ministers, admirals and generals, and their dates, saved from aridity only by some thread of epic that guided the reader through the maze. More recently, these great personages have been almost written out in the rise of social history. We do not want to go so far in the history of science, but there also the common man and woman must receive much more of the attention formerly lavished upon the great discoverers. The nineteenth century was the Age of History as well as the Age of Science: both became academic disciplines, crucial to understanding humanity and nature. They have in common the problems of deciding what, from masses of possible evidence, is relevant; of being careful not to throw out babies with the bathwater as new insights become infectious; and of interpreting an unfamiliar world, a project that can never be completed because continually revised in the light of observation, fashion and brilliant perceptions of a world suddenly more coherent. These last come from geniuses, which to Davy and his contemporaries were spirits, genies, djinns that might function as a muse for the receptive and talented,[47] and to later more prosaic generations came to mean the person so inspired.

So, whereas history of science has been seen variously as a tale of triumphs, a branch of intellectual history, investigation of a flimsy superstructure erected upon economic reality, a series of individual careers, a study of equipment and machinery, or a matter of human interactions, academic historians of science have become more self-conscious about their approach and reluctant upon the whole to paint big pictures.[48] Like scientists in positivist times, they keep their heads down, paint more or less exquisite miniatures and tend to leave grand vistas to others.[49] So in this book, necessarily using a broad brush, I shall be an opportunist, picking and choosing among these historiographic traditions to suit the subject matter and context. For others, a closer focus upon narrower themes, such as the making of scientific instruments, the rise of romantic biology in Germany or the publication of a notorious book in England, where many factors can be weighed and given full credit, has cast bright light upon science and culture.[50] Natural history is nowadays at last being given its due, rather than being portrayed as an awkward stage like adolescence,[51] and there are bibliographies accompanied by essays to help those who are new to the history of sciences.[52] There are also companions of various kinds,[53] and dictionaries, both of concepts[54] and of scientists. The multi-volume *Dictionary of Scientific Biography* is being updated, with entries for people previously left out, while for readers of this book the *Dictionary of Nineteenth-century British Scientists* will prove a most valuable resource indeed.[55] So will the *Oxford Dictionary of National Biography*, available online, and rewritten a century on from its Victorian predecessor (which is not superseded when Victorian perspectives are wanted), and many other national biographies. Published letters such as the *Faraday Correspondence*, and that huge international enterprise, the *Darwin Correspondence*, which will go on for many years more, are wonderful ways into nineteenth-century networks in the electrical and the life and earth sciences.[56] Two great multi-volume compilations are also in process of coming out: the *Cambridge History of Science*, published by Cambridge University Press, and the *Storia della Scienza*, published by Enciclopedia Italiana.

The history of science, like science itself, has its societies and journals. Some are general, like the History of Science Society, based in the USA, which publishes *Isis* and *Osiris*, and the British Society for the History of Science, whose journal is the *BJHS*, along with similar societies in other countries. There is also a commercially published general journal, *Annals of Science;* the Royal Society publishes its historical *Notes and Records,* as well as obituaries of its Fellows (British and foreign), which are a valuable source, and *Interdisciplinary Science Reviews* usefully augments the sometimes very narrowly focused papers published elsewhere. More specialized are the Society for the History of Alchemy and Chemistry, which publishes *Ambix;* the Society for the History of Natural History, publishing *Archives of Natural History;* the Society for the History of Technology (SHOT) and its journal *Technology and Culture;* and the Newcomen Society and its *Transactions,* for the history of engineering, while for history of medicine there is the Society for the Social History of Medicine, publishing *Social History of Medicine.* All these societies also hold meetings and conferences, and are international in their membership; their journals have, like scientific ones, been a subject of scholarly study.[57] Scientific societies, such as the Royal Society of Chemistry and the American Chemical Society, have their historical groups or sections. Anyone with a serious interest in the history of science, technology or medicine will need to join one of these groupings. There are also various local societies and institutions which have strong interests in the field.

1

SCIENCE IN AND AFTER 1789

PREAMBLE

Electricity a hundred years ago had been a matter of parlour tricks: now
it was the key to understanding physics. Science had been transformed,
and life with it. So Arthur Balfour (1848–1930), President of the British
Association for the Advancement of Science, and Prime Minister, told
the world in 1904.[1] We cannot fathom the nineteenth century if we leave
out the science. At that same period the German-educated polymath John
Theodore Merz (1840–1922), having retired from chemical and electrical
engineering in Newcastle, was writing his magnificent *History of European
Thought in the Nineteenth Century*.[2] Half (the first two of its four volumes) is
devoted to science, in thematic chapters. There was no doubt in his mind
that 'our century was the scientific century', and that the place to begin
was France. He coined the useful term 'research school'. He believed that
in later life, from 'personal knowledge and experience', one could handle a
hundred years or so as contemporary history, checked against memories: he
would have met many people born and brought up in the eighteenth century, as I did people from the nineteenth. He was writing the history of his
own times, and I am not. Today, our indirect personal experience cannot
take us further back than about the 1880s. Nevertheless, we have some
advantages: we know how stories turned out, what happened next. We have
witnessed the test of time. Our perspectives, and hence what we seek and
find in the past, are different.

Merz, in focusing upon scientific thought, was following the footsteps of Whewell, of Trinity College, Cambridge.[3] Writing against the current view that science meant open-mindedly accumulating facts until generalizations emerged, Whewell saw it as a matter of getting the right perspective, in an imaginative leap, and then filling in this broad picture by directed observation and experiment.[4] His *History of the Inductive Science* was meant to demonstrate this, with the characteristic Fundamental Idea or essence of each science as his clue: chemistry was analytical, geology dynamic, botany classificatory, physics mechanical. Merz, writing at a time when Idealism was triumphant in philosophy, looked similarly at the astronomical, the atomic, the mechanical, the genetic, the statistical and other views of nature. While Merz is a wonderful guide, and our later writings are in a sense footnotes to that amazingly footnoted work, his was not the last word. He quoted Johann Wolfgang Goethe (1739–1842):[5] 'History must from time to time be rewritten, not because many new facts have been discovered, but because new aspects come into view, because the participant in the progress of an age is led to standpoints from which the past can be regarded and judged in a novel manner.' Science is an intellectual activity, but practical and social as well, and scientific life and practice must be our themes also.

We accept that science includes more than established facts, but also that it includes more than ideas. Whewell, Merz and their contemporaries took it for granted that science was progressively finding out truth about the external world, through observation and experiment. In writing its history, they were also evangelists, promoting science. We live in more pessimistic or suspicious times, careful about interpretations, reading between the lines. Nowadays, experts are distrusted and historians must write about Inquisitors, Nazis and others they do not admire. Science is no longer innocent. In the twentieth century, with fresh interest in organized science as a part of culture, some portrayed it as a social construct, projecting on to nature the structures and assumptions of societies.[6] After all, the phrases 'the struggle for existence' and 'the survival of the fittest' came into evolutionary thinking from the political economists Thomas Malthus (1766–1834) and Herbert Spencer (1820–1903) – it was 'social Darwinism' from the start.[7] But that does not mean it was no more than a reflection of 'Victorian values'. Models and metaphors are crucial, but I believe there is a real world, that we can find out more and more about it, and that the sciences are methods of doing this, patiently worked out over time – the nineteenth century being especially important. They are provisional (Balfour noted that much of what he had learnt forty years before was false) but with luck self-correcting. There is indeed no royal road to truth, no ready way to detect 'pseudo-science', and we must not forget instructive examples both of neglected truth and of unsuccessful sciences like phrenology and animal magnetism in our story. History is not only concerned with winners. And just as medical historians have come to look from below, at patients as well as doctors and nurses,[8] so we must also remember those on the receiving end of the benefits and hazards of science (in both the industrialized and the colonial world), and the great number of people who were not involved in startling innovation but who carried on and adapted older, tried-and-tested practices, crafts and ideas.

'Science' is an abstraction: it is something carried on by scientists. Its history is more than the sum of biographies,[9] but looking at lives wonderfully illuminates this human activity we call science.[10] Apostles of science preached prosperity, the alleviation of toil and disease, and the pleasures of teamwork, as well as liberation from superstition and ignorance. Societies, publications, museums and exhibitions proliferated, peripatetic associations visited provincial cities like scientific circuses, transport and communications were revolutionized, religious practices such as calling days of prayer to alleviate plagues were challenged, and governments found themselves having to regulate science-based industries that were polluting the environment. Science's twentieth-century connections with the military-industrial complex go back to the nineteenth, when scientific education and professions emerged, and science needed more and more money if it was to be carried forward. The English word 'science', which had meant organized knowledge of any kind, was restricted in the later part of the century to its present sense, and the word 'scientist' gradually came into use. What had began as a leisure activity (perhaps mildly comic) for European and North American men had by 1914 become a number of increasingly specialized disciplines in which people were educated and trained for careers in research, teaching and administration, or as technicians.[11] Women, Asians, Arabs, Africans and Latin-Americans had played important but backstage parts; by 1914 they were also breaking through into the limelight as science became a global activity. Our story starts in the age of revolutions.

In April 1789 George Washington had been inaugurated as the first President of the USA, and in July the storming of the Bastille marked the beginning of the first French Revolution. And in August 1914 the Great War began. The 'long' nineteenth century begins and ends with political cataclysms. Politics has never been the same since 1789, but it was also an important scientific date: Lavoisier's *Elements of Chemistry* and Antoine

Laurent de Jussieu's (1748–1836) natural system of botanical classification were published. In France, 'two cultures', scientific and humanistic, were perhaps becoming discernible; and a second scientific revolution was under way, in which science was becoming specialized, a demanding vocation (requiring accuracy, precision and numeracy) rather than a hobby. Men of science were mobilized to defend the infant French republic. They did oversee the conversion of church bells into cannons, but science did not yet seem very potent or threatening. In the opening years of the twentieth century, the tumultuous intellectual changes were in physics, genetics and medicine, and by then science-based industries were vital in national economies. The war of 1914–18 was not really the first world war – there had been such lamentable things in the eighteenth and nineteenth centuries[12] – but it came to be described as the 'chemists' war', notorious for the use of high explosives in bombs, artillery and torpedoes fired from submarines, and of poisonous gases, as well as the serious mobilization of scientists in the war effort. This meant the break-up of the international scientific community that had been a feature of the nineteenth century after two decades of warfare ended in 1815: in 1914, enemy aliens were soon expelled from scientific societies and academies, and amid stories of atrocities the free and open communication of knowledge again ceased.

FRANCE IN 1789

The science of what came to be called the Enlightenment was a matter of detecting and imposing order in and upon nature. This might be through mathematics and experiment, or description and classification: natural philosophy or natural history. Voltaire (1694–1778), in exile in England, encountered Newton's physics, and popularized it among his countrymen, while Jean-Jacques Rousseau (1712–78) similarly got across the botany of Carl Linnaeus (1707–78). The great *Encyclopaedia* of Denis Diderot (1713–84) and Jean le Rond D'Alembert (1717–83) began as a project to translate Ephraim Chambers's *Cyclopedia* (1727) but turned into something much bigger and more important.[13] The superbly illustrated[14] scientific and technical articles were augmented by other notorious ones, eluding censorship, on politics and religion, helping to undermine the *ancien régime* with its absolute royal government, feudal privileges for aristocrats and persecution of dissenters from its established Roman Catholic Church. In the hands of these 'philosophes' and their successors, some of them avowed materialists,[15] it seemed as though a modern, scientific world-view must challenge

established traditions in politics and religion, assisting in the emancipation
of the plebeians. Although Lavoisier was beheaded (as a 'tax farmer') in the
Reign of Terror of 1794, and there was unease about elitism, science in
France was to be associated with revolution and secularity from 1789 on.
The life of a man of science who was also religious was not always easy.[16]

Whereas in England the Royal Society, dating from the 1660s, was
in effect an intellectual gentlemen's club, supporting and publishing the
researches of a minority of active Fellows, its contemporary Académie des
Sciences in Paris was an elite salaried body of forty-four eminent men of science.[17] As civil servants, they were required from time to time to undertake
practical investigations into gunpowder or street-lighting, and to adjudicate
on perpetual-motion machines and Mesmerism. They enjoyed prestige and
authority, and it was possible in France for a bright boy to envisage a scientific career: elsewhere, he would also need another profession, such as medicine, the Church or the law.[18] Academicians were elected for life, and some,
like Michel Eugène Chevreul (1786–1889), lived to a great age: there were
set numbers for different sciences, and the total was fixed. After a death,
the survivors would nominate – in effect elect – their next colleague, who
stepped into the dead man's shoes. Would-be academicians therefore had
to make themselves known and respected in Paris in the hope of a vacancy;
it was a competitive world, with canvassing and block voting. Sometimes
the categories might be stretched a bit so that a bright man could get in, the
ham-fisted applied mathematician Denis Poisson (1781–1840), for example,
as an experimental physicist.[19]

Associated with the Académie was the Paris Observatory, founded like
its Greenwich counterpart with the improvement of navigation in mind,
because we can tell where on Earth we are only by looking at the heavens.
Important projects had included work on the exact shape of the Earth; and
following the Revolution, and the adoption of the new metric system in
place of the various and arbitrary-seeming feet and inches in use, the metre
was defined as a fraction of the Earth's circumference, which entailed new
measurements.[20] Also in Paris was the Jardin du Roi, renamed Jardin des
Plantes in the revolutionary period, the splendid botanic garden on the left
bank of the Seine[21] where several generations of the Jussieu family reigned.
Associated with it from 1793 was a Museum of Natural History, and after
the Revolution the royal menagerie was brought from Versailles and the
zoo set up adjoining it. Zoologists were appointed – Georges Cuvier (1769–
1832) and Jean-Baptiste Lamarck (1744–1829) – and gave lectures open
to the public. In these heady days of liberty and equality, medical lectures

Figure 1 *Dens for the wild beasts at the Paris Zoo: J.P.F. Deleuze,* History and Description of the Royal Museum of Natural History, *Paris: Royer, 1825, p. 364*

were also readily available and Parisian hospitals became great centres of medical learning and activity.[22] A former abbey was converted in 1797 into a museum of science and technology, the Conservatoire des Arts et Métiers, including models submitted to the Academy for approval as new inventions. Here lecturers held forth in due course to working men.

At all these places, scientific careers were opening up by 1800. The Revolution provided opportunities for able young men and new patrons on the lookout for protégés. The educational system, which had depended upon privilege and patronage, was also revamped, and especially important was the foundation in 1794 of the École Polytechnique to train engineers and scientists, where entry was by competitive examination. There were innovative medical schools, as at Leiden and Edinburgh, but most universities were conservative places, giving students very much the same education, humanistic and liberal, as that of their fathers and grandfathers. A new idea, embodied in the French Hautes Écoles, was to ally teaching and research: everybody was learning, and it was the duty of professors both to add to knowledge and to pass it on to students. The École Polytechnique under Napoleon became increasingly militarized, but the École Normale, for training teachers, was also involved in scientific education, and the university, reorganized under Napoleon with branches in the provinces, began to adopt the ideal of research and teaching too. Thus training and opportunities became available to bright young men attracted to the sciences.

Unfortunately, the salaries involved were low, and especially after 1815 a practice (*le cumul*) became widespread whereby prominent men held several jobs, and got much of the work done by juniors to whom they paid a pittance – rather like dignitaries in the contemporary Church of England.[23]

The mathematician Pierre-Simon Laplace (1749–1827) and the chemist Claude-Louis Berthollet (1748–1822) flourished under Napoleon, becoming senators and peers. This involved a good income with few duties, and they acquired country houses in what was then the rural suburb of Arcueil. There they founded a select society, inviting their protégés to discuss their research before bringing it to the Academy.[24] Papers presented there were published, with greater rapidity than the Academy could achieve. Most of these disciples became academicians in due course: Laplace and Berthollet were responsible for selecting, training and encouraging the next generation of elite men of science, such as Poisson and the great chemist Joseph-Louis Gay-Lussac (1778–1850), in this semi-formal way. The meetings might be described as residential research seminars, or workshop conferences, in attractive surroundings.

This concentration of resources, coupled with government approval for science, meant that Paris, with its large critical mass of qualified manpower, was the world centre of excellence across all the sciences for thirty years or so. There were extremely able men of science in other countries: Priestley and Henry Cavendish (1731–1810) in England, Joseph Black (1728–99) and James Hutton (1726–97) in Scotland, Tobern Bergman (1735–84) and Carl Wilhelm Scheele (1742–86) in Sweden, Luigi Galvani (1737–98) and Alessandro Volta (1745–1827) in Italy, Benjamin Franklin (1706–90) in America, and Karl Friedrich Gauss (1777–1855) and Alexander von Humboldt (1769–1859) in Germany, for example. But they had to look enviously to Paris and its institutions for standards and recognition as their nations were, from the scientific point of view, in a different league from France, where the main stream of science flowed. Hence it was the French government which asked two academicians, Jean-Baptiste Delambre (1749–1822) of the Observatory, and Cuvier, to report in 1808 on the progress of science in Europe since 1789.[25] The reports were larded with praise for Napoleon, but the centrality of France was appropriate. The two men were joint secretaries of the Institut, for 'mathematical' and 'natural' sciences. It had already become necessary to make this division as, in the years Britain was going through the first Industrial Revolution and building the economic power that ensured the defeat of Napoleon, France retained its lead in science. And even before the Revolution, France meant Paris:

those exiled from it, for one reason or another, pined in relative isolation. Napoleon's victories depended upon concentrating a mass of troops and in science, too, numbers and nearness counted.

During the long nineteenth century, rival centres of power and influence grew, but people still made great advances in places remote from them: John Dalton (1766–1844) in raw Manchester, Justus von Liebig (1803–73) in Giessen, Gregor Mendel (1822–84) in Brno, Dmitri Mendeleev (1834–1907) in St Petersburg, Willard Gibbs (1839–1903) at Yale, and, at the very end of the era, William Henry Bragg (1862–1942) in Adelaide and Ernest Rutherford (1871–1937) in Montreal. In science as in technology innovations often come from such individuals or small groups, then are taken up by those in metropolitan cities – where talent may be stifled as easily as fostered. But as in opera, a chorus to carry the work forward is needed as well as the soloists. The consolidation of science, its transformation into public knowledge and the research programmes that follow upon new insights come from the 'scientific establishment' of learned societies, prestigious journals and institutions with up-to-date facilities and large numbers of able people.

Science could be a wonderful vehicle for social mobility, as the medieval Church had been for boys from humble backgrounds. But comparing Gay-Lussac with Davy, his exact contemporary, involves a tale of two cities: Paris, where a high-powered scientific education led to a career as a professional scientist, and Regency London, snobbish yet meritocratic, where Davy's talents as a lecturer attracted crowds of the nobility and gentry to the Royal Institution. Turning his professorial skills into a performance art, he competed successfully with the entertainment industry to attract patronage, and his research on electrochemistry and safety lamps was done in the laboratory beneath the lecture theatre, where he trained Faraday. Nevertheless, his income was uncertain and like Dick Whittington in the nursery-tale he secured his status and position by marrying an heiress, enabling him in due course to perform the role of President of the Royal Society.

MATHEMATICAL SCIENCES

This group within the Académie, as we now turn to the specific sciences in the *Reports*, included mathematics in its various branches, astronomy, geography, mathematical physics, mechanics and manufacturing. There was no hard and fast line, in France or anywhere, between what came to be called 'pure' and 'applied' science, or science and technology. Knowledge was power and science would be useful: the École Polytechnique was after all an engineering school and the Académie had been founded with utilitarian aims – that's why governments continue to support science, tolerating some blue-skies research at the taxpayers' expense for cultural prestige and long-term hopes. The great achievement of French mathematicians in the later eighteenth century had been to build upon the work of Newton and Gottfried Leibniz (1646–1716) on the differential and integral calculus, developing it into a powerful system of mathematical analysis. Newton's method of 'fluxions' had been slavishly adopted by his disciples in Britain, and his notation, where $\dot{x}$ (with a dot above it) indicated the fluxion of x, was taught at Cambridge up to 1810 and beyond, while one exam paper set was headed 'Newton'. This provincialism cut off British mathematicians from continental Europe, where Leibniz's more convenient notation of dy/dx for the differential of x was preferred, and Newton and Leibniz's work carried forward. A leading mathematician there was Joseph Louis Lagrange (1736–1813) who, before moving to Paris in middle age, worked in Turin and then in Berlin (where under Frederick the Great French was the language of the learned and cultured). He published his formidable *Mécanique analytique* in 1788. He was proud that this book, unlike Newton's geometrical *Principia* of 1687, contained no diagrams or constructions; it was wholly algebraic. This abstract and exact mathematics was becoming the language of science, clear and definite.

The most eminent mathematical physicist in the 1790s was Laplace, who worked especially in astronomy and in probability, using and building upon the work of Lagrange. Whereas Newton had been able to demonstrate that under gravity a planet's orbit about the Sun would be elliptical, his theory indicated that planets would also attract each other, producing wobbles that might accumulate. For him, God was a painstaking governor, not an absent prince, who had both created the system and would from time to time regulate the planets. Leibniz had been scornful: Newton's idea was derogatory to God, who would surely devise clockwork that never needed to be reset.[26] Laplace, a former seminarian, used the powerful methods of analysis in the five volumes of his *Mechanique celeste* (1799–1825) to show that the wobbles cancelled out in the long run: Leibniz had been right. But Laplace's conclusion was different: in a famous reply to Napoleon about God, he declared that he had no need of that hypothesis. For three decades, under various regimes, he dominated the physical sciences[27] and his vision of a world made up of bodies immense and minute interacting under gravity, or central forces akin to it, became all-powerful.

Laplace also improved understanding of the Moon's motions, accounting

for them more exactly than Newton had been able to do, and in more popular and accessible vein speculated about the origins of the solar system. The
orbits of the planets lie almost in a plane: they all follow the same track,
the zodiac, across the heavens – which had made possible the science of
astrology. Like his older contemporary William Herschel (1738–1822),
who had come to England as a musician in a German band and to general
astonishment identified the planet Uranus (the first addition to the list since
ancient times),[28] Laplace suggested that they had solidified out of a mass of
nebulous material that had been spun into a disc shape by the rotating Sun.
Bigger and better telescopes were disclosing minor planets – asteroids –
which had perhaps simply not, or not yet, coalesced into a planet to fill the
gap between Mars and Jupiter. Laplace's evolutionary nebular hypothesis
was to excite astronomers and physicists right through the century.[29]

Laplace also wrote mathematical and popular works on statistics. Life
is messy, but science seemed appropriate only to the tidily causal realm.
To bring chance under the rule of law would be a great achievement.
Eighteenth-century mathematicians had begun analysing games of chance,
and Laplace took this further by extending the study into human affairs.
Making assumptions about how frequently witnesses told the truth, how
sensible jurors were in coming to conclusions, and how acceptable it was to
have a few innocent people convicted and a few guilty acquitted, he came
up with the best size for a jury. This highly a priori approach was rapidly
superseded by what were called 'vital statistics' as governments began to
collect vast quantities of data about populations from which inductive conclusions could be reached, crucial in improving public health. But in physics Laplacian deductive probability remained very important as statistical
thinking became one of the great scientific achievements of the nineteenth
century.[30]

Mathematical sciences included algebra, where unusually the German
Gauss rather than a Frenchman was the star, and (perhaps curiously)
geography, where voyages were the 'big science' of the time, requiring a
large investment and government backing. One or more ships with their
crews were committed for years on surveys and exploration of previously
unknown regions, yielding specimens for natural history, charts and maps,
the claiming of territory, and the settling of colonies. The perfecting of
chronometers meant that longitude could be found with greater accuracy.[31]
There were international agreements that warships on such peaceful voyages were not fair game for enemy vessels even in wartime: science was
above the quarrels of kings or peoples and its devotees should be assisted.

Thus the French expedition under Nicholas Baudin (1754–1803) was welcomed in Sydney, though Britain and France were at war.[32] This could be
problematic. The charts James Cook (1728–79) made of the St Lawrence
river had made possible the capture of Quebec, and in Rio de Janeiro the
Portuguese were terrified that his arrival might end the same way.[33] Skilled
surveyors from a foreign power seemed very like spies.

Magnetism was also a key to voyaging and its study formed a part of
'physics', becoming in the hands of Charles Coulomb (1736–1806) a mathematical rather than simply experimental science, with parallels to electricity
but giving no reason to suppose any close connection. Optics, mechanical
devices like clocks, telegraphs and pumps, and the arts and manufactures
were also there. Especially important was the new metric system, in principle operating both in the sciences, where Delambre saw it as a great bonus,
bringing standardization, and in ordinary life, where it was long and widely
resisted. It was not until the 1870s that its use in science in Britain became
general, so that conversions to and from grains, troy ounces and cubic
inches into centimetres and grams remained necessary for many years,
bringing possibilities of confusion.[34] British industry continued with inches
and pounds (carefully reconstructed after the standards were destroyed in
the fire at the Houses of Parliament in 1834) for well over a century. The
'perfection' that the French saw in their system did not persuade everyone:
the metre, derived from the Earth's dimensions, was supposed to be 'natural' whereas inches, deniers, toises and pounds were arbitrary – but those
brought up with them found them natural enough, and John Herschel
(1792–1871) defended them in a lecture in Leeds in 1863.[35]

NATURAL SCIENCES

Divisions between sciences are clearly social constructions, but it is curious to us to find in Cuvier's volume that chemistry (which included heat
and electricity) appeared along with natural history and the 'applied science' of medicine. They were all accessible without advanced mathematics.
Astronomy was sublime, and those who could afford a good telescope could
and did take it up and make discoveries right through the nineteenth century. Chemistry was much cheaper, practical, and required all the senses.[36]
Solids might be unctuous or gritty, liquids mobile, viscous or viscid, gases
nauseating in various ways: chemists were trained to think with their fingers,
nose and tongue – so that even after a long lapse the smell of the laboratory
is instantly evocative. Chemistry was, and remained until the second half of

the twentieth century, essentially hands-on, looked down on by physicists as akin to cookery, an art as much as a science, where distinguished practitioners were proud like surgeons of their manual skills.

Laplace indeed worked with Lavoisier, trying to measure the heat emitted in chemical reactions by carrying them out in vessels surrounded by ice, and seeing how much was melted. The results were hard to reproduce and inexact, and Lavoisier instead made his new chemistry a science of weights and their changes, balancing the books as he did in the tax office. An ambitious and energetic man, indecently rich, systematic and logical, not much liked by colleagues who found him pushy, his greatest achievement was to establish by careful experiments that in burning, something was absorbed from the atmosphere.[37] Combustion was the classic chemical reaction. The old explanation had been that everything combustible contained 'phlogiston', named from the Greek word for flammable, and that this was emitted into the air on burning. When the atmosphere was saturated with phlogiston, no more could be absorbed and the fire went out. Lavoisier demonstrated that phlogiston would have to have negative weight, and argued that all sorts of ad hoc hypotheses had to be made about it – it was incoherent. Conservatives suggested modification rather than abandonment, proposing a range of phlogiston theories to account for the phenomena.[38]

In 1774 Lavoisier met Priestley, accompanying an embassy from Britain to Paris, who told him about the new, eminently respirable, vital kind of air that he had recently prepared by heating the red 'calx' produced when mercury is roasted in air. Priestley called it dephlogisticated air, because it would support combustion much longer than ordinary air before getting saturated with phlogiston. He came to see that this differed from other airs he was busily isolating, just as metals and salts differed from each other, making chemistry a science of gases as well as liquids and solids. For Lavoisier, however, Priestley and everyone else had got things the wrong way round: this vital air was a component of ordinary air, absorbed on burning (or when the mercury was roasted), and he called it 'oxygen'. Work on gases, as the airs came to be called, helped to make chemistry exciting, so that it became and remained the most popular science of the nineteenth century. Astronomy was sublime but chemistry was hands-on, promising to be useful in all sorts of ways but also fun to do.[39]

Lavoisier thought the idea of indivisible atoms 'metaphysical', but published a famous list of simple substances, the limits of analysis, replacing the ancient elements (earth, water, air and fire) with iron, copper, oxygen, sulphur and other substances which themselves were soon called 'elements'.[40]

Nevertheless, he considered them simple only in the solid state – unknown for oxygen. Lavoisier believed that the change of state from solid to liquid, and from liquid to gas, was due to chemical combination with the simple substance 'caloric'. Black had shown that definite, often large, quantities of heat were required to bring about these changes (familiar to all who have melted ice or boiled water). Whereas the usual view had been that heat was the motion of particles, for over twenty years the caloric theory became orthodoxy. Caloric was thought of as a weightless kind of fluid, 'subtle' or 'imponderable'. It could be combined chemically like other elements, being absorbed or emitted in reactions, or free, merely mixed with matter, swelling things as more was absorbed and they got hotter. Its affinities with light, which also featured on Lavoisier's list, became clearer in these years. William Herschel, studying radiant heat, investigated which colour in the Sun's spectrum had the greatest heating effect, and found that invisible rays beyond the red end were most powerful. Johann Ritter (1776–1810) and William Hyde Wollaston (1766–1828) both thought of looking beyond the other end, the violet, and found there 'actinic' rays that would set off chemical reactions such as the blackening of silver chloride. Light and heat thus fell within chemistry in Cuvier's story.

Electricity, Balfour's parlour tricks, had meant what we call electrostatics: tremendous in lightning, but when generated by Franklin, Priestley and others by friction on a glass plate or globe, indeed entertaining. The hair of someone standing on an insulating block and charged up would stand on end, and sparks would fly from his fingers. Galvani showed that frogs' legs twitched in the presence of electricity; Cavendish, Humboldt and Davy studied torpedo fish and electric eels.[41] Galvanizing the listless became a fringe-medical activity, with the corpses of executed criminals being made to grimace horribly when prodded with electrified terminals.[42] Mary Shelley invoked the 'spark of being' in *Frankenstein* (1818);[43] it was clear that electricity was, like oxygen, a key to life. But in 1799 Volta demonstrated that electricity could be generated by immersing two different metals in water: Galvani's frogs' legs had detected rather than generated it, and his kind of low-voltage 'galvanic' electricity was not simply an organic phenomenon. As Davy put it, Volta's discovery (communicated in French to the Royal Society in London) was an alarm bell to chemists all over Europe.[44] William Nicholson and Anthony Carlisle dipped the wires from a cell like Volta's into water, and found that oxygen and hydrogen bubbled off.

The word 'battery' that had been applied to electrostatic condensers, 'Leyden jars', soon came to be used for Volta's cells. These, when water was used, soon ceased to work; but with acid instead, went much better. All over Europe, Cuvier, Humboldt, Davy, Ritter, Jacob Berzelius (1779–1848), Wollaston and many more took up the investigation. Volta thought mere contact generated electricity; others that a chemical reaction was necessary. Puzzlingly, around the pole where oxygen was emitted, the water turned acid; and where hydrogen came off, alkaline. Was electricity, like light and heat were widely thought to be, a kind of substance that might combine with ordinary matter?

Davy's attention was diverted to urgent technical questions: how to improve agricultural productivity and the tanning of leather. In 1805 these researches earned him the Royal Society's highest award, the Copley Medal. His reputation made, in the autumn of 1806 he distilled water in a silver vessel and passed an electric current through it in cups of agate and gold, believing rightly that glass apparatus was far from inert. He concluded that the acid and alkali came from dissolved nitrogen reacting with the emerging oxygen and hydrogen, and duly demonstrated that pure water is decomposed into its elements, with everything else being a side-reaction due to impurities – always something very important but difficult to demonstrate in chemistry. His conclusion was momentous: that chemical affinity and electricity were manifestations of one power. The Académie in Paris (more interested in theory than the Royal Society) awarded him the prize it had proposed for advances in galvanism. In the following year, he used an electric battery to decompose fused caustic potash, liberating in a spectacular reaction (bright sparks flew about the room) a highly reactive, soft, silvery-white substance that floated on water, decomposing it so violently that the hydrogen emitted caught fire. Despite its anomalous properties, he identified this 'potagen' as a metal, and called it potassium. In his report Cuvier was still uncertain about its status, but soon London and Paris competed to build the biggest battery. Huge troughs held banks of metallic plates which could be dipped down into the acid, but, as is often the case, expensive (and dangerous) equipment did not lead to very much that could not be done with the pocket-sized batteries of which Wollaston was proud. Bigger and better apparatus can make original thinking harder.

Burning and breathing both used up oxygen, and for Lavoisier the secret of life really was the vital flame, the conversion of oxygen to carbon dioxide. Human bodies were thus a kind of low-temperature soft-shelled chemical works, as were plants, which as Priestley and others demonstrated absorbed carbon dioxide and emitted oxygen, allowing life to go on. But conditions in the furnace and the lungs were very different. The surgeon John Hunter

(1728–93) and the anatomist Johann Blumenbach (1752–1840) argued that a vital force must supervene in organisms, controlling development and to some degree suspending ordinary chemical reactions. In a corpse, the digestive juices may attack the stomach, which during life is immune; in decay, ordinary chemistry asserts itself. French surgeons and physiologists (making the running in the early nineteenth century) were more materialistic, but in Britain the Animal Chemistry Club, a subset of the Royal Society in which Davy and the surgeon Benjamin Brodie (1783–1862) were prominent, were vitalists. William Lawrence (1783–1867) was denounced for blasphemy when he disseminated French ideas at the College of Surgeons, though this didn't in the long run hinder his career.[45] What we would call biochemistry had an uneasy beginning but its development forms a very important line through the nineteenth century, promising a genuine applied science of medicine.[46] Our devotion to the 'natural' and the 'organic' is perhaps vestigial and continuing unease from two centuries ago.

For Cuvier, studies of the composition of organic matter, and fermentation and putrefaction, marked the frontier territory of chemistry and his own science of natural history, where his despised colleague Lamarck had coined the term 'biology'. Natural history included meteorology, but Cuvier was unaware of the classification of clouds by Luke Howard (1752–1864), being more interested in the electrical and chemical states of the atmosphere. Much easier to classify, because fixed in their shape, were minerals: in France crystallography had begun with René Haüy (1743–1822), publishing in Lavoisier's journal, *Annales de Chimie*. But with geology and the functions and structure of animals, Cuvier came into his own, though only alluding briefly and in the third person to his own work.

Cuvier had the chance, with the zoo adjacent to the museum, to compare living animals with the fossils that around 1800 were being found in the quarries of Montmartre as Paris was rebuilt in imperial splendour. There was no longer serious doubt that they were, as the surgeon James Parkinson (1755–1824) described them, 'organic remains of a former world'.[47] Cuvier was the pioneer of reconstructing the skeletons of extinct creatures. That meant deciding from bones found all mixed together which ones came from the same sort of animal, and how they had fitted together. He, like Aristotle, was a teleologist: all the parts of an animal cohere and conduce to its mode of life. Thus herbivores have big grinding teeth and sideways-moving jaws so that they can chomp up vegetation, and they are likely to have eyes placed to give them wide vision, and long legs with hoofs, that enable them to spot and escape predators. Carnivores have canine teeth, strong jaws that move up and down, powerful shoulders and claws, and eyes closer together to focus upon prey. What he called the principle of correlation led him to claim that he could identify and reconstruct an animal from a single bone. It meant he could demonstrate that there had been different kinds of elephants and other creatures near where Paris now stood: in fact a whole series of former worlds, with different faunas and floras now extinct. His vision dominated geological thought for a generation and his *Discourse* about it came out in a new edition as late as 1877.[48]

In the eighteenth century, it was still plausible that animals like the huge elks whose skeletons turned up in Irish peat bogs were only locally extinct. After all, why would God allow species of animals that He had created to die out? They might be found flourishing in the unknown southern continent, in the almost equally unknown Wild West of America, in Siberia or in Central Africa. By Cuvier's time, this no longer looked likely, partly because there were so many such fossil animals, and partly because of the progress of geography. Travellers from Russia beyond the Urals could find no living mammoths, but reported dead ones found frozen, fresh enough to be eaten by sledge-dogs even if not by less-than-intrepid travellers. For Cuvier, this proved that they had been overtaken by a catastrophe. A country with enough vegetation to support elephants must have suddenly turned into a deep freeze, entombing them while still fresh. The history of the Earth was thus a succession of periods of little change, with periodic cataclysms in which the populations of huge regions were destroyed.[49] Cuvier, a good Protestant, was not worried by the implications of this: humankind was recent, appearing only after the last great convulsion, and the Bible not concerned with mammoths and mastodons.

Through physiology of plants and animals, Cuvier's survey moved on to taxonomy, where the French had produced natural methods based upon recognizing the affinities, homologies and correlations that went with family groups. The language implied common ancestry, but for Jussieu as for Linnaeus this was a metaphor. The natural method was superior but more difficult to learn and apply: the great appeal and value of Linnaeus's more arbitrary and artificial system (based on external features) had been that it was straightforward and easy. From here, the survey's last step was into the applied sciences of medicine and surgery: once again, the lead that had been held by Leiden and then Edinburgh had passed to Paris. In pathology and in surgery, the hospitals of Paris were centres of innovation in the revolutionary and Napoleonic years. Bold and innovative, proud of their manual skill, ready to use the vivisection that earned François Magendie (1783–1855) the epithet 'murderous' from British animal-lovers, they undertook new surgical operations in days before anaesthetics or antiseptics. Daring and dashing, heroes in their native land, they aroused awe and horror among foreign visitors.

Cuvier finished his report with agriculture, and then 'Technologie', the study of arts and techniques. It seems to us a curious placing, in a different volume from physics and engineering, and the word did not enter the English language for many more years. The term included researches on baking and on wine, on tanning, bleaching and dyeing, on making soap and cement, and on paper suitable for money, the 'assignats' associated with the inflation of the revolutionary years. It is a miscellaneous list, but it shows that in France as in Britain there were high hopes in these opening years of the century that scientific understanding of processes would lead to their improvement, and that scientific discoveries must have useful applications of some kind. Such optimism was characteristic of men of science and those who flocked to their lectures, but it did not please everyone. These were the years of the Romantic movement, and deep unease (especially outside France) about some kinds of science and their effect on nature and the minds of those studying them.

DISENCHANTMENT?

William Blake wrote of dark Satanic mills, doubtless having new industry in mind as well as diabolical machinations within the human mind and society, while for John Keats (1795–1821) (natural) philosophy would clip an angel's wings, erasing with its cold and withering touch all the magic of the rainbow. Everyday experiences that had held a religious dimension, the rainbow recalling Noah's Flood and God's comforting message, and the butterfly reminding us of the Resurrection, were demythologized into examples of refraction and metamorphosis. Blake's picture of Newton, the basis for the statue in the forecourt of the British Library, has him looking down at his geometrical figures, where a real seer would have looked up into heaven. William Wordsworth's Newton on the other hand, with his prism and silent face, is forever voyaging in strange seas of thought, alone – a properly romantic hero. The Romantic poets and artists in Britain and Germany, reacting against the Enlightenment, were not simply anti-scientific and certainly not irrational – Reason was extremely important for Samuel Taylor Coleridge, for example, though it should not be divorced from the shaping spirit of Imagination. In their delight in the sublime rather than just the calmly picturesque, in the works of God rather than of man, in their use of the language of ordinary people and ballads rather than high-flown poetic discourse, and in their love of history, ruins and myth, they turned against mechanical explanations, determinism and what we can call scientism – the idea that reductive scientific explanations are the only genuine kind.[50]

Thus Goethe, like his contemporaries in the German Romantic movement,[51] delighted in chemistry, making use of alchemy in *Faust* (1775–1832) and of more up-to-date theory in *Elective Affinities* (1809), where the characters are learning chemistry as their own relationships break up and reform, disastrously. We are made to think how like or unlike we are to the chemical elements whose affinities are not really 'elective' at all but are determinate. Coleridge relished chemistry, which his friend Davy, revealing the underlying forces that modify matter, had made 'dynamic', and Percy Bysshe Shelley was also a devotee, seeking to account for vitality through chemistry.[52] At the beginning of the century, Coleridge had written to Davy for advice when he and Wordsworth were intending to set up a chemical laboratory in the Lake District. Atomic theory was for Coleridge materialistic, a world of little things, incapable of accounting for the big things; and in later life as he suspected Davy of atomism, their relationship cooled. Coleridge's own 'magnum opus', the product of a lifelong interest in medicine, was to be a theory of life: a slim volume written when he was living in Highgate at the end of his life, *Hints Towards . . . a Theory of Life* was published posthumously.[53]

Keats had had medical training, and ideas from medicine were also prominent in Romantic writing: the charnel house and the twitching corpse in gothic-horror novels, but also a real interest in how the body and mind work, and in the writers' own health and its effects on their state of mind. Coleridge coined the useful word 'psychosomatic'. Voyages were also sources of excitement and interest: the 'noble savage' was an invention of the later eighteenth century, and 'The Ancient Mariner', published in the mould-breaking *Lyrical Ballads* of 1798, was inspired by reading narratives.[54] Cook was accompanied on his voyages by painters who depicted people, landscapes, flowers and animals, and this example was often followed on later scientific voyages.[55] Engraved and published with the usually plain and unvarnished prose of the sailors' stories, these worked readily upon the imagination: this kind of science was not reductive or disenchanting. Back home, painters found the new mills, foundries and quarries sublime, especially when lit up at night;[56] Caspar David Friedrich made careful studies of trees for his moody canvases;[57] John Constable was fascinated by Howard's

meteorology for his own cloud studies; and J.M.W. Turner later showed
how thoroughly romantic steam tugs and railway trains can be.[58] The message of *Frankenstein*, that science is an activity of sorcerers' apprentices playing God, was not nearly as resonant in its own day, when readers were used
to gothic novels, as it has become in ours.

Sir Joseph Banks, who had accompanied Cook on his first voyage,
was President of the Royal Society from 1778 to 1820.[59] As the French
Revolution turned into terror and violence, and the war with France into
a war on terror, so philosophes like Voltaire, Diderot and Rousseau were
seen as its ideologues, which from a British perspective made science a
prime suspect as the cause of social and moral collapse. Some Fellows of
the Society, notably Priestley, were indeed left-wingers (the seating in the
French assembly introduced the idea of political left and right) who looked
forward to an end of privilege and deference in Britain. These were heady
days and presented a problem for Banks, who can be seen as a major figure
in an 'English Enlightenment'. His vast international correspondence and
archive is a valuable route into his times.[60] He had no doubts about law
and order, was a great landowner, appreciated the economic importance of
science, loved natural history and was suspicious of hypotheses, theorizing
and systems. He saw that the Society's Royal Charter and elite membership would, with care, protect it from suspicion in the perilous days when
habeas corpus (protection against summary imprisonment) was suspended,
illicit gatherings dispersed and a fifth column of revolutionary sympathizers
was dreaded. He cold-shouldered Priestley, whose house and laboratory in
Birmingham had been looted by a mob shouting 'Church and King', and he
opposed the founding of new or specialized scientific societies. Partly, no
doubt, this was because they would not have been part of his learned empire,
but also he saw that if men of science did not hang together, they would hang
separately. A Privy Councillor, he persuaded governments that science was
important, politically safe and patriotic, with an important role in agriculture
and industry in Britain, and in the new empire – notably in Australia whose
settlement he had promoted.[61] He set up botanic gardens in the colonies and
sent out collectors to bring home potentially valuable species.

To Keats, the rainbow's significance and beauty had been reduced by
Newton's optics to a matter of refraction and reflection, but to Wordsworth
the loss, glory passing from the earth, was due to leaving childhood behind.
For everyone who found science disenchanting, there were several who
found it entrancing. Paley's *Natural Theology*, published in 1802, encouraged
its many readers to wonder at all the adaptations in a world in which God

had added pleasure to what might have been mere existence.[62] Discoveries in astronomy and natural history were exciting, indeed marvellous, while Priestley's forecast was verified – that work in chemistry, electricity and optics would go beneath the surface of things and reveal the underlying forces, eclipsing even Newton. The popularity of Davy's lectures, delivered at the Royal Institution to the affluent in London's West End, confirmed that science was not a threat to the ruling class, but promised a cornucopia with benefits to everyone – the message of his famous inaugural lecture of 1802.

CONCLUSION

Thus in Paris a critical mass of people working right across the sciences, invigorated and brought into prominence with the Revolution of 1789, had achieved world leadership. At the apex was the Académie des Sciences, re-founded in 1794 as the First Class of the Institut de France, with its salaried elite of researchers. Otherwise, there were educational institutions, museums, government-supported industries and groupings such as those who met at Arceuil, in which academicians were prominent, and in which young men could hope to make careers as they aspired in turn to become full members of the Académie. Laplace, Lavoisier, Berthollet, Delambre, Cuvier and their contemporaries had little patience with what they saw as the airily optimistic 'systems' of their predecessors, in what we call the Enlightenment. Instead, they saw real science as based upon complex mathematical calculations, exact experiment with precision apparatus and rigorous taxonomy. Their work can be seen as a 'second scientific revolution', bringing in new standards of what would count as science and demonstrating the value of diverse groups and a competitive ethos. Abroad, their dominance was recognized, often ruefully, though this kind of reductive and seemingly materialistic science was also opposed, notably in Germany and in Britain where the Romantic movement was under way. But in both these countries, occupied by French armies or engaged in war with them from 1793 to 1815 and suspicious of everything French, it was an age of wonder, with wide enthusiasm for science of a dynamic kind. The coming of this 'Romantic science' can also be seen as a second scientific revolution, different from but complementary to the one going on in France.[63] Nineteenth-century scientists were the heirs both of the exact French tradition and of the Romantic one, where polar forces underlay apparently solid matter.

Thus Davy in his lecture demonstrations made science exciting and accessible, a source of metaphors for Coleridge and of conversation and stimulated imagination for his audiences. So at least it seemed, though some hearers noted how difficult it was to remember what he had said unless one followed up the lecture with study at home.[64] Some science was more formidable. The Latin names that Linnaeus introduced could be made palatable and even entertaining to learn, as Erasmus Darwin showed in his scientific poem *Loves of the Plants* (1789).[65] His verse, mixing instruction with amusement, was very popular until *Lyrical Ballads* marked a new tradition, and the parody 'The Loves of the Triangles' by William Pitt, George Canning and Hookham Frere, in their conservative *Anti-Jacobin*, made Darwin a joke. The parody showed incidentally that real mathematics could not be learnt painlessly in the same way through heroic couplets, like natural history, or by watching experiments, like chemistry. Indeed, mathematics remained (and remains) for most people a barrier to much science. And gradually scientific jargon began to grow also, with scientific communities. It was becoming more difficult for nature's interpreters to get their message across. But that is a topic for the next chapter.

2

SCIENCE AND ITS LANGUAGES

We are the talking and laughing animal. Other creatures communicate, and hyenas, jackasses and kookaburras make guffawing noises, but speech distinguishes us from every other species. We are informal, we joke, we are vague accidentally or on purpose, we mislead and we gossip. But scientists are not like that. Their enterprise depends upon terse and sober truth-telling. Children taking science are taught to write that a test-tube was taken and a result was obtained: the passive voice gives weighty authority, a veneer of objectivity over what would otherwise be personal, subjective and doubtful. If children find something really surprising, they are made to do the experiment again, better. Then they are taught jargon: they must use ordinary words like 'energy' or 'atom' with care, and new ones like 'ion' with exactitude. They should eschew ordinary terms like 'alcohol' or 'salt' in favour of 'ethanol' and 'sodium chloride', and are instructed to call the Peacock butterfly *Nymphalis io*. They learn about laws, like Boyle's, and lawbreakers, like carbon dioxide, that do not 'obey' it. As they get older, they have to 'unlearn' over-simplified explanations drilled into them earlier, and find science a puzzling and disconcerting mixture of dogmatic certainty and probability. Unlike history textbooks, rival scientific ones normally do not differ in interpretation, only in clarity and arrangement – but different editions will disagree over facts and theories, which therefore seem provisional; so it is essential to be up to date. Proof in science may seem timeless and logical, as in the mathematical ideal; really it is legal, meaning beyond reasonable doubt given the evidence at a particular time, and like everything else it won't endure for ever. Scientific language reflects the aim of objectivity and the fact of instability.

Things were different in the past. Galileo wrote expansively, in fine literary style – indeed, that was partly what got him into trouble with the Inquisition. Founders of the Royal Society like Robert Hooke (1635–1703), naturalists like Gilbert White (1720–93) and Erasmus Darwin, and even chemists like Davy, were natural philosophers and did not write like scientists in compressed style abounding with formulae and tables. The science they described was interesting and inviting to anyone, not arcane and forbidding, and their plain style was admired and followed in pulpits where witty eloquence had previously held sway.[1] Nowadays, there is a great deal of what Karl Pearson (1857–1936) called the grammar of science to learn.[2] It was in the course of the nineteenth century that the language and culture of science became thus inaccessible to those without formal training, as arts turned into sciences, and knowledge picked up by working with a practitioner gave way to book-learning.[3] How did this happen, and why?

Answering that question will take us into realms of professional organization, specialisms, experts, education and publishing, as the growth of knowledge by 1800 made it impossible to be a walking encyclopedia or renaissance man. The word 'scientist', coined in the 1830s, did not come into use for half a century and by then there were not just 'two cultures', humanities and sciences, but more 'ologies' in what we call social sciences with their jargon, as well as the old closed worlds of law, Church and medicine. Ordinary language comes loaded with values: in mastering a foreign language, seeing how words convey approval or disapproval, praise or derision, is crucial – and over time words like 'specious', 'portly' and 'gay' have evolved. Languages are unstable: science may no doubt be progressing towards a permanent and value-neutral state, but its terms and categories do become obsolete and inevitably it employs metaphors and models. Seventeenth-century natural philosophers were struck with the way that we all read numbers 1, 2, 3 and so on, though making different noises in England, France and Germany, and by Jesuit missionaries' reports that in China speakers of mutually incomprehensible dialects and languages could read the written characters and get the same message. John Wilkins (1614–72), another founder of the Royal Society, devised an exact and artificial written language for science in symbols like Chinese characters, but it never caught on. It was much too restrictive and allowed the imagination no play.

Beginning with language itself, we shall note its role in various sciences, botany, chemistry and physics, and in human sciences, as description, classification and explanation went forward in the nineteenth century with new words coined for new sights and new thoughts. We shall look at visual language, as scientific illustration came to look less like fine art, and at the gradual fading of the language of natural theology, praising (or trying to prove) the goodness and wisdom of God, in which science in Britain and America was saturated. These things triggered debates about liberal, as opposed to technical, education that have never stopped – ancestors of our 'science wars'.

LANGUAGE AND LANGUAGES

Gradually, in the course of the eighteenth century, Latin ceased to be the learned language in which all well-educated people felt at home. In seventeenth-century Italy, Galileo had published his inflammatory *Dialogue* in Italian; soon afterwards, Descartes published in French, and Robert Boyle (1627–91) in English. Nevertheless, it was many years before men of science wrote everything in their vernaculars: very dry and recondite works like Newton's *Principia Mathematica* (1687) were in Latin because the international market was important. Newton's more accessible *Opticks* (1704) was published in English; later, a Latin translation was prepared for sale on the continent. In Germany (a mass of different states), and in Russia, Latin continued much longer, but by 1800 vernacular languages were the norm, French being often used by those from smaller nations. Conversing and corresponding with foreigners in Latin ceased to be commonplace. The effects of this were contradictory. On the one hand, it made science much more accessible to the surgeon, the merchant, the sailor or the craftsman whose education had not included ancient languages – and to educated women, who were more likely to learn modern languages than dead ones. On the other hand, it brought back the curse of Babel: few people were at home in more than one or two foreign languages, and men of science could remain in deep ignorance of what was going on in places remote from them.

If language distinguishes us from animals, the orthodox view was that God both gave us language and created through his word: he spoke and it was done.[4] Adam, created in God's image, named the animals in the Garden of Eden – his vision, unclouded by sin, had enabled him to see their essential character, and his names were significant, not just conventional noises. At Babel that divine language was lost, and now ordinary language was too vague for scientific purposes. A 'sycamore' in the Holy Land is a kind of fig; in Britain a maple, and in North America a plane tree. In the eighteenth century, words like 'force' and 'energy' were loosely used in ordinary speech, and metaphors from family relationships were freely employed in natural history and chemistry. The use of deliberate obscurity to deter the uninitiated, for example, in alchemy, was frowned upon in the Enlightenment but precision combined with clarity was hard to achieve. For the great philosopher Johann Gottfried Herder, language was not a divine gift but a human accomplishment acquired in our 'apprenticeship to nature'.[5] Diverse languages were expressions of diverse cultures in different environments.[6] The study of languages was therefore crucial to understanding the past and indeed the present.

Natural historians put humans at the top of a scale of nature that, when the missing links were found, would have no great gaps or leaps. Thus in Linnaeus's genus *Homo*, the orang-utan was at our heels but could not speak.[7] Language made us what we were. Revolutionary years brought euphemism and political correctness into European speech.[8] Most of those who recorded exotic languages had noted vocabulary rather than structure and thus came up with thoroughly fanciful connections, 'philological co-incidences', such as the Mandans in the USA being supposed to speak Welsh.[9] But Jesuits in China and Japan had been struck by the very different way that languages worked there. Cook and Banks learned Tahitian, and saw its likeness to other Polynesian, and difference from Melanesian, languages.[10] William Jones (1746–94), in India as a judge in the 1780s, presided over Warren Hastings's Asiatick Society of Bengal and with a pundit studied Sanskrit, recognizing its affinity with most European languages and with Persian, and different structure from Semitic languages like Hebrew and Arabic. In 1799 the Rosetta Stone was brought to England in the wake of Nelson's victory on the Nile. Here, Thomas Young (1773–1829), a great beginner of investigations, notably in optics, began the deciphering of its hieroglyphs, and François Champollion (1790–1832) in France achieved it. After Jones died in India, the study of linguistics became a particularly German science in the nineteenth century. The pioneers were Herder[11] and Wilhelm von Humboldt (1767–1835), founder in 1810 of the University of Berlin (and brother of Alexander the explorer) who studied Basque.[12] Max Müller (1823–1900) brought philology to Britain, as Professor at Oxford from 1854, and demonstrated the fluidity and evolution of languages.[13] Linnaeus had classified races of mankind using his customary external characters. In the early nineteenth century, language rather than colour became the key to human variety;[14] later, physical attributes became crucial again.[15]

NAMING AND PLACING

It had been the great achievement of Linnaeus to perfect and propagate a kind of terse Latin to describe species, accompanied by a two-word key, the generic and specific (or 'trivial') name.[16] Through his collections and publications, he (almost like Adam) gave the definitive names to animals and plants, and his successors, with more sophisticated taxonomies, continue to do so.[17] Clear, systematic description became possible. Newton was the model for explanatory natural philosophy, Linnaeus for descriptive natural history, and both reflected the 'order and beauty' in the world.[18] The now-dead languages of Latin and ancient Greek, mainstays of the classical, liberal education that European universities imparted, became and remain the source of scientific jargon, freed, it was hoped, from misleading values and vulgar associations. The important task of naming was a matter for the scientific elite, mandarins at home in the ancient world as well as in the botanic garden or the laboratory. Thus by 1800 scientific language had become both more straightforward, being English, French, German or Italian, but also more technical and theory-laden with specialized vocabularies. Science is open, public knowledge, but like secret societies scientific communities shared rituals and language that excluded outsiders. Just as the French had an academy to regulate their language, the Académie française, so national and then international bodies codified and approved nomenclature.

It took time for Linnaeus's work to catch on, and in Britain a major influence was the poetry of Erasmus Darwin. His *Loves of the Plants* was published in 1789.[19] Conveying science in verse, and taking up Linnaeus's idea that plant reproduction, and therefore classification, was sexual, Darwin played with the idea of 'vegetable love' amid the numbers of stamens and pistils in the flowers, here of broom and balm:

Sweet blooms GENISTA in the myrtle shade,
And *ten* fond brothers woo the haughty maid.
Two knights before thy fragrant altar bend,
Adored MELISSA! and *two* squires attend.

The book also contained copious notes, covering not just botany but also electricity, chemistry, the mythology of the Portland Vase and anything else which attracted its polymath author's attention.[20] This was a time of great voyages, of colonial settlement in Australia, of developing love of gardens and of increasing leisure, and all these things brought botany to the fore. Linneaus's sexual system, with its girlish Latin names, made science palatable and fun, though some contemporaries and later the straitlaced critic and sage John Ruskin saw it as prurient.[21] It also made elementary botany quick to learn, so that ship's surgeons could collect and make some sense of the flora of distant regions (provided that the plants were in flower) before bringing specimens back for formal naming to the experts in Europe – maybe in Paris, London, Leiden or Uppsala, otherwise at the 'physic gardens' of medical schools elsewhere. Under the guidance of Banks, a friend of George III, the grounds around the royal residence at Kew were transformed into a botanic garden, which after various ups and downs became a national institution with a wider focus on plants of economic importance – and an aesthetic appeal to the general public, admitted for a penny.[22]

The Botanic Garden was a great success, and Darwin became one of the best-known poets of the time. But as a sympathizer with the French Revolution, he was satirized in 'The Loves of the Triangles' in the conservative *Anti-Jacobin*,[23] and the *Lyrical Ballads* made his kind of instructional poetic diction with its footnotes and endnotes seem old-fashioned. His huge popularity waned, but he went on to write *The Temple of Nature*, published in 1803 soon after his death, where (in striking contrast to Linnaeus) he presented an evolutionary world-view. Progress, which had meant travel (as in Royal Progresses around the country), was coming to imply movement onwards and upwards,[24] and while conservatives deplored the loss of 'Merrie England', Thomas Jefferson's rural America or Biedermeier Germany, 'you can't stop progress' became a slogan generally accepted in the nineteenth century. The slow emergence of civilization from barbarism was for Darwin the last stage in a process beginning with the formation of the Solar System, and continuing through the progressive evolution of plants and animals.[25]

THE CHEMICAL REVOLUTION

Just as the French Revolution was bringing in a new political vocabulary, Lavoisier published his *Elements of Chemistry* incorporating and justifying the new language for the science – a project he had taken over from Guyton de Morveau (1737–1816) and others, and made his own.[26] The old language had been disorderly, with terms like 'flowers of sulphur' and 'sugar of lead' based upon appearance and taste, 'Glauber's salt' being named after its discoverer, 'spirit of hartshorn' from its origin, and 'orpiment' and 'alcohol' from Arabic. Names gave no clue about composition. They might mislead: 'lunar caustic' was named from the supposed affinity of silver and the

Moon. Lavoisier's ideal science was algebra, wholly abstract, value-free and immune from metaphorical thinking. Chemistry should follow that route, avoiding speculation ('metaphysical', which among men of science meant pointless) about atoms and elements. Simple bodies were those which could not be further analysed – necessarily an unstable list, but corresponding more or less to our chemical elements. Apart from familiar metals, they were given names derived from ancient Greek and Latin in the hope that the new language would, like Linnaeus's, be international. Thus 'Glauber's salt' became sulphate of soda, subsequently sodium sulphate. Like Linnaeus's, expanded and amended in detail, this language has survived into our day as the basis of its science.

This did not happen effortlessly. Just as Linnaeus's language depended upon realizing the importance of vegetable sex, so Lavoisier's involved more than renaming. Priestley had after all named the vital air he isolated 'dephlogisticated' because he held the phlogiston theory. For Lavoisier, this gas was 'oxygen', supporting combustion and yielding acids (hence its name, from Greek). With 'hydrogen' it formed water, and both were gases because they were combined with 'caloric', the cause of heat – emitted, for example, when they combined explosively. Lavoisier's sober Enlightenment rhetoric (very different from the language of revolutionary politics) convinced his peers, his powerful position in the Academy enabled him to enforce his views, and with his associates he founded a journal, *Annales de chimie*,[27] which would publish only papers written in his new language. French power, intellectual and military (soft and hard, in our political jargon), spread it over continental Europe,[28] though pharmacists and doctors were sceptical even in France,[29] and 'antiphlogistic' remedies to reduce fever and inflammation were still being prescribed forty years after 1789 (to the dying Davy, for example). Priestley never accepted it, seeing it as speculative and unnecessary, a throwback to the make-believe world of Descartes. But to younger chemists especially it seemed logical and attractive – even to Davy, Priestley's chemical heir.[30]

This is perhaps the episode which best fits Thomas Kuhn's analysis of 'scientific revolutions', where 'normal science' in an accepted paradigm in which some anomalies have inevitably built up is challenged by someone with a different world-view.[31] If the worrying anomalies disappear in the new perspective, not too many babies are thrown out with the bathwater, and the proponent gains disciples especially among the younger generation, then the new paradigm will prevail and normal science will resume within it. A new language will replace the old one and full translation between them will not be possible. To those of us chafing under the dogmatic teaching of science, which goes back to scientific catechisms of the early nineteenth century,[32] which (for Kuhn) had been evolved to help students avoid errors like an Earth-centred cosmos, or phlogistic chemistry, Kuhn's schema was liberating.[33] In the 1960s, revolutions seemed to be one-way: there was no turning back. Since 1989 we have known that this is not so: Leningrad can become St Petersburg again, and the history of science also looks less fractured by revolutions than it did, less dominated by its Lavoisiers and Newtons. But Kuhn helped everyone to see that while truth may be mighty and will prevail, it needs assistance. Scientists are sceptical, have their own agendas and are conservative, so that innovators, if they are to make headway, must see the importance of what they have done, get it across clearly and work through institutions to convince their colleagues.

Kuhn drew attention to the way that translations are always imperfect. Languages, as Herder had perceived, involve world-views. Scientific language is inevitably theory-laden: there will be losses when old science must be expressed in the new manner. It can mislead too. Lavoisier had wrongly but reasonably concluded that oxygen was the cause of acidity, and named it accordingly. Bloodless precision may be the ideal in algebra, but in chemistry analogy and geometry have proved essential. Oxygen has even given rise to metaphor: 'the oxygen of publicity'. Davy demonstrated that the choking green gas from sea salt was an element, analogous to oxygen but not a compound of it, 'oxymuriatic acid' as the French had believed. Avoiding theory, he called it chlorine, from its colour.[34] But what was it? Since Newton, natural philosophers had seen everything as made up from particles or corpuscles of undifferentiated matter, differently arranged. John Dalton, however, imagined each simple body or element as composed of identical, irreducible atoms.[35] This enabled him to explain why elements combined in simple ratios by weight. But the language of the science thereby became confused: 'atom' was used to mean corpuscle, or the smallest particle of elements like iron or chlorine, or (by Dalton among others) the smallest particle of a compound like nitrous oxide. The defining property of Dalton's atoms was weight, but many contemporaries thought him too imaginative, and preferred simply to treat the weights as 'equivalents' needed in recipes for chemical reactions.[36] Thus eight equivalents (ounces, grams or tons) of oxygen combine with one of hydrogen to form water. Specially calibrated slide-rules, devised by Wollaston, were sold to facilitate calculations.

Lavoisier admired algebra, but Dalton's imagination was geometrical. He advocated circles shaded or with letters inside to symbolize his atoms in compressed chemical language, and arranged them in various ways to indicate structures. But lacking direct evidence, he was guided by a simplicity rule: where only one compound of two elements was known, it must be 'binary' – so water was what we would write as HO. Contemporaries, including Amadeo Avogadro (1776–1856), André-Marie Ampère (1775–1836), Gay-Lussac and Davy, were struck by the way that two volumes of hydrogen combined with one of oxygen, and saw its composition as H_2O. For them, the atomic weight of oxygen was 16 rather than 8 (hydrogen being 1). When Davy, as President of the Royal Society, presented a medal to Dalton he was very careful to say that it was for the laws of the proportions in which bodies combine rather than for his atomic hypothesis.[37] Dalton's symbols were difficult to remember and to print; they gave an illusion of structures, and to his chagrin the letters devised by Jacob Berzelius (O for oxygen, C for carbon and so on) prevailed when the question was considered at the British Association for the Advancement of Science meeting in Dublin in 1835.[38] The initials were taken from Latin names, so iron is Fe (*Ferrum*), and gold Au (*Aurum*). A second, lower-case letter was used for metals and to prevent muddle where two non-metals, like carbon and chlorine (C and Cl), begin with the same letter. Because of uncertainties about formulae even for things like water, chemical equations (which reflect Lavoisier's insight that constituents and products must, like his tax accounts, 'balance') were at first tentative, and did not become a central feature of chemical language until there was agreement about H_2O in the 1860s. This followed the first major international chemical congress, at Karlsruhe in 1860, where despite agreement that equivalents were more empirical than atomic weights, delegates afterwards read the handouts that Stanislao Cannizzaro (1826–1910) had prepared, arguing for his countryman Avogadro's ideas, and were gradually convinced.

Since then, successive meetings of International Unions for sciences have regularized vocabulary and even spelling, so that 'sulfur' rather than 'sulphur' is now the official usage. The status of the simple bodies or elements remained uncertain: could there really be so many distinct building blocks making up the world? Dalton's atomic theory had come out in the wrong order:[39] first propagated in 1807 in the third edition of a very successful textbook by Thomas Thomson (1773–1852) in Scotland; then supported by analyses done by Thomson and Wollaston, published by the Royal Society;[40] and finally fully set out by Dalton himself in his book in 1808.[41] Thomson's textbook was translated into French, bringing to an end a period of one-way traffic, and his *First Principles* (1825) was an attempt to base chemistry on analytical experiments done by him and his students in the medical school in Glasgow.[42] He concluded that, as suggested in a paper by William Prout (1785–1850) in Thomson's journal, *Annals of Philosophy*, the atomic weights of elements were exact multiples of that of hydrogen.[43] The book's introduction, in a rhetoric of progress and improvement in chemistry, praised Gay-Lussac for his law that the volumes of reacting gases are in simple ratios, but criticized Dalton for pig-headedness and Berzelius (whose 'analyses, in point of accuracy, infinitely surpass all those which had preceded him') for excessive empiricism, giving weights to implausible numbers of decimal places and failing to spot simple ratios. Thomson and his students knew which results were best: the ones that gave the whole numbers. Berzelius's retort in a review was withering: 'science will derive no advantage whatever. Much of the experimental part, even of the fundamental experiments, appears to have been made at the writing desk.'

There were real problems here about precision and accuracy: underlying simple laws in science are hard to spot beneath a host of particular circumstances. To be accused of fabricating results is very serious indeed. Berzelius was felt to have gone too far. The famously irascible mathematician Charles Babbage (1791–1871) wrote on precision in observation, and the frauds of observers,[44] distinguishing hoaxing, forging, trimming and cooking, and adding with sexist rhetoric that 'the character of an observer, as of a woman, if doubted is destroyed'. Astronomers had realized how observations cluster around the true value, and Gauss with his bell-shaped error curve and method of least squares demonstrated how to find it. In other sciences, estimates of accuracy were trickier and slower in coming. The language of science is meant to be 'parliamentary', the way the gentry would converse and disagree without having to challenge each other to a duel, but it could become ruder when passions were aroused. The controversy set off series of further analytical experiments over the next half-century, never quite managing to set all minds at rest. As is often the case, a hundred years on there was seen to be truth on both sides: Berzelius's analyses were more accurate, but the existence of isotopes vindicated the ideas of Prout and Thomson. In quarrels, participants are often right in what they affirm but wrong in what they deny, as Coleridge somewhere remarked.

THE LANGUAGE OF MATHEMATICS

Equations seemed to promise to chemistry the precision and prestige of mathematics, which was becoming the language of natural philosophy.

This was not just a descriptive language: mathematical relations might point towards a real physical relationship or analogy. They might however be misleading. Thus Sadi Carnot (1796–1832) investigated the efficiency of steam engines on the assumption that caloric passed through from source (the boiler) to sink (the condenser), just as water flowed through a water-wheel, and discovered what we call the Second Law of Thermodynamics.[45] Because the caloric theory was passing out of favour, nobody saw the importance of what he had done until after his death – when the result had to be proved to be compatible with a different theory. Carnot's result did not prove the truth of his basic assumptions. This kind of story has always encouraged those who see scientific theories as scaffolding, needed while the edifice is under construction but to be taken down when it is completed. Joseph Fourier (1768–1830) devised a system of equations for heat flow without making any assumptions about the nature of heat; they proved valuable in other contexts also.[46] Comte in his 'positive philosophy' saw positive, mathematical knowledge freed from theology and metaphysics as the only real kind.[47] For some, science wasn't and isn't there to explain, but to yield equations that predict.

In Cambridge, a small group including Whewell, Babbage, George Peacock (1791–1858) and John Herschel had seen in the 1810s how behind-hand and insular the university was, and set about jerking it into the nine-teenth century. In 1816 they translated a standard French textbook of 1802 by Silvestre Lacroix (1765–1843), and in 1817 when Peacock was appointed an examiner he set questions using the French notation.[48] At last a genera-tion of English-speaking mathematicians was able again to keep abreast of the French. As the science we call physics emerged, it was not enough to be experimental; as Herschel wrote (in italics for emphasis):[49]

A sound and sufficient knowledge of mathematics, the great instrument of all exact enquiry, without which no man can ever make such advances in this or any other of the higher departments of science, as can entitle him to form an independent opinion on any subject of discussion within their range.

There was no longer room for the dilettante or indeed for the interested outsider; they must believe what they are told on authority. Speaking the right language can lead to what looks like arrogance or at least impatience; learning it was like learning Latin had been, the key to knowledge. Mary Somerville (1780–1872), the group's protégée, not only had problems about gender, but also had great trouble in getting right the level at which to communicate physics to a wider public ignorant of mathematics.[50]

The Cambridge men were, however, uneasy about pure mathematics, which could divert its adherents from the real world. For Whewell, it was purely a matter of deducing consequences from axioms, and 'deductive reasoners, those who cultivate science of whatever kind, by means of mathematical and logical processes alone, may acquire an exaggerated feeling of the amount and value of their labours'.[51] This offended Charles Babbage (1790–1871), for whom mathematicians were not like logic-choppers, but Whewell's position at Cambridge ensured that applied mathematics remained the main focus, and his version and John Herschel's of inductive philosophy preferred.[52] As new mathematical tools were developed, 'quaternions' and then 'vectors' for dealing with forces having strength and direction, the vocabulary available to physicists increased. But the mainstream view in Cambridge was that a theory ought to be an expression of the real processes of nature producing the phenomena, rather than just a model.[53]

Faraday consulted Whewell when he wanted language free from preconceptions suitable to describe and account for his experiments in electrochemistry and electromagnetism. He knew no Latin, Greek or higher mathematics, and found the language he inherited misleading. Like Leibniz, he thought that the Sun's action upon the Earth across void space remained mysterious and unexplained, and found anyway that analogies with gravity from Newtonian physics did not work well in his new world. In electrical contexts, empty space seemed oddly to be both a conductor and an insulator. Terminals where the wires from the battery dipped into the fluid were called 'poles', implying that like magnets they attracted substances – he was uneasy with that. Polarity and polar opposites were pervasive in German *Naturphilosophie*,[54] where their clash led to new and higher syntheses, and apparently solid substances endured only as we and waterfalls do, through the ceaseless flux of particles in dynamic equilibrium. Faraday asked for neutral terms, and in the end he and Whewell came up with anode and cathode for the terminals, and ion for a charged particle. Faraday also conceived the idea of a 'field' in which atoms were mere points, centres of the forces that we actually encountered.[55] Faraday's lecture audiences may have felt that they got an intuitive grasp of what he meant; his scientific peers were bewildered by his new language and saw him as a genius whose notions led to wonderful and revealing experimental results. This was typical of the chemist rather than the physicist, which indeed was how Faraday was seen by contemporaries. It was only when William Thomson (1824–1907, later Lord Kelvin), and then James Clark Maxwell 1831–7, gave mathematical form to his ideas that physicists appreciated him – when his work was at last translated into their now-extended language.[56]

Chemists were unsure about the real existence of those unobservable entities, atoms, and debated the question twice in London in the 1860s. There, to free the science from hypotheses, Benjamin Brodie (1817–80) proposed using an abstruse operational calculus, based upon the algebra of classes that George Boole (1815–64) devised in which x+y=xy.[57] But in chemistry as in physics, realism was and is most people's position, even though they may be 'mitigated sceptics' like Thomas Henry Huxley. To eschew theory, or use one but not believe it, is awkward. It is not just associated phenomena or mathematical models which interest us, but the comprehensible world, and highly abstract ideas do not have the door-opening heuristic value of more tangible and earthy conceptions. Thus the language of atomism prevailed over that of thermodynamics, even though it was opposed by both Wilhelm Ostwald (1853–1932), pioneer of physical chemistry and Nobel Prize winner, and Marcellin Berthelot (1827–1907), Minister of Education and Permanent Secretary of the Paris Academy. Chemists were not at home with higher mathematics, whether Boolean algebra or differential equations: they practised a craft as well as a science. Atomic theory worked and it could be taught through models, making theory hands-on in the same way that an experiment was.

VISUAL LANGUAGE

Wollaston made elliptical billiard balls, strictly oblate and prolate spheroids, to illustrate ideas about crystal form,[58] and boxes with wooden crystal shapes were made for sale. But molecular models entered mainstream science when August Hofmann (1818–92), brought to England by Prince Albert (1819–61) who realized the educational backwardness of his adopted country, demonstrated 'glyptic formulae' at the Royal Institution in 1865, not long after formulae were agreed. Seeking to make tables of formulae arranged in 'types' more vivid and intelligible, he had begun with little boxes stacked up, labelled N, H or O, combining to form equivalent bigger boxes labelled ammonia and water. He then moved on, a 'happy experimentalist', to croquet balls and rods so that actual structures could be imagined and constructed. He concluded optimistically:[59]

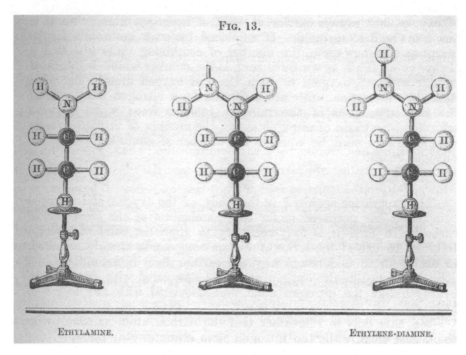

FIG. 13.

ETHYLAMINE. ETHYLENE-DIAMINE.

Figure 2 *Molecular models, A.W. Hofmann:* Proceedings of the Royal Institution
4 (1863): 421

Modern chemistry is not, as it has so long appeared, an ever-growing accumulation of isolated facts . . . A sense of mastery and power succeeds in our minds to the sort of weary despair . . . For now, by the aid of a few general principles, we find ourselves able to unravel the complexities of these formulae, to marshal the compounds which they represent in orderly series; nay, even to multiply their numbers at our will, and in a great measure to forecast their nature ere we have called them into existence . . . It is a movement as of light spreading itself over a waste of obscurity, as of law diffusing order throughout a wilderness of confusion, and there is surely in its contemplation something of the pleasure which attends the spectacle of a beautiful daybreak, something of the grandeur belonging to the conception of a world created out of chaos.

Hofmann's rhetoric was suited to the occasion, and models became and remain a huge asset to the chemist, both the expert and the student. Curiously, because he was bringing diagrams to life for his audience with boxes and then balls, he still thought two-dimensionally and did not exploit his models fully. Models, in wood or in the mind, are like metaphors. The

Dutch chemist Jacobus van't Hoff (1852–1911) took Hofmann's toys seriously and realized their full implications in 1874: instead of being at right angles in a plane, the four carbon wires would point to the corners of a tetrahedron.[60] This allowed him to explain the different optical behaviour and crystalline form of various compounds (notably tartrates, investigated by Louis Pasteur (1822–95)). Chemists now had to think in three dimensions, and the spiralling DNA models are a contemporary example of this.

Hofmann's models had been easy to depict but it was harder to draw van't Hoff's clearly, and just as crystallographers like William Hallowes Miller (1801–80) had fallen back on abstruse geometrical notations,[61] so chemists found themselves illustrating structure in new ways not always easy for the outsider to understand. Chemists had moved steadily away from the realistic pictures of apparatus, laboratories and works which were a feature of Lavoisier's time, when spherical or cylindrical vessels were carefully shaded (as indeed in Jacques Louis David's magnificent portrait of the Lavoisiers, with apparatus at the great man's feet[62]). By the mid-nineteenth century, anyone taking up chemistry knew that retorts were spherical and test-tubes cylindrical; thus they could be shown in uncluttered linear style. Nevertheless, Hofmann's paper has a picture of apparatus showing the grain of the wooden bench, while the molecular models and their stands are also fully detailed, because they were new. Depicting exactly the apparatus used must have seemed akin to recording uncooked data: it would make it as easy as possible for anyone to repeat the experiment.[63] And often the disembodied hands of the experimenter were shown, indicating how things should be manipulated.

LIKENESSES

We also have portraits of chemists, sometimes showing the tools of their trade just as admirals had ships in the background, landowners their mansion and Banks a globe: thus Davy's at the Royal Society shows his safety lamp.[64] These were often engraved for publication and may be powerful (almost speaking) likenesses. Together with the eminent portrait-painter Thomas Lawrence (1769–1830), Wollaston wrote a paper for the Royal Society on why the eyes in portraits follow you around.[65] From the middle of the century, we also have photographs: there are a few of scientists in laboratories, but usually they are studio portraits (often done for *cartes de visite*, to be circulated to friends and colleagues) which give no indication of the sitter's profession or achievements. Photography meant that artists need no longer strive for topographical exactitude but could concentrate upon light, mood or shock value, and those illustrating science could try to depict chemical structures and processes, using symbols rather than straightforward pictures.[66] The range of illustration in physical sciences was extended by the imaginative use of tables and graphs.[67] These things were made possible with the emergence of trained people who could read them, unlike ordinary members of the public, for whom they made science a closed book.

The long nineteenth century was also the great age of illustrated natural history, when drawings and paintings of plants and animals made for scientific purposes were also works of art.[68] For purposes of classification, it was necessary to show flowers dissected, and seeds as well as magnificent blooms, and to show the life cycle of an insect (although caterpillars, pupae and adults would not normally all be about at the same time). Birds and animals are hard to spot in the wild, so again the pictures did not do justice to their camouflage.[69] At first, exotic birds were drawn in a studio in Europe by someone who had never seen them alive, and who had only the skins of dead specimens, rubbed in arsenic then sent home packed in a barrel, to work from, and they look correspondingly stiff.[70] John James Audubon (1785–1851) in the field in America, Edward Lear (1812–88) in zoos, and John Gould (1804–81) and his very talented wife Elizabeth in Australia, were among those who sketched such birds from life. They may then be shown eating, nesting or in movement: the visual language conveys behaviour and not just appearance. John Constable the painter said 'we see nothing truly till we understand it'.[71] In the magnificent Australian flower paintings of Ferdinand Bauer (1760–1826), we see how fruitful was the cooperation of a talented artist with the great botanist Robert Brown (1773–1858), with whom he was sailing on Matthew Flinders's (1774–1814) epic voyage, while at Kew and other botanic gardens a similar tradition of informed illustration developed.[72] Visual language was a necessary complement to text.

In natural history, a picture is worth a thousand words indeed, but it must be printed for circulation. Audubon had to come to Britain for his to be engraved on enormous copper plates (at great expense). The publication of the double-elephant folios, a metre high, of his *Birds of America* was prolonged: the parts came out at intervals, receipts from one paying for the printing of the next, and the costs were great.[73] Thomas Bewick (1753–1828) at the turn of the century had used the technique of engraving on the very hard end-grain of boxwood for his charming studies of British birds.[74] This was a relief process: the wood was cut away so that what would print black stood up from the block, which could then be set

Figure 3 *Blue and yellow macaw: E. Lear*, Illustrations of the Family of Psittacidae or Parrots, *London: Lear, 1832, pl. 8*

in with type, and the illustrated page printed in one operation. For really long runs, a cast called a cliché could be made, like the stereotyping used for books. Curiously, this high technology gave us metaphors for stability. These processes made possible well-illustrated books at reasonable prices. To take a noteworthy and recently analysed example, engraving was used for the superb plates by Henry Carter (1831–97) in the classic textbook *Anatomy* (1858) by Henry Gray (1827–61).[75] Then, in the 1820s, lithography was perfected: when a drawing was made on stone with a waxy crayon, wetted, and inked with oil-based pigment, prints could be made. These soon became as detailed as engravings but at a greatly reduced cost, being much less laborious to do. Engraving is an intaglio process, where what is to be printed is below the surface, which is inked and then wiped, which meant both this technique and that of lithography needed different printing presses from those used for typography, and preferably different paper. With both, the illustrations have normally to be separate from

Figure 4 *Pattern plate, annotated lithograph: W. Swainson,* Zoological Illustrations, *London: Baldwin Cradock, 1820–3*

the text. Engraving is a painstaking art, where the sharp 'burin' is pushed into the copper to cut out the line; in lithography the line is much more flowing and the vocabulary thus greater. Colour printing (long practised in Japan) was tried in various experiments, but did not become the norm for scientific purposes until the twentieth century, so natural history books remained hand-coloured and therefore subject to some variation. With lithographs, the stone is wiped and reused; there is no 'original' and the artist would colour one as a pattern plate for colourists to follow as best they could, perhaps over a period of years as buyers turned up. In popular works, the process of chromolithography brought bright but less-subtle colouring relatively cheaply.

THE LANGUAGE OF DESIGN

Charles Darwin's works were illustrated, sometimes even with photographs; but the *Origin of Species* (1859) has only one illustration, a chart indicating how divergence and extinction happen. When diagrams and dissections replace pictures, they give biological writings look less attractive to the untutored eye, but the visual element is still essential to the scientific language. God also had a walk-on part in the *Origin*, where by contrast the words 'probably' and 'perhaps' are very frequent.[76] And for much of the century, especially in the Anglo-Saxon world, religious language in science was very important. Much scientific popularizing was done in the form of an unrigorous theology of nature, where the language of wonder and enjoyment could be artlessly employed, and God praised for the intricacy of his designs and the promotion of happiness throughout the animal kingdom. Natural theology, using nature to make a case for the benevolence and wisdom of God, was more stringent. The classic author here was Paley, whose *Natural Theology* (1802) was written at the end of a life devoted to promoting Anglican faith.[77] He recommended that his books should be read in the reverse order to their publication, so that after being convinced that the world had a maker, one would be confronted with the evidence for the truth of the Christian revelation. The first chapter of *Natural Theology*, describing finding a watch, is often anthologized: nobody could doubt that the watch had a maker. The rest of the book, in fine clear style, was a cumulative argument to establish beyond reasonable doubt that we and the whole cosmos were designed and made just like the watch, where every component had its purpose. The book, with revisions and updatings by various editors, went on selling right through the century.

In the 1830s it was joined by the Bridgewater Treatises, commissioned under the will of the clergyman Earl of Bridgewater, heir to a canal fortune. Eight authors received the substantial sum of £1,000 each to demonstrate the goodness and wisdom of God using evidence from their particular science. Whewell's volume on astronomy, making his points about mathematics, and William Buckland's (1784–1856) on geology, were particularly noteworthy. Paley's design argument went well with Cuvier's reconstructions of extinct animals and his Aristotelian emphasis upon function. Buckland was fascinated by dinosaurs and conveyed his enthusiasm, arguing that they were not poor designs tried out and abandoned in favour of mammals but were exactly suited to the Earth's state at the time, when God was making it ready for humans.[78] God had not, like a doctor, buried his mistakes. The Bible, literally understood, indicated that Creation had taken place about six thousand years before; the date computed by the Irish archbishop James Ussher (1581–1656) was 4004 BC.[79] Buckland did not read it literally, believing as people always had that it contained allegory, poetry and stories rather than being a kind of textbook. His frontispiece showing the Earth's history folded out, and was more than a metre long, covering millions of years. Believing in the recent appearance of mankind, he suggested that everything before that was contained in the line 'In the beginning', and that there had been a succession of extinctions and creations through deep time. This made the Earth's history into an epic.[80] Buckland's language, polished by his wife Mary (née Morland), conveyed optimism, purpose and delight. Many followed in his path in writing and lecturing, although they might differ in their interpretations of Genesis, with some like Hugh Miller (1802–56) preferring to make the days of creation into epochs.[81]

The shocking but hugely popular *Vestiges*, the anonymous publishing sensation of the 1840s, was denounced as atheistic but was really deistic, with God as the remote First Cause rather than loving father.[82] By the time the *Origin* (1859) made development respectable, there were many who no longer accepted the God of Paley and Buckland, for example, Huxley (to the distress of his fiancée Henrietta, 'Nettie').[83] Denying design made talk of function more difficult but then, for Huxley, Darwin saved the day because questions that seemed to involve purpose became safely scientific when answered in terms of the struggle for existence and survival of the fittest – phrases borrowed from political economists. Similar language, marvelling over adaptations, was thus available to Darwinians and to Paleyans. That should not surprise us because Darwin had relished Paley, and struggled to find a way round his argument for design in a world that seemed too red in tooth and claw to be the work of a benevolent creator.

In 1860 a group of Oxford liberals published *Essays and Reviews*, urging that the Bible be interpreted like other ancient texts, that geology was the key to the Earth's history and that miracles were impossible.[84] William Wilberforce (1759–1833) denounced them, but he was no literalist insisting upon a six-day creation, a universal flood, Balaam's speaking donkey or the Tower of Babel as matters of fact. Well-educated and widely read, he chose to take issue with Darwin mostly over the hypothetical, probabilistic nature of his arguments – as sceptics did with atomic theories. Religious language gradually disappeared from scientific rhetoric, though there are many echoes of the King James Bible in the writings of Huxley and his contemporaries. Huxley was at one with Wilberforce in insisting that all humans were brothers and sisters under the skin, united in opposition to slavery and to the belief that black people belonged to a different species so therefore might be treated differently.[85] Whereas Jesus had deplored classifying baddies as irredeemable publicans and sinners, many in the nineteenth century distinguished the deserving and undeserving poor, and by the end of it scientists had a taxonomy in which idiots, imbeciles, cretins and degenerates were carefully tabulated.[86]

LIBERAL EDUCATION OR TWO CULTURES?

Darwin's education at Cambridge had been unspecialized, suitable for gentlemen who after graduation would join the elite and perhaps specialize in a profession or hobby. The majority of his contemporaries expected like him to be ordained, serving as clergy or teachers.[87] All shared a common culture and language. It was much the same at Harvard, Yale or Princeton. In France ever since the Revolution there had been institutions, the École Polytechnique and the École Normale, for example, giving a more specialized education to the engineers and teachers who would hold the highest professional posts. After Waterloo, German universities followed the model that Wilhelm von Humboldt had devised for Berlin,[88] with teaching and research united, and students encouraged in their *Bildung*, or development, to follow their own inclinations and go deeply into narrower fields. In medicine, in Edinburgh, Göttingen, Paris and London, students studied for their profession, and from the 1850s pressure mounted on Oxford, Cambridge and the Ivy League to run degree courses in new subjects, including history, theology, languages and sciences. There was great anxiety amongst conservatives that liberal education would be replaced by training: a real and persistent fear. But after 1870 and the Prussians' victories over the Austrians and the French, the arguments of Huxley and others that education was the key to national progress and even survival seemed justified.

Inevitably, specialization meant a weakening of common culture, but it would be wrong to see the language(s) of science as value-neutral and soberly descriptive, quite unlike that of history, literature, religion and politics. Scientific rhetoric takes different forms – in the public lecture theatre as Davy was succeeded by William Thomas Brande (1788–1856), Faraday, Huxley, John Tyndall (1820–93), and Bragg; in popular writings; in textbooks; and in the scientific paper, addressed soberly to peers – but it needs to be there if the ideas are to be got across. Scientists disagree, not only, like everyone, about politics and religion but also about their science. They are not dispassionate, like desiccated calculating machines. Thus chemical debates could become vehement: Berzelius denounced as incompetent not only Thomson, but also Liebig and Jean Baptiste Dumas (1800–84) over their work on the substitution of chlorine for hydrogen in organic compounds, which did not produce the radical change in properties that his electrochemical theory predicted. Later, Hermann Kolbe (1818–84) became abusive against Van't Hoff's structure theory. Huxley's denunciation of Richard Owen (1804–92) in their Darwinian debates about apes and men became notorious. The risk of polemics or invective in science were and remain high, and indeed the reputation of their authors in these cases suffered whatever the outcome. Nevertheless, great rows between scientific prima donnas were a feature of the nineteenth century.[89] People pour themselves into their research and strong feelings are inevitable.

In principle, scientists prefer controversies to be coolly and rationally conducted within the confines of academies and societies; thus the great and good deplored and played down Huxley's bad-tempered public confrontation with Wilberforce.[90]

More peaceable scientific rhetoric, in the tradition of Lavoisier denouncing phlogiston (rather than a colleague directly) as a 'veritable Proteus' for its undefined vagueness, can be seen in the French chemist Charles Gerhardt (1815–66) demanding a clear convention when certainty seemed impossible – he favoured the 'one volume' HO system and saw formulae as no more than condensed recipes which could be arranged rather as in a cookery book. His associate Auguste Laurent (1807–53) came out eloquently against him, in favour of a hypothetico-deductive approach in the face of chemical confusion, and wittily wrote of his discovery of a new metal which turned out to be hydrogen, displaced from acids by metals just as a more reactive metal would displace another from its salts.[91] Rutherford, on the other hand, when exasperated by the chemist William Ramsay (1852–1916), who was hogging the country's stock of radium, is supposed to have said that the word 'chemist' meant 'damn fool'. There is a long history of suspicion between, for example, chemists and physicists: scientists are not a monolithic group but have intellectual territory to preserve.

CONCLUSION

Science has been written in many languages, Latin giving way to vernaculars by the later eighteenth century, when for a time French was dominant. Meanwhile, philologists cast new light on the nature and relationships of languages, as language was perceived as the essential characteristic of humans. From the late eighteenth century, physicists increasingly turned to mathematics as the way to express their science, while chemists sought new, systematic, and as far as possible theory-free, terms in a convenient and necessary jargon. Natural historians had always needed illustrations, but gradually their visual language diverged from that of fine art as more anatomical detail of taxonomic importance was required. Physicists and chemists, too, had needed good pictures of apparatus. As equipment became standard these were no longer so necessary, but chemists began to depict hypothetical molecular structures, using different kinds of illustration for different purposes. Elsewhere in science, diagrams became an important means of communication in visual languages that had to be learned. In fact, analogy and metaphor have proved irresistible and necessary to augment these various but austere languages, especially (but not only) in popularizing and teaching, making science palatable. Portraits of scientists lightened nineteenth-century presentations, and so did ways of placing science in context, such as theology of nature or the improvement of society. And at this point, we leave words in order to look at actions, in what came in the nineteenth century to be called technology, and perceived as applied science.

3

APPLIED SCIENCE

On 21 January 1802, Davy gave an inaugural lecture to his course on chemistry, in which he sketched the power of science to transform the world:[1]

> We do not look to distant ages, or amuse ourselves with brilliant, though delusive dreams concerning the infinite improveability of man, the annihilation of labour, disease, and even death. But we reason by analogy from simple facts. We consider only a state of human progression arising out of its present condition. We look for a time that we may reasonably expect, for a bright day of which we already behold the dawn.

This vision of applied science, especially powerful in Britain, the first industrial nation, galvanized his audience and will be the theme of this chapter, which will perforce be focused especially upon Britain.

Knowledge is power, and Bacon's vision two centuries earlier of a world in which natural knowledge would enable us to conquer poverty and disease (and enemies too) had been very attractive to our ancestors all over Europe. Those who could interpret nature could command her (by obeying her). There was no snobbery about useful devices among men of science in the eighteenth century, but it was clear that very few of them owed anything to the latest science. When natural philosophers turned their attention to machines, processes and practices, and farming, they generally hoped to understand what was going on when rules of thumb were applied, to vindicate best practice, and maybe to identify the active agent or component in natural products. Interpretation followed invention, as in the Wild West survey followed settlement. Thus in medicine quinine was isolated from the febrifuge 'Jesuits' bark' imported from Peru, and morphine from opium – and could therefore be given in controlled doses and properly assessed, though that took time.[2] Davy's researches on tanning, for which he was awarded the Royal Society's Copley Medal in 1805, fit this pattern: he found tannin in various trees and shrubs, and confirmed that slow tanning with dilute solutions yielded the most flexible, durable leather, as leading tanners knew already.

The great agricultural innovations of the eighteenth century, and the inventions that set off the early Industrial Revolution, depended upon practical common-sense reasoning and boldness rather than sophisticated theoretical deduction. It is a commonplace that science owed more to steam engines than they owed to science.[3] The term 'science' then meant a logically organized body of knowledge, distinguished from arts, fine (as in the new Royal Academy where Joshua Reynolds (1723–92) presided and in the French salons that Diderot reviewed) or useful. Testable explanation within a connected framework of theory did not really apply to techniques: the proof of a pudding is in the eating. And certainly many innovators and projectors had little formal education, while some of their achievements seemed impossible to their more philosophical contemporaries. It is still sometimes the case, after all, that inventions take scientists, just as much as ordinary folk, by surprise: 'they said it couldn't be done'. The nineteenth century is important in our story because it was then that science caught up with what came to be called technology, and Bacon's hopes thus began to be realized: some inventions really were the application of new discoveries, and were guided by novel or long-accepted theory.[4]

COLOURS

Bleaching had previously been done by spreading out cloth in sunshine but was entirely transformed by the new chemistry, when Berthollet (a director of the Gobelins dye works from 1784) and others found that the poisonous, choking gas they called oxymuriatic acid was a powerful agent. In Lancashire, with its rapidly expanding cotton industry and damp cloudy climate, this was a particularly valuable discovery, though the use of formidable chemicals (notably chlorine) presented problems. Textiles once bleached were usually dyed, and by the early nineteenth century there was a considerable body of knowledge about natural colourings, animal,

Figure 5 *Mr Walker's improved steam engine, 1802: A. Walker*, A System of
Familiar Philosophy, *London: Walker, 1802, pl. 28*

vegetable and mineral. Their effect upon wool, cotton, linen or silk was
different, and to make them adhere mordants were required, alum being
most important. Making alum was an industry where rule of thumb, involving urine and kelp, was beginning to give way to what we might call process
control, requiring knowledge of chemistry. New pigments for painters
or dyers were suggested by chemists, including Davy, who analysed pigments from Roman paintings in Pompeii. In 1835 Thomas Thomson
wrote a series of papers on dyeing and colouring, illustrated with patterned
swatches, some of them spectacular, pasted on to the page.[5] Here was a
professor of chemistry, like Davy, elucidating and promoting industrial
best practice.[6] Thus when synthetic dyes were first discovered two decades
later, and illustrated again with swatches by another professor, Hofmann,[7]
there were many chemists experienced in dyeing so that these could be
rapidly exploited and added to the repertoire of available colours. The
Royal College of Chemistry, Hofmann's laboratory, where William Henry

Perkin (1838–1907), trying to make quinine, happened in 1857 upon the intense purple he called mauve, was founded by Prince Albert to promote applied science, although textile firms were already employing chemists and providing them with laboratories.[8] Mauve caught on as a fashionable colour and the new industry was launched. We think of Victorians as in mourning much of the time, like Victoria for Albert, swathed in black: a real black had been hard to get with natural dyes, but new technology in the form of 'aniline black' made mourning ritual much easier.

The highly competitive dye industry was one of the first where rule of thumb experience was not enough, and trained scientists were needed to work full time, rather than being called in as consultants only when things went wrong. Although Britain continued to dominate the coal and textile industries, the British and the French (magenta was a French discovery, named after recent victory at the Battle of Magenta) gradually dropped out of the coal-tar dye industry, which became increasingly German and Swiss. Amid intense competition, huge companies like Hoechst, AGFA and BASF emerged, employing teams of chemists. In 1869 Heinrich Caro (1834–1911) at BASF patented a commercial synthesis of alizarin just one day ahead of Perkin; they came to an agreement to produce it and this led to the collapse of the cultivation of madder. From 1877, a comprehensive patent law in Germany protected chemical inventions. Synthetic indigo was produced at BASF and Hoechst from 1897, leading to the decline of the trade in natural indigo. Here, the power of modern science was displayed as products of the laboratory entered ordinary life. Determinations of structure and composition went hand in hand with dye production as the pure and applied sides of chemistry were brought together. By 1914, nearly 90 per cent of the world's synthetic dyes came from Germany and 10 per cent from Switzerland, with Britain having around 2 or 3 per cent from its surviving works.[9]

PROTECTING PEOPLE

A trumpeted example of scientific consultancy, in this case unpaid, had been Davy's invention of the safety lamp for coal miners. In his inaugural lecture of 1802, when he was twenty-three, he had electrified his audience by invoking a world transformed by applied science rather than political revolution.[10] His previous research on nitrous oxide and other gases had been of the 'suck it and see' kind, like Priestley's, but when in 1806 he returned to electrochemistry, he was guided by a theory and its consequences in demonstrating that chemical affinity was electrical. Having isolated potassium and other extraordinary metals, and shown that chlorine was an element, he was famous and perceived as Mr Fixit.

In 1815, knighted and married to a wealthy widow, Davy hurried back from the continent and Napoleon's 'hundred days' comeback (that ended at Waterloo) to be met by a request that he do something about explosions in coalmines. Demand for coal meant deeper mines, drained now by steam pumps but menaced by gas. He had samples sent to London and found they were methane, and that methane and air would only explode at high temperatures. In a burst of activity he tried limiting access of air to a lamp, and egress from it, using narrow tubes where it would cool down. Then, after announcing success to the Royal Society, he hit upon the idea of using wire gauze, which dissipates heat so fast that it never gets hot enough to detonate the gases. His classic lamp was therefore a cylinder of gauze surrounding the wick: there was no glass. Banks was delighted, claiming the invention for the Royal Society: a scientific genius in a laboratory in the metropolis had devised a simple cure for a dreadful ill, and when tried in mines it really worked. A lamp is a seductive metaphor for intellectual illumination, but the case was not quite so simple. A rival device, similar to Davy's, was invented by George Stephenson (1781–1848), using trial and error and restricting access of gas (which he thought hydrogen). Looked at closely, both men seem to have gone about the investigation in that way, rather than using profound deduction.[11] Scientists writing up experiments often make their discoveries look more logical than they were. In this case, the propaganda of Banks, Davy himself, the Royal Society and the Royal Institution prevailed: science was seen to work. Davy received presents and accolades, British and foreign, and was the clear candidate for President of the Royal Society when Banks died in 1820.

Miners often carried on using candles and smoking pipes down mines where so far there had been no accidents. And with better ventilation of mines, even where Davy lamps were used a flame might be blown through the gauze to ignite methane or ambient coal dust, another explosive agent. There were still terrible disasters, though far less per ton of coal extracted. The lamp had therefore to be modified into its modern shape, with glass again surrounding the flame, but the air passing in and out through gauze. Then, alongside electric lighting, the lamp survived as a device for detecting 'fire damp' rather than for illuminating the mine. Mining remained a dangerous trade. In our complex and changing real world, practical problems are rarely solved for all time, even by eminent scientists.

Whereas fire damp was one natural hazard, industry, economic change and a growing population were by 1800 generating other noxious substances. Nobody would ever have wanted to live downstream from a tannery, or a shambles where animals were slaughtered, yet the burning of coal was creating smog in cities such as London. The ventilation of buildings was recognized as a problem, and in London the House of Lords and Newgate prison both received scientific attention from Davy, who caught 'gaol fever' (typhus) and was at death's door in 1808 following visits to Newgate. Because the prevailing winds in Western Europe are from the west, the wealthy left the smoky east ends of cities. In London and Newcastle, these were also downstream so the water was dirtier, too. Water closets, which replaced the earth closet or privy emptied by night-soil men at intervals, were an environmental disaster because they discharged straight into the brick sewers, built to carry rain water, that ran into rivers. Where they were tidal, like the Thames, the Tyne and the Elbe, the sewage could slosh to and fro rather than be carried out to sea and out of mind.[12] It was accompanied by the effluents from the various works and wharves along the rivers and increasingly (as in the Rhine and the Mersey) these were chemical by-products or impurities for which there seemed to be no use, discharged into the air or water. Henry Luttrell in 1820 hoped for a clean Thames:[13]

O Chemistry, *attractive* maid,
Descend in pity to our aid!
Come with thy all-pervading gasses,
Thy crucibles, retorts and glasses,
Thy fearful energies and wonders,
Thy dazzling lights and mimic thunders!
Let carbon in thy train be seen,
Dear Azote and fair Oxygene,
And Woolaston and Davy guide
The car that bears thee, at thy side.

In the event, it was the provision of clean water and efficient sewers, one of the great achievements of the engineers of the later nineteenth century, that saved the day: notably in London, by Joseph Bazalgette (1819–91).

By the end of the twentieth century, chemists had a bad name as polluters but in the nineteenth the public turned to them.[14] Berzelius had devised repeatable methods of analysis for mineral compounds, and Thomson began training cohorts of students in analysis in what would later be called a research school. In the little University of Giessen, Liebig devised apparatus (now depicted as the logo of the American Chemical Society) for the analysis of animal and vegetable products, and trained graduate students in its use.[15] These techniques helped to confirm atomic theory but they were also useful in other ways. The excise laboratory was set up in Britain in 1842 to check the strength of beer and wines for tax purposes, but by then chemical analysts were also testing food and drink for adulteration and were being called as expert witnesses in poisoning trials, notably in France.[16] Following campaigns in *The Lancet*, in 1860 the British government passed a Food and Drugs Act, and ten years later local councils began to appoint public analysts. Meanwhile, in 1863 the Alkali Act laid down limits to the pollution, notably hydrochloric acid vapour, from works making soda and sulphuric acid, and Liebig's former pupil Robert Angus Smith (1817–84) was appointed Inspector. The old idea of laissez-faire government was coming to an end. Control of industry in the public interest, an aspect of 'reform', was beginning, though it was often ineffective.[17] In 1894 Thomas Edward Thorpe (1845–1925), trained in Heidelberg and Bonn, was appointed Government Chemist in Britain, and director of the government chemical laboratory in London, setting standards in food, beer, tobacco and other trades.[18]

FEEDING THE WORLD

Davy's recommendations for agriculture, in lectures delivered in hungry years of war (including the blockade that brought the USA into the war in 1812) and published in 1813, were conservative, and vindicated best practice through experiment.[19] He showed that manure was most effective when ploughed in fresh, indicated how to determine the acidity of the soil, and supervised experiments on various meadow grasses done on the Duke of Bedford's estates at Woburn. Charles Daubeny (1795–1867), professor of both chemistry and rural economy at Oxford, took a strong interest in agricultural chemistry, and one of his pupils, John Bennett Lawes (1814–1900), experimenting on his estate at Rothamsted, found that ground-up bones treated with acid were beneficial as a fertilizer. He called the product, patented in 1842, superphosphate, and inaugurated the fertilizer industry, becoming a prominent manufacturer of other chemicals also. In 1843 he employed Joseph Henry Gilbert (1817–1901), who in 1840 had got a PhD under Liebig, on the experimental farm he set up to test fertilizers. Liebig, facing the problems of the 'hungry forties' was radical: he analysed soil and crops so that any missing mineral components could be added as fertilizer.

Hoping to eliminate crop rotation and the practice of leaving ground fallow, he saw no merit in the notion that soil quality in the form of humus, vegetable mould, was important. In contrast to Davy's recommendations, this was theory-driven and in practice it did not always work well.

Experiments at Rothamsted, which became a pioneer agricultural research institute, showed that Liebig's mineral fertilizers were not enough: phosphate was necessary, but Davy had been right in emphasizing ammonia. On his farm in Alsace, Jean-Baptiste Boussingault (1801–87) found that clover enriched the soil with nitrogen, but puzzlingly when he tried growing it in the sterile conditions of his laboratory in Paris, it did not work (bacteria are responsible). These various researches led to the importation of nitrate and of guano from Chile, and eventually in the twentieth century to the synthesis of ammonia from nitrogen and hydrogen under enormous pressure in the Haber-Bosch process, just perfected by 1914 (when its products were used instead to make armaments for the war). Lawes and Gilbert were involved in cooperation and controversy with Liebig over the possibility of using sewage from cities as fertilizer. It was said that in the neighbourhood of London sewage was thus being transformed into strawberries and cream, but schemes to make money from effluent always proved speculative. Like Liebig,[20] Lawes and Gilbert also investigated animal nutrition and physiology, to find out how plants nourished cattle and what promoted muscle and fat.

The food problems of the 1840s were largely solved by the importation of cheap grain from the American prairies as the West was won, while in Britain one of the cataclysmic political events was Robert Peel's repeal in 1846, partly in response to the Irish potato famine, of the corn laws that had protected agriculture from competition. This split his party, the Tories, and was seen as a triumph of the town over the country as Britain became an urban nation, depending upon free trade. Food prices duly fell. By the 1870s, cheap imports led to decades of agricultural depression. Market gardening flourished, but there was little money to invest in improvement for farming. Previously, chemistry was not the only science to be applied: stockbreeders had carefully improved herds long before evolution or genetics were accepted as science, while steam engines for ploughing, pumping, threshing and other activities made processes less backbreaking. It is indeed the inventions of the engineers that we most commonly think of when dealing with applied science, but it was not until well into the nineteenth century that the majority of people even in Britain lived in towns and cities rather than in the country, and peasant customs survived around

Manchester after Waterloo.[21] Agriculture was a crucial part of the economies of all the countries where modern science was being carried on, and it was a sphere to which chemistry and other sciences could be effectively and profitably applied.

FACTORIES

Spinning and weaving took place at home, while other industries were done in small workshops, forges and foundries. Arsenals and naval dockyards, the defence industries of the eighteenth century, were very unusual in their scale. The word 'factory' in the early eighteenth century had meant a warehouse abroad where a factor had looked after goods awaiting shipment. Richard Arkwright (1732–92) set up the first modern factory in 1771, using water power to drive cotton-spinning machinery, and was soon followed by others, especially in Lancashire. In 1790 he was one of the pioneers in the use of a steam engine, making factories independent of rainfall. All this required much capital and factories began to adopt shift-working so that the expensive machinery would not stand idle. The system spread from textiles into other industries, such as making steam engines.[22] Notable here was Matthew Boulton (1728–1809) and James Watt's (1736–1819) factory in Soho, Birmingham,[23] which William Murdock (1754–1839) 'illuminated' outside with coal gas, a by-product of the making of the smokeless fuel coke, to celebrate the Peace of Amiens in 1802. By the following year, the factory was lit inside: the flaring gas jets were cheaper and brighter than candles, too smelly for use in small rooms but good for factories, public rooms and street lighting. Initially, a factory would have its own gas apparatus, but soon municipal gas works were set up, an omnipresent and smelly chemical industry, with networks of pipes laid beneath streets. London streets were gaslit from 1807.

Gas lighting made possible the development of Mechanics' Institutes, literary and philosophical societies, and libraries, because it was now easier to study in the evenings. In effect, it lengthened the day, especially important in Northern Europe. Faraday, in the course of experiments on whale oil and the gas prepared from it as a rival to coal gas, isolated and analysed benzene, recognizing that its ratio of carbon to hydrogen was the same as in acetylene, later used in carbide lamps. After his patron Davy died, he turned from work on glass and steel to the study of electricity and magnetism, and his researches made possible the first electric telegraph, devised by Charles Wheatstone (1802–75), professor of experimental philosophy in London and inventor of the piano accordion.[24] The introduction of railways taking passengers in the 1830s, with trains soon running at 30 miles per hour or more, meant that signallers with flags were no longer adequate to prevent accidents. Telegraph lines laid along the tracks meant that information could be sent to signal boxes and stations, quick as a flash. As timetables became more exact, time signals were sent, too, and in December 1847 'railway time' prevailed over local times based on the sun as the clock became ever more controlling of life in Britain.[25] The railways also opened their systems to others wanting to send urgent messages. Telegraph poles ranged at regular intervals enabled travellers to gauge the speed at which they were travelling.

STEAM AND SPEED

Engineers in Cornwall, where tin was mined but coal expensive, built efficient high-pressure engines, and even put them on wheels,[26] but it was in the north-east that railways had begun as wagon-ways, and there that they were adapted for steam power. The self-taught George Stephenson, and then his son Robert (1803–59), whom he had sent to study in Edinburgh for a year, built the locomotives that proved on the Liverpool and Manchester railway (1830) to be reliable and faster than horses. Railways soon spread throughout the industrial districts of England, and then Robert was appointed engineer of the line to link London to Birmingham.[27] In 1838 this line was opened, with locomotives to Camden Town where a stationary engine waited to haul the carriages up and down to the terminus at Euston. Railway mania followed, and in the 1840s a national network came into being in Britain. Lines were also built in continental Europe and North America, as transport was revolutionized. Railways made possible fast postal services as well as cheap travel, and meant that cows need no longer be kept in cities. In France, there had been lumbering diligences connecting towns, while in Britain, a system of fast mail-coaches and turnpikes was seen by the romantic Thomas de Quincey (1785–1859) as the acme of speed and efficiency.[28] But many people might hardly have gone beyond the borders of their county, whereas now visits by rail to London, Paris and other metropolitan cities for meetings and exhibitions were easy, and holidays by the seaside caught on.

The standard gauge was four feet eight and a half inches, as it had been for wagons since time immemorial, but when a railway was planned for the opulent and genteel parts of England, from London to Bristol, the engineer appointed chose to do things on a grander scale. He was Isambard Kingdom Brunel (1806–59), like Robert Stephenson the son of a distinguished engineer.[29] He believed that a seven-foot gauge would allow bigger loads and higher speeds, and from the splendid giant folios of John Bourne, illustrating the construction and early running of this and the London and Birmingham railways, we can see how a line serving Windsor and Bath might differ from one to the Black Country.[30] Building railways was a huge enterprise, involving armies of navvies digging and building embankments, cuttings, tunnels and bridges. In Western Europe, where cities were close together and land fully occupied, this involved great capital expense but proved very profitable. In North America, railways were built more cheaply and helped to open up the West. Americans, used to steam boats on their great rivers, spoke of boarding the train, and modelled their carriages upon saloons rather than Europeans' stage-coaches bolted together into compartments. The great surveys commissioned by Congress for the projected Pacific Railroad took the 'topographical engineers' into virgin territory. Their reports were published in lavish volumes: railways attracted aesthetic attention on both sides of the Atlantic, making people simultaneously aware of the beauties of nature and of technological advance.[31]

Brunel was an imaginative, indeed visionary, engineer, and he saw his line not as stopping in Bristol but as continuing to New York, again by steam power.[32] Such an idea would not have seemed possible earlier in the century. Steam engines were indeed adapted for ships, but burned coal so fast that ocean-going was out of the question. River boats, and tugs that could haul sailing ships in and out of harbour so they didn't have to wait for a wind, were developed. The Royal Navy had its first tug in 1823,[33] and the new Seaham Harbour in the Durham coalfield was built to be worked by paddle-driven steam tugs, one of which is preserved at Greenwich. Tugs increased the reliability of sailing ships, which reached the peak of their development in the later nineteenth century, with the 'down-easters' from Maine and the clippers from London racing to get home first with tea or wool and get the highest prices for their cargo. Better-fed (tinned meat was available by 1815), armed with good charts (the job *HMS Beagle* was doing) and assisted by the work of Matthew Maury of the US Navy on currents, they followed great circle routes much more direct than traditional passages from one landfall to another, though they were exposed to storms and icebergs in the high latitudes these routes often entailed. Their safety near coasts was improved by lighthouses and Faraday advised about electric lighting,[34] but in his lifetime it was not reliable enough to replace oil.

The first steamships were not expected to use their engines most of the time. Once at sea, they relied on their sails when there was a breeze. In the doldrums or other calms, or on a lee shore, the engine would get them out of trouble, but it was not possible to carry the coal required for a long voyage. The warships with which Commodore Matthew Perry (1794–1858) of the US Navy 'opened up' Japan in 1852–4 were like that.[35] Brunel, realizing that bigger ships would not need much more fuel, saw that true steamships must be much larger and designed the *Great Western* for the Atlantic passenger and mail trade. In due course, she, her larger sister the *Great Britain*, and their successors made the voyages of immigrants to the USA much more tolerable, and could even give them the prospect of returning to or keeping up with their country of origin – something usually impossible for those who had left for America in earlier times. For the more affluent, steamships made transatlantic travel increasingly easy, speedy and comfortable, following definite schedules like trains. Brunel went on to design and find backers for the much larger *Great Eastern*, for voyages to Australia. She was never an economic success, though in the end she played a vital role in the laying of undersea cables. Her engines drove both paddles and screws, a rather belt-and-braces arrangement. Francis Pettit Smith (1808–74) had invented the screw, and with advice from Brunel built *HMS Rattler*, which passed a series of tests culminating in 1846 in a public tug-of-war against a similar paddle steamer, *Alecto*; and thereafter screw propellers gradually became standard in navies. Screw steamers could mount a more formidable broadside and were harder to damage, but paddle steamers continued in civilian use into the 1950s.[36]

SETTING STANDARDS

Nelson's navy was built of wood, but Brunel's great ships were made of iron. Shipwrights, like stonemasons building great cathedrals and builders putting up terrace housing, had considerable scope for adapting designs. Natural materials like wood could not be exactly standardized. By the late eighteenth century, naval architecture, like bridge design, was more advanced in France than elsewhere; plans were becoming more detailed, and those executing them were expected to stick closely to them. The coming of iron, cast and wrought, and then after 1855 cheap steel (following the invention of the 'converter' by Henry Bessemer (1813–98)) for ships and bridges speeded up the process: these were materials of definite and predictable properties. At the École Polytechnique, Gaspard Monge (1746–1818) had taught descriptive geometry, the foundation of modern engineering drawing. Patent specifications demanded exact descriptions and drawings.

The drawing office became an essential part of the engineer's practice, and craftsmen were expected to follow instructions, drawings and blueprints exactly.[37] In effect, they were being deskilled by technical advance – a continuing consequence of science. Expertise learned on the job was no longer enough: some book-learning, formal science, mathematics and engineering were needed for those who were to do the designing and give the orders, although British industrialists long remained sceptical about paper qualifications and expensive apprenticeships remained the norm for engineers. We should also remember, in concentrating upon scientific and technical advances, how much industry remained traditional everywhere, with small workshops and rule of thumb continuing to flourish despite the coming of high technology.

Faraday in his *Chemical Manipulation* (1827) promoted the bright idea that one should make the screw pitches on different pieces of apparatus the same, so that they could be fitted together in different ways. But there were no standards generally followed, so that nuts and bolts from one maker were incompatible with those from another. Also, guns or watches made even in Birmingham, a centre of precision metalworking since Boulton's time, were hand-finished by skilled workers from parts made to loose tolerances. Faced with damaged weapons on the battlefield, artificers had to work hard with files to make parts from one musket fit another. At the Great Exhibition of 1851 at the Crystal Palace in London, an event made possible by the development of railways, Britain did indeed project itself as the workshop of the world, but the French impressed visitors with their industrial design and the Americans (though not ready on time) with their six-shooters. The revolvers of Samuel Colt (1814–62), crucial in winning the West because reloading a musket would give an Apache ample time to dash in for the kill, were assembled from accurately machined interchangeable parts. Mass-production paid off because skilled labour was harder to come by and more expensive in the USA than in Britain, but here was something actually better suited to conditions on the battlefield, mass-produced in a vast armoury.[38]

The eminent Manchester industrialist Joseph Whitworth (1803–87) visited the USA for the New York Industrial Exhibition of 1853 as one of three commissioners officially dispatched by the British government, reporting to parliament on the 'American System of Manufactures'.[39] Their report carried weight and the American system duly spread, particularly in the armaments industry. Such technical transfer was a great feature of the times. Whitworth and his Newcastle contemporary William Armstrong (1810–1900) improved guns with rifling in the barrels, and breech-loading mechanisms. The nineteenth century was a great period for the military-industrial complex. As steel replaced bronze gunmetal, ever-bigger guns gave a completely new role to long-range artillery (until then, targets were always in sight) and transformed warships. The old principle of getting in close and giving them a broadside was no longer possible, with battles fought between ironclads at ever-increasing range. There were wars in the Crimea and Baltic, in the USA, in the German lands as Bismarck brought about their unification and humbled Austria, in France and in South Africa, as well as in other colonies (notably India). In Britain, people trembled at the prospect that Napoleon III might be tempted to invade, and signed up as riflemen in auxiliary forces. At the Royal Institution, lecturers regularly held forth on military matters – notably Frederick Abel (1827–1902), inventor of cordite and an expert in this era of gun-cotton and high explosives. These explosives – like nitro-glycerine and the dynamite that made a fortune for Alfred Nobel (1833–96) – also had important civilian uses. And the Japanese bought a navy from Armstrong, using it to defeat the Russians at the outset of the twentieth century. As an imperial maritime power, Britain had taken the line that she required a navy as large as the next two most powerful put together, and this led to a naval arms race with Germany.[40] For shots from a moving battleship to hit an enemy vessel, several miles away and also moving, calculations were required that could be both transferred to instruments and reduced to rules of thumb for ease of application when fighting.

Whitworth hit upon the idea of drawing up specifications for the pitches of screws and bolts that could be standardized, so that everything could be readily fitted together. This was a huge step, but unfortunately British industry still worked in feet and inches, which meant the eventually agreed standards on the continent of Europe, based on metres, were different. The USA kept feet and inches, but adopted a different (Smithsonian) convention so that British and American pitches were also incompatible. In the interests of national unity and economic efficiency, the French had abolished the various units of length and volume that had prevailed in different cities: Britain was slower to change, so coal, for example, was sold well into the nineteenth century in 'chauldrons' of various sizes in different parts of the country. Steamships and telegraphs were making the world smaller, and standards that had been satisfactory in a district now needed to be national, or even global. This applied to railway gauges as well as screw pitches. Brunel's seven-foot lines covered the west of England, but met with the rest of the network in places such as Birmingham, where transferring

goods and passengers was a great nuisance. After some years of having a
third rail attached to its sleepers, the Great Western line went over in 1892
exclusively to Stephenson's gauge. Russia, however, kept to its larger gauge,
and Australian states had different standards. By the 1870s, the British
Association went at last for the metric system devised by the detested revolutionaries of eighty years before, bringing British science into line with
that elsewhere in Europe, as a genuinely international scientific community
was forming. But just as Britons stuck with pounds, shillings and pence
despite pressure for decimal currency, British industry stuck with the older
units for another hundred years. Publishers of ready-reckoners and makers
of slide-rules did well out of it: we have seen some disasters because of mud-
dles over units, notably on a rocket to Mars, but our ancestors seem to have
been luckier on the whole.

DISASTERS

Naturally however, things often went wrong. Disasters make people look
for causes. The opening of the Liverpool and Manchester railway in 1830
was marred by the death of the local Member of Parliament, William
Huskisson, who wandered across the lines to make up a quarrel with the
Duke of Wellington and was frozen with horror on seeing a train bearing
down on him at fifteen mph. That upset the celebrations completely, but
it took many years of development before the brakes on trains were made
continuous, using vacuum links through the carriages, despite increasing speeds. When Babbage was experimenting with safety devices on the
Great Western railway on a Sunday, expecting to find no other trains on
the line, one passed going the other way at fifty miles an hour, driven by
Brunel. Their engine was to go at forty, and a shaken Babbage asked what
would have happened if either had been on the wrong line. Brunel's answer
was that he would have accelerated in hopes of pushing Babbage's off.[41]
Perhaps drivers sometimes followed this method. Certainly, on occasions
less momentous than official openings, trains collided or became derailed,
boilers blew up and accidents happened at level crossings. It was generally
believed that engine drivers should be exposed to the elements like coach-
men so they would have a good view and keep awake; proper cover for the
footplate, or enclosed cabs, came much later. Whenever a serious accident
happened, there was a public inquiry, and from such occasions much was
(and is) learned about appropriate procedures, strength of materials and
other valuable engineering data.

Major explosions in coalmines occasioned such inquiries, in which prominent men of science might be called upon as investigators. Thus Faraday and the geologist Charles Lyell (1797–1875) were appointed to find the cause of one such disaster:[42] they discovered that safety precautions were lax (Faraday had been invited to sit on a sack of gunpowder) and showed that in mines with powerful ventilation (rather than small boys opening and shutting doors), safety lamps could ignite ambient coal dust. Although the report received little publicity, precautions and lamps were gradually improved. Bridges also were all too liable to collapse, sometimes spectacularly and fatally – as with the Tay Bridge, which fell with a train on it, on a stormy Sunday in 1879.[43] Sabbatarian pressures were weaker than they had been, and trains now ran on the only day off that most working-people got, but the celebrated Scots versifier William McGonagall (1830–1902), describing the event in lugubrious doggerel, duly noted what day it was. The pillars from the collapsed bridge can still be seen from its replacement. The subsequent inquiry showed that the contractors had botched the job, but also raised questions about design, appropriate use of materials and safety. This bridge was not a particularly high-tech project. Although it was long, it traversed a broad estuary in a series of spans rather than a breathtaking sweep. Others were much more impressive: the Stephensons' Britannia Bridge that took the railway across from mainland Wales to the island of Anglesey, from which mail to and from Ireland was shipped; Brunel's Saltash bridge, bringing the railway into Cornwall; and John Fowler (1817–98) and Benjamin Baker's (1840–1907) Forth Bridge.[44] All these were high so that ships could pass beneath, and were successful projects, demonstrating their engineers' mastery. In the USA, John Roebling (1806–69), born and trained in Germany, designed the astonishing Brooklyn Bridge, the first of many great spans around the world supported by steel cables.

EDUCATING ENGINEERS

Just as people who work in pharmacies may be called chemists, so those who repair appliances or drive trains may be called engineers, sometimes provoking indignation from those who have laboured to attain high qualifications in chemistry and engineering. The French had begun the professional training of engineers even before the Revolution, and more systematically after it in the École Polytechnique. But the course there, with its competitive examinations and advanced teaching of science and mathematics, could produce only a small elite and was also militarized under Napoleon. In Britain, practical men learned by apprenticeship. Just as formal courses in chemistry for medical students seemed desirable (and became compulsory after 1815), so a year at Edinburgh University completed Robert Stephenson's training. The Scots were a better-educated nation than the English, with more (and more democratic) universities, and exported not only doctors and chemists – on steamships, the engineer was often 'Mac'. Formal courses in universities were slow in coming. Examinations had traditionally been oral, but from about 1800 written papers began to be set – for example, for mathematics in Cambridge from which one sometimes comes across papers annotated by a tutor coaching pupils.[45] The infant University of Durham, which had recruited a Cambridge mathematician, Temple Chevallier (1794–1873), began a course in engineering in 1840, and the examination papers make fascinating reading. Some are highly practical, to do with pricing and specifying for earthworks, and asking about sparrable nails and how to graith buckets.[46] Others deal with mathematics, natural philosophy and chemistry, and the candidates were also examined in a modern language. At the foundation of the university in 1832, the Bishop of Durham had checked that other bishops were prepared to ordain its graduates, and they were. Unfortunately, industrialists were less impressed by paper qualifications. It may not have helped that the university felt unable to take the novel step of granting degrees to its engineers (though their course looks more formidable than that in arts) and gave only a certificate. So the graduates were taken on as apprentices on the shop floor, having to pay the usual premium on top of what they had paid to go to university. It was in effect a slow track (though no doubt their acquired skills may have helped later), and the course failed.

Meanwhile, Mechanics' Institutes in industrial towns in Britain and America had long been assembling libraries and holding evening meetings and classes. In the later eighteenth century the Manchester and then Newcastle Literary and Philosophical Societies had opened, and so had Glasgow's Andersonian Institution; in 1799 London caught up, with its Royal Institution. This was ritzy, and soon others, such as the London and the Surrey Institutions, aimed at white-collar families, were also founded. Faraday, a bookbinder's apprentice, attended a humbler society; for mechanics, or skilled artisans, sought education just as avidly as their social superiors. 'Improvement' was the order of the day and industrialization was creating new opportunities. In Albany, New York, Joseph Henry (1797–1878) studied and then taught mathematics at the Albany Academy before going on to direct the infant Smithsonian Institution in

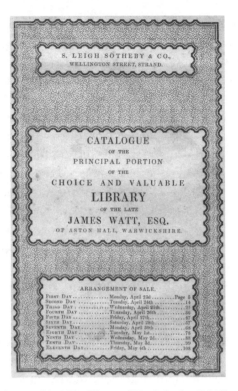

Figure 6 *Sale catalogue of 2,665 of James Watt's books (1849); the extensive library of a well-read engineer, and member of the Lunar Society of Birmingham*

Washington.[47] In London, the physician George Birkbeck (1776–1841), who had begun mechanics' classes at Glasgow's Andersonian, was soon involved in similar work, culminating in the founding of the London Mechanics' Institution, and of London University. Radicals, associated with the new *Mechanics Magazine*, were suspicious of such *de-haut-en-bas* activity, and while some Institutes were run for artisans by members of the middle classes, others were more democratic. Gas lighting and steam heating made evening classes comfortable, and in the quarter-century after 1827 the price of books was about halved with entrepreneurial publishers like Henry Bohn (1796–1884) taking advantage of new technology for paper-making, printing and case-binding. Books therefore ceased to be a luxury, and even plebeian institutions could afford libraries.[48]

In Germany, once again many states after the fall of Napoleon in 1815, men of science responded to Lorenz Oken's (1779–1851) call for annual meetings, each year in a different city. The various authorities were at first nervous about such a pan-German group, but soon perceived that *Naturforscher* members were harmless. The states began to compete in their universities as in their opera houses, and show off to visitors. The meetings attracted attention in Britain, and in 1831 a meeting on similar lines was called in York, and the British Association for the Advancement of Science was formed.[49] Its next meetings were in Oxford and Cambridge, but then it branched out and soon cities were bidding for its meetings, held each summer and attracting considerable publicity, while rousing local enthusiasm for science and industry. Part of the bid might be a promise to build a museum, a library or an institution, several of which in due course formed the nucleus of a technical college and then a university. Like the Royal Institution, the BAAS was dedicated to the view of science as useful knowledge, in the belief that knowledge was power. Each year, the President would boast of new discoveries, and demand more money and influence for science and technology.

'Technology' was a new word in English, and when in 1855 George Wilson (1818–59) was appointed as professor of it in Edinburgh (and simultaneously curator of a museum), he had to spend much time explaining what it meant.[50] In London, the Society of Apothecaries, one of the City Companies that had preserved its original objectives when the others embraced chiefly bankers and lawyers, ran a pharmaceutical laboratory at its Hall in the later eighteenth century and on into the nineteenth. Here, with continuity little interrupted by upsets in chemical theory, the main business was preparing the traditional drugs medical men were prescribing, strictly following inherited recipes and methods. There were some innovations in apparatus, such as ceramic mortars for grinding small samples, and then in 1819 steam power for this job that had previously been the exhausting bane of apprentices. Gradually, new drugs entered the pharmacopoeia, but doctors were suspicious of innovation. And even in this industry, chemical theory played little part, the most important factor being that Apothecaries Hall drugs could be relied upon as genuine – and things were similar in the provinces.[51] While in Britain industrial skills were seen as crafts, Liebig at Giessen was from the 1820s training students in chemical analysis, especially of organic compounds, for PhD degrees, and helping them to find jobs not only in pharmacy but also in other fields (like dyeing). Chemists steadily became ubiquitous, with railway companies, for example, by 1900 having chemistry laboratories, testing rails, fuel and dangerous consignments. Companies in Manchester recruited trained Germans (along with Scots), and set up works laboratories for them. And with the coming of synthetic dyes in the second half of the century, companies in Britain, Alsace, Germany (united after 1870) and Switzerland needed trained chemists to produce new dyes, investigate competitors' products and oversee the transition from pilot-plant to full-scale production.[52] These practices spread into other branches of the chemical industry, prominent figures in Britain being Ludwig Mond (1839–1909) and (Sir) John Brunner MP (1842–1919), 'the chemical Croesus', who had come from Germany as part of the brain drain. America similarly benefited from its immigrants, as well as from citizens who studied in Germany.

German universities produced a mandarin elite, including a formidable array of scientists. But in technology and science a chorus of technicians is needed as well as these soloists, and technical high schools were the German answer, though their status remained uncertain as they began to compete with universities, demanding the right to grant degrees at all levels. In America, Johns Hopkins University from 1874 followed the German model,[53] and land-grant colleges, both public and private (like the Massachusetts Institute of Technology, 1861), had begun teaching applied science. In Britain, following alarm at German, American and French industrial progress, a Royal Commission on scientific instruction was established (chaired by William Cavendish, Duke of Devonshire, and with Norman Lockyer (1836–1920) as its secretary) and provided a report in 1872–5, in two enormous volumes containing all the evidence submitted, with tables and illustrations as well as their conclusions and recommendations.[54] In the wake of this, civic 'redbrick' universities were founded, generally emphasizing useful knowledge,[55] and by the end of the century they were receiving government grants.[56] William Garnett (1850–1932), who had assisted James Clerk Maxwell at the Cavendish Laboratory in bringing experimental physics to Cambridge and had hoped to succeed him, left when Lord Rayleigh (1842–1919) was appointed instead. In 1884 he went to the industrial city of Newcastle to direct the Durham campus there, the 'College of Physical Science'. The college on its new site on the edge of the city was later joined by the long-established medical school, and in the twentieth century the institution became a separate university. Having overseen its move, set it on course and even helped with the building work, Garnett went to London in 1893 to take charge of the system of polytechnics, well established by 1914 as the equivalent of the technical high schools.

These had arisen piecemeal, but an important progenitor was Finsbury Park Technical College, funded by some of London's City Companies, originally craft guilds but by the mid-nineteenth century mostly indecently rich clubs of City gentlemen no longer deserving charitable status unless they did something socially useful.[57] The result was investment in schools and colleges, and the setting up of the City and Guilds Institute which drew up qualifications and set examinations for technicians. At the opening of the twentieth century, several such institutions were fused into Imperial College in South Kensington, a great centre for applied science. It is often said that Britain lost industrial pre-eminence and declined through its neglect of scientific and technical education in favour of the classics education supposed to favour generalist empire-builders. But that is the view of scientists fervently demanding more funds and ignores what actually happened in the forty years following the reports of the Devonshire Commission. Pioneers inevitably get caught up by those copying their successful practice – the French in science and the British in technology – but decline is another matter.[58] Certainly there were snobberies, as indicated curiously enough by a controversy in the mid-nineteenth century over whether the craftsman Watt or the aristocrat Henry Cavendish (1731–1810) had discovered the composition of water decades before.[59]

Doctors and lawyers had a long tradition of elite societies that admitted candidates to the professions, setting standards of competence and ethics. Engineers similarly formed professional, chartered Institutions before the end of the eighteenth century, and during the nineteenth the profession divided into civil, mechanical, electrical and other branches. When universities began to teach engineering, these Institutions were prepared to grant exemption from their own examinations to graduates if and only if they had approved of the course offered. Others entered the profession directly from the shop floor. The Chemical Society of London was, by contrast, a learned society, dedicated to the advance of knowledge and to publishing papers. Some members, working in industry, added FCS after their name and pushed for the Society to demand that jobs such as public analysts should be confined to members of the Society, and not open to doctors and others – arguing, in fact, for professional activity. In the end, this led to a split that endured for a hundred years, with the new (Royal) Institute of Chemistry setting examinations that enabled technicians to demonstrate levels of knowledge up to the equivalent of a degree, and to qualify themselves for suitable posts.[60]

APPARATUS AND EQUIPMENT

Technicians were necessary as apparatus, equipment and machinery became more complicated. Josiah Wedgwood (1730–95) had devised methods for process control in the kilns of his pottery, including a 'pyrometer' indicating temperatures above the melting point of glass that he described to the Royal Society. A standard lump of clay from a particular pit was put into the very hot kiln for a standard time, when it would contract. Then it was removed and dropped into a calibrated groove, and the further it fell the higher was the temperature. Wedgwood put them on sale, and the 'degrees' were used in metallurgy and chemistry in the early nineteenth century, but it was not clear how this scale could be integrated with the Celsius or Fahrenheit degrees of ordinary thermometers. In the event, electrical devices – the platinum resistance thermometer, and the thermocouple – came to the rescue, and high and low temperatures could be properly expressed.[61] Wedgwood's device replaced the experienced foreman estimating temperature by the glow, and the sizzle that spit made, but it was pragmatic and was in its turn superseded by a scientific instrument. Through the century, there was interchange between industry and science in this way. The great instrument makers of the eighteenth century used vernier scales, had devised ways of dividing circles so that angles could be very accurately measured, and added telescopic sights.[62] These made possible the great triangulations linking Britain to France and to Ireland, and similar surveys elsewhere, notably that carried out under George Everest (1790–1866) in India. The quest for greater precision in optical and electrical devices, and in chemical balances, continued, while new manufacturing processes made them more available and cheaper.

This helped to transform laboratory practice. In his *Chemical Manipulation*,[63] Faraday told readers how to cut circles out of bibulous (blotting) paper for filtering and how to bend a glass tube into zig-zags to carry out fractional distillations. He was proud of his manual skills and indeed all chemists were, well into the twentieth century. There was much tacit knowledge of manipulation that had to be learned hands-on and then became intuitive. Chemists had to make apparatus because it was not commercially available and therefore glass-blowing was a skill required. Tricky or complex pieces, like vacuum flasks and fractionating columns, were made by technicians who also mixed 'lutes', sticky but inert cement to join glassware together, until rubber tubing became standard. Glass had to be thick to stand up to mechanical stresses and thin so as not to crack when heated – the laboratory could be frustrating as well as dangerous. Gradually, a few years behind the pioneer researchers, equipment became commercially available. Wedgwood sold apparatus like that he made for Priestley, and later Frederick Accum (1769–1838) published a book on chemical analysis which is also in effect a catalogue, its illustrations showing his stock and his address.[64]

Chests of apparatus and reagents, called portable laboratories, were on sale and were used by surveyors, then by itinerant lecturers, or at home by enthusiasts. Institutions and forward-looking schools that taught science also acquired them. Additionally, they might have some apparatus that we could call conceptual, like the handsome brass and mahogany apparatus for demonstrating the principles of mechanics made for the young George III,[65] wooden geometrical forms for crystallography from Haüy's time, and the ball and wire models introduced by Hofmann. More exacting analyses and syntheses needed a dedicated room fitted out as a laboratory, and the number of these grew steadily through the nineteenth century, especially as chemistry became a major feature of scientific education and industrial practice. This demanded standard equipment, available from entrepreneurial makers' catalogues, including not only stands and clamps to hold apparatus, with glassware and racks to store it in, but also new and more costly purpose-built devices.

Robert Bunsen (1811–99) and his physicist colleague Gustav Kirchhoff (1824–87) at Heidelberg realized in 1860 that when the light emitted from substances in the hot flame of a Bunsen burner is passed through a prism, its spectrum shows bright lines characteristic of the elements present. People before had used flame tests, but colours were often masked by the brilliant yellow of omnipresent sodium. The spectroscope provided a way of doing chemical analysis without using blowpipes, charcoal blocks, test-tubes, smelly reagents like hydrogen sulphide, and powerful acids and alkalis. It marked the beginning of an instrumental revolution, depending upon specialized and expensive equipment, that culminated in the later twentieth century when analysis became a matter of reading dials and print-outs as the technology of 'physical methods' deskilled chemists along with other craftsmen.[66] Spectroscopes were largely responsible for raising the standards of purity with which chemists worked, being so sensitive; the human nose is very good for smelly things, but not for everything. Whereas chemists used to spend much time purifying their reagents, by the end of the century a wide range of substances suitable for immediate use in research (industrial or academic) was available commercially through a branch of the much-diversified and no-longer-crude chemical industry.

The spectroscope was more than a tool for clean chemical analyses, or even a step in the reduction of chemistry to physics. It raised all sorts of questions in physics, too, about atoms and the spectra associated with them. A finely ruled grating could replace the prism, producing a spectrum by diffraction rather than refraction, and then the frequencies of the spectral lines could be calculated. The American Henry Rowland (1848–1901) became particularly well known for his very accurate gratings, and was one of a number of Americans working with the high-precision apparatus that had become essential for making the accurate measurements required in the last decades of the nineteenth century. Human error was avoided in self-registering apparatus such as the barograph that records variations in pressure in weather stations, making a line on a slowly rotating drum. The beam of light in the very sensitive electrical apparatus made by Kelvin for telegraphy could similarly be recorded on a photographic plate, and so could spectra. Optical instruments became better and cheaper. The achromatic microscope (cutting out coloured fringes round the image), vital in industrial analyses, became available in the early nineteenth century, long after the principle had been applied to the larger lenses of telescopes. Opticians used diffraction gratings as test objects. The best microscope was the one which could reveal the lines as separate – and this stimulated makers of gratings to make them even finer. The optical glass of Joseph Fraunhofer (1787–1826) had been the best there was in the 1820s, enabling him to be sure there really were dark lines in the Sun's spectrum, but Faraday strove in vain to make really clear glass of high refractive index using various additives. Half a century on, good glass was a product of industry rather than craft and easy to obtain.

TRANSFORMING SOCIETY

The electrical industry had begun with telegraphs, but electric light was by the 1890s competing with gas. In 1885, Auers von Welsbach (1858–1929) invented the gas mantle, in which a hot (pale-blue) gas flame plays upon a web of asbestos fibres impregnated with thoria. This meant that gas lights could be as bright and white as the electric bulbs introduced by Thomas Edison (1847–1931) and Joseph Swan (1828–1917) in 1879, which were much more convenient than the spluttery carbon arc lamps, or the limelight in which an oxy-hydrogen flame was directed on to chalk or limestone. It is common for improvements in devices to hold off challenges from new inventions: water-wheels were made much more efficient in the eighteenth century as steam engines came in; steam engines for road vehicles looked a better bet than early internal-combustion engines; and balloons in the form of Zeppelins looked as promising as the new-fangled aeroplanes slowly developed from the Wright brothers' hazardous first flights. It is easy with benefit of hindsight to say that in these cases the newer technology was bound to prevail, and an unalloyed 'good thing': in fact, things might have worked out differently in our complex and open-ended world.

Electricity was used for traction too, with trams in cities moving vast numbers of people, suburban trains like those on which Merz worked, and underground metro or subway systems. The first of these, London's Metropolitan line, was in its early years steam-powered and must have been unpleasant, but it linked London termini and in due course brought 'Albertopolis', the cultural zone in South Kensington, within easy reach of the city and West End. The later, deeper 'tube' lines, with their narrow tunnels, would have been impossible without electric locomotives. Electricity was the motif of the Paris Exposition of 1900, with its technological optimism about the new century. The electrical and chemical industries, global transport networks, telegraphs (and now telephones and radio), high farming, cheap publications and penny post had transformed society. The dream Davy had shared with his audience in 1802 was becoming more like reality, as the power of applied science became obvious. Typewriters, sewing machines, gramophones and motor cars were by 1914 common-place, while Victorian advertisements let us into a consumer society, with prosperity spreading downwards.[67] Photography was well established, and films (even documentaries about science) were there to entertain.[68] We know, as Merz and his readers did not, just how significant the explosives, rifled guns and battleships were to be in the twentieth century.

Brunel built grandiose structures and sought speed and power in the locomotives and ships he designed, going for size rather than economy. But not all engineers were like him, and those beloved of investors were often more modest in making things that worked without too much bally-hoo. Bicycles were one of the great and highly influential inventions of the nineteenth century, as the 'ordinary' or penny-farthing that required a steady nerve and a head for heights gave way to the safety bicycle, in models suitable for both sexes, with light wheels having wire spokes in tension. On a bicycle, people could get out of the towns cheaply; indeed, bicycles have been seen as posing a greater threat to religion than any scientific theory because it was more fun to go for a ride on a Sunday than to go to church. The physics of dynamic equilibria behind cycling – working out how it is possible – followed the practice, as so often with technology. Rather similarly, the eminent Scottish physicist Peter Guthrie Tait (1811–1901) devoted considerable time to the physics of knots, and then of golf, in which his son was a champion. He might have hoped to improve his game, but one of the big attractions of science, more plausible in this case, has always been intellectual excitement. It is to this, an aspect just as important as utility and application, to which we shall turn in the next chapter.

CONCLUSION

It is difficult, and often artificial or actively misleading, to separate science and technology. The classic pattern, from cannon to steam engines and on to bicycles, was for invention, based on common-sense trial and error, to precede science. Whereas for science an intellectual structure of generalization and theory is required, in technology the tests are practical and economic. But in the course of the nineteenth century, beginning with French work in chemistry and engineering, the science started to come first: and the most trumpeted case was Davy's invention of the miner's lamp. Here, his vision that applied science could transform society seemed to be realized. In railways, running was made possible by electric telegraphs, the result of scientific research, while dyeing and the making of fertilizers demanded more and more chemical knowledge and theoretical input, and new standards were set for equipment used in discovery and in manufactures. Applied science could be carried on only by well-educated, trained people; so scientific and technical colleges (where chemistry was especially central) augmented and even came to replace apprenticeships. Universities followed the German model of research and teaching carried on together by offering science degrees, taken by many who made careers in industry, where research laboratories blurred the line between scientists, who publish their work, and technologists, who patent it.

4

INTELLECTUAL EXCITEMENT

The chemistry lab can be an exciting place, not only because of the bangs and smells but also because of the insights achieved when thinking with fingers, nose, eyes and tongue as well as more abstractly. In acquiring this knowledge, both tacit and formal, there is great pleasure. We love science not only because it is or promises to be useful, but also because it is beautiful.[1] My physics teacher used to growl, 'You're not here to understand, boy, you're here to learn', and much elementary science is taught dogmatically. Probably, like spelling and grammar, it has to be learned that way: Kuhn's account of the student of 'normal science', drilled in the accepted 'paradigm' and filling in the big picture sketched by some great scientist, is like other caricatures a vision of truth, distorted.[2] If it were the whole truth, science would be both very dull and more subjective than in fact it is. But there is a real danger that dogmatic teaching and specialization can obscure the sheer excitement of interpreting nature, finding out how the world works, picking up diagnostic clues[3] and connecting phenomena that had seemed wholly distinct.

Our ancestors in the first half of the nineteenth century might have learned their science from a catechism, just as children learned Church dogma.[4] But they would have had a less specialized schooling, although this was a time when the expert was beginning to be valued. Davy had time to moon about on the seashore and to tell tall tales to his schoolfellows, and Charles Darwin to collect beetles, when they might have been learning more Latin and Greek. Medicine and mathematics were taught at universities in Britain and elsewhere, and the former included some chemistry, comparative anatomy and botany, the latter some physics, so that both these courses were important in training men of science. Others, like Darwin at Cambridge, took more general and less-demanding courses, allowing them time to go to extra-curricular science lectures, and he thereby became excited by botany and geology. Others again served apprenticeships and educated themselves by using libraries, Mechanics' Institutes and other institutions, as Edward Frankland (1825–99) did.[5] In professional courses there was inevitable drudgery, textbooks heavy with facts to read and examinations to pass. But where science was not a matter of grinding through a syllabus and could be learned unsystematically and enthusiastically, those who opted in were full of curiosity and excitement. Their knowledge might be, as Berzelius said of Davy's, a matter of brilliant fragments rather than a firm edifice:[6] 'he was never made to work diligently in all parts of chemistry as a whole'. But the minimization of drudgery is and ought to be attractive, and often it is those slightly outside the mainstream who make the interesting discoveries, as indeed the opportunistic Davy did. They haven't been taught what is impossible or what is the only right road to follow, and if they are very lucky their road leads somewhere good.

As had happened in France in the second scientific revolution, the sciences everywhere in Europe became more specialized so people began to know more and more about less and less. The process was not confined to science but was particularly obvious there. Specialist societies and journals were started, while general societies began to split into sections during their meetings because different groups were working on different kinds of topics by different means. Fragmentation became a real possibility. Banks was very worried about the weakness of a divided scientific community and opposed the setting up of the Geological Society in 1807,[7] believing that specific sciences should be the concern of groups within, rather than distinct from, the Royal Society over which he presided. In France, the powerful centralizing Academy kept things together, but everywhere the pressures of making careers and pursuing research on an increasingly distant frontier of knowledge were potent – perhaps oppressive – though many welcomed the chance to develop their particular talent. By 1830, John Herschel was like Humboldt becoming unusual in his refusal to specialize, and Mary Somerville's high-level popularizations were being welcomed because they revealed to men of science what their fellow experts in other spheres were up to, and pulled it together.[8] A Baconian ideal of fact collection and suspicion of hypothesis was widely preached, so that the sciences began to seem like masses of information only very loosely connected together. If science were to be real natural philosophy rather than technicality, however fascinating, acting as an exciting and major component of the Western worldview and culture, then it needed great overarching generalizations.

The middle years of the nineteenth century saw two of these grand syntheses – the idea of energy and its conservation, which brought into the new and fundamental science of classical physics a whole range of previously unconnected sciences,[9] and evolution, which united geology, zoology, botany, anatomy, physiology and psychology in a powerful synthesis. Lavoisier had made the conservation of matter explicit, and the Academy in his time had refused to examine any more claims for perpetual motion machines, which were supposed to get work done for nothing. Following upon Erasmus Darwin's poetic sketch of 1803, his contemporary Lamarck published his evolutionary theory in 1809. Lamarck, and Gottfried Trew (1776–1837) in Germany, coined the word 'biology', bringing together botany and zoology in an updating of the old 'animal, vegetable and mineral' kingdoms of natural history. In Germany, around 1800, Romantic thinkers associated with Friedrich Schelling (1775–1854) saw a dynamic world in which indestructible polar forces in shifting equilibria lay behind all phenomena:[10] we humans, and indeed more solid-seeming things, endure like waterfalls through the flux of our component particles. The task of the *Naturphilosoph* was to reveal the clash of opposites that resulted in new syntheses; the attempted objectivity of Newtonian physics with its inert 'inanimate brute matter' was a sham. And, in natural history, the researches of Goethe revealed all plants as developments of a simple Ur-plant, petals being modified leaves, and the skulls of vertebrates as developments of vertebrae.[11] These were exciting, frothy times.

THE FORCES OF NATURE

Liebig saw *Naturphilosophie* as an intellectual Black Death, and Cuvier condemned Lamarck's wild speculations. The exacting demands associated with that second scientific revolution ruled such unverifiable generalizations out of court. In his very influential *Preliminary Discourse* (1830), John Herschel required in Newtonian vein that scientific explanations involve a *vera causa*, a genuine rather than a hypothetical mechanism. 'Energy' was in the early nineteenth century a vague, ordinary word, much less scientific than 'force', and belief in evolution was discreditable and ideological. Far more respectable was the science of phrenology, a mechanistic psychology in which the development of the brain, and hence mental faculties, was revealed in bumps on the head.[12] That promised to be directly useful in education and social policy, and seemed empirical and testable: dynamical and evolutionary notions were not. It is one thing for a big idea to be in the air, part of the zeitgeist; it is another for it to become a part of science. It may well involve an outsider rather than someone in the mainstream of a discipline and at the centre of things. There also seem to be times and places where analytical, reductive science flourishes, like Revolutionary and Imperial France, and others where a synthetic approach, in terms of systems rather than individual entities, is relished, as in Germany in those same years.[13] So, how did energy conservation and evolution, both proposed by numerous people over the years, become central features of established science, with implications fascinating to everyone? It is important, but tricky in practice, that we do not see these things as rather like the South Pole, there to be found, with expeditions pushing nearer and nearer to the goal. It was not apparent to contemporaries that these were inevitable or even useful organizing principles. And we know that such theoretical structures are not timeless truths, but always require modification, as indeed happened around 1900 to both these great syntheses.

Around 1800, chemistry was exciting because its theories were new and uncertain. The tyro might stumble into something important, and lecture audiences pick up the drama of a mind gaining ground upon the dark. When Volta perceived that mere contact of wet metals generated a current, and electrified contemporaries with the news of his battery, chemists entered a field that had been largely dominated by medical men, trained or quack.[14] The problem was to connect this phenomenon with others. Priestley had hoped that optics, electricity and chemistry would be the key to a new world, of explanations deeper than Newtonian mechanics.[15] In a leap of faith, several chemists inferred that the 'Franklinic' electricity of clouds and frictional machines, the 'Voltaic' electricity of batteries and the 'Galvanic' or 'animal' electricity of frogs' legs, torpedo fish and electric eels were essentially the same, rather than just similar. But there were curious, puzzling effects: when an electric current passed through water it decomposed it, but not into exactly the right proportions of oxygen and hydrogen, and moreover acid was generated along with the oxygen, and alkali with the hydrogen. In 1806, Davy, in his prize-winning research, announced that these were generated from dissolved nitrogen and that with apparatus of gold, silver and agate freshly distilled water gave the expected ratio, and no side reactions.[16]

His explanation was that chemical affinity and electricity must be mani-festations of one power, and he confirmed this hunch by isolating potassium and other metals in the following year. This caused enormous excitement. Davy had drawn previously distinct sciences together in a new synthesis, but could not explain in detail what went on in the passage of electricity through the substances, fused or in solution, that it decomposed. Faraday (who eventually proved that all electricities were indeed one) later wrote that a dozen different and incompatible theories based upon this insight had been proposed.[17] Meanwhile, the systematic, industrious and wide-ranging Berzelius saw Davy's work and his own as a key to organizing chemistry into a system, dualism: each compound had for him a positive and negative part, held together by electrical attraction. This helped to transform chemistry into a stable but growing body of knowledge, although no longer so exciting to the outsider looking for new hypotheses and striking experiments – in Whewell's words,[18] as a science becomes established 'all must be left far behind who do not come to it with disciplined, informed and logical minds, the cultivators are far more few, and the shout of applause less tumultuous and less loud'.

Priestley was a pioneer in studying photosynthesis, the process in which plants absorb carbon dioxide and emit oxygen (using Lavoisier's terms), when and only when light shines on them. Chemists knew that both heat and light were effective in stimulating chemical reactions in the laboratory, and in 1802 Thomas Wedgwood (1771–1805), assisted by Davy, had done some pioneering experiments in photography, using the effect of light on silver salts (but unable to fix his images) – researches taken up by John Herschel.[19] The discovery of infra-red and ultra-violet radiation, radiant heat and 'actinic' rays, made Priestley's forecast more plausible: light, elec-tricity and chemical affinity were somehow connected.

ELECTROMAGNETISM

After Davy's death in 1829, Faraday could move out of his shadow and leave behind the humdrum investigations into steel and glass that he had been doing, thus moving into 'blue skies' research that turned out to be of infinitely greater usefulness. A great experimentalist, untrained in and uneasy about mathematics as armchair and a priori science, he saw the labo-ratory as a place where God's creation was revealed to the humble inquirer. He sought coherence and connections. Sorting out what was happening when an electric current passed through a liquid, he found that he needed a new vocabulary. The language of 'poles', positive and negative, went with Romantic notions of polarity and the dialectical clash of opposites, and implied that it was the attraction of the terminals that broke up chemical compounds. Faraday was anxious, as Davy had been about naming chlorine, to avoid uncertain or misleading implications in scientific terms. Having, in correspondence with Whewell, coined 'anode' and 'cathode' for the terminals, and 'ion' for charged particles, he then moved into the new field of electromagnetism.

Electricity and magnetism were both polar phenomena, but beyond that there had been little to connect them. Magnetism was to do with iron and steel, with geography and navigation; electricity with thunderstorms, physiology and chemical affinity. Hans Christian Ørsted (1777–1851), a Dane who had studied in Germany and picked up *Naturphilosophie* there, believed that such polar forces must be linked and tried to find magnetic effects from electric currents. Eventually, to the surprise of everyone, it worked: during a lecture in 1820, when the electricity was switched on and off, the compass needle kicked.[20] This discovery was an alarm bell comparable to Volta's. In France, Ampère saw the experiment repeated before the Academy of Sciences and set to work to account for it, using Newtonian physics to found electrodynamics.[21] Others joined in, in a flurry of excitement, including Faraday who was woundingly accused of trespassing on intellectual territory staked out by Wollaston. Wollaston had died not long before Davy, another reason why Faraday felt free to take up research again. Unlike Newton's, Ampère's equations seemed to yield no unexpected predictions, and anyway Faraday could not follow them. Instead, he resorted to experiments guided by ideas that contemporaries found hard to understand. He discovered that a changing electric current in a coil of insulated wire wrapped around one side of an iron anchor-ring induced currents in a coil wrapped around the other side – making the first transformer. Simultaneously with Joseph Henry in the USA, he made an electromagnet, giving the prospect of enormous magnetic power that he demonstrated in his lectures, flinging up poker, tongs and then coalscuttle to be suspended high above his audience.

He built the first dynamo, a coil of wire spun between the poles of a magnet, in which an electric current was generated from mechanical movement (as static electricity had been from friction on glass globes or plates in electrical machines), and then its opposite, where initially a star-shaped wheel dipping into mercury rotated sparkily when a current was passed through it. This was the ancestor of the electric motor. Electricity and magnetism together would produce motion. These discoveries make Faraday the godfather of the electrical industry, though there were no power stations by the time of his death in 1867. Delivering two unusually informal lectures at the Royal Institution in 1844 and 1846, the first on electric conduction and the nature of matter, he challenged the accepted billiard-ball notion of atoms, and speculated (his word) that they were really point centres of force – an idea going back through Priestley to the eighteenth-century Jesuit Roger Boscovich (1711–87).[22] The world was therefore largely empty space, filled with what in the subsequent (perhaps impromptu) lecture of 1846 on ray vibrations he described as lines of force, which may be illustrated in the patterns iron filings fall into when sprinkled on a sheet of paper with a bar magnet beneath it. Faraday came to see space as a 'field' of force, more interesting than the particles (mysteriously acting at a distance) on which previous men of science had concentrated. The inverse-square central forces of Newtonian mechanics, used in that very year to predict the new planet Neptune as the cause of wobbles in the orbit of the planet William Herschel had identified, Uranus, did not seem applicable to electromagnetism. Faraday may have got his vision across to his lay audiences as far as they needed to understand it, but his peers, puzzled by his lack of mathematics, were uncomprehending.

Meanwhile, in November 1845, Faraday had astonished them in a paper, read before the Royal Society on 20 November, on the magnetization of light. The plane of polarization of a ray passing the poles of a magnet was rotated. Faraday's paper began:[23]

> I have long held an opinion, almost amounting to a conviction, in common I believe with many other lovers of natural knowledge, that the various forms under which the forces of matter are made manifest have one common origin; or, in other words, are so directly related and mutually dependent, that they are convertible, as it were, one into another, and possess equivalents of power in their action. In modern times the proofs of their convertibility have been accumulated to a very considerable extent, and a commencement made of the determination of their equivalent forces.

Faraday's ideas about lines of force were appreciated by Maxwell, who saw him as a great intuitive mathematician, and translated his understanding into equations which formed the basis for subsequent field theory.[24] But while Faraday grasped what he called the correlation of forces, he lacked the means to work out the rates of exchange between different forms of what came to be called energy, set out in his last sentence.

ENERGY

Faraday occupied a position in a hub of the scientific world, but his audiences at the Royal Institution were lay men and women interested in science as a branch of culture rather than as a profession, and the lectures of 1844 and 1846 were therefore popular in tone. It is striking that in the development of thermodynamics many of the other crucial insights were presented in public lectures, by James Joule (1818–89) and Hermann Helmholtz (1821–94), for example, or by books intended to be popular such as Sadi Carnot's, and we shall see something similar happening with evolution. 'Money' is an abstract idea only realized as pounds, euros or dollars – we need to know what each is worth in terms of the others. 'Energy' is a bit like that. There were many of Faraday's contemporaries, at least a dozen, formulating something like conservation of energy: but to both clarify that point and establish rates of exchange between its different forms (heat, light, mechanical work and so on) was very difficult.[25] Two crucial figures were Julius Robert Mayer (1814–78) and James Joule (1818–89). One story is a sad one of neglect; the other of triumph, illustrating the role of luck in science. Both were beautifully told by Tyndall, Faraday's successor and obituarist, and the translator of Helmholtz's crucial lecture.[26]

Mayer was ship's doctor, trained in Tübingen, who bled a patient in Java and noted to his surprise how bright red the venous blood was, like arterial blood – in temperate Europe it was always much darker. He reasoned that in an ambient temperature about the same as blood heat, the body needed no fuel to keep warm and therefore used less of the oxygen in the bloodstream. He then thought of a way to calculate the conversion rate of heat and work. The specific heat of anything is the amount of heat required to raise its temperature by one degree, but for gases there are two figures. The gas may be heated in a sealed vessel at constant volume, or in a U-tube with mercury keeping it at constant pressure, when it will expand. The latter measure is always higher; and Mayer reasoned that this is because the gas is doing work against atmospheric pressure. It was straightforward to calculate the mechanical equivalent of heat, and in 1842 Mayer, by then town physician in Heilbronn, sent Liebig a paper for his journal, and elaborated his conclusions in 1845. Unfortunately, a paper by an obscure physician in a chemical journal, with a derivation of the equivalent that must have seemed convoluted, did not arouse interest and Mayer had a breakdown. But he did not die in utter obscurity: late in life, in 1871, the Royal Society awarded him the Copley Medal.

Joule was also an outsider, a wealthy Manchester brewer and a private pupil of John Dalton, used to the careful control of temperature in his business, and to keeping exact numerical records for tax purposes – like Lavoisier, he had an accountant's eye for balanced books. He was very interested in electric motors and their efficiency as possible alternatives to steam engines, and he tried to quantify his electrical researches, publishing papers in 1841 and 1843 where conservation is implicit. He perceived that if heat were motion of particles, then stirring water should make it warmer, and on his honeymoon in Switzerland he tried measuring the temperature at the top and bottom of a waterfall to test this idea. More formally, he devised in 1845 an experiment in which water was churned by a clockwork with a falling weight. The vessel was carefully insulated, and the thermometer was exceedingly accurate; he duly found a direct relationship between the work done by the weight in falling, and the rise in temperature, the 'mechanical equivalent of heat'. As an outsider, he was fortunate that William Thomson (later Lord Kelvin), of Cambridge and Glasgow, met him at a British Association meeting, subsequently collaborated in research on heat, and brought Joule into the inner circle of men of science: his refined and definitive experiment was published by the Royal Society in 1850. Meanwhile, in May 1847 he had lectured in Manchester on 'Matter, Living Force and Heat', concluding:[27]

> I do assure you that the principles which I have very imperfectly advocated this evening may be applied very extensively in elucidating many of the abstruse as well the simple points of science, and that patient inquiry on these grounds can hardly fail to be amply rewarded.

Joule was duly awarded the Royal Society's Copley Medal in 1870, and has a unit of energy named after him, as we see on yogurt pots (kJ).

Despite Faraday, Mayer and Joule, the scientific world did not take the conservation or correlation of force or energy very seriously until Helmholtz awoke his hearers and readers from their dogmatic slumber.[28] First in Berlin in 1847, and then at Königsberg in 1854, he gave lectures pointing to the importance of what had been done, and drawing out its wide implications. Inclined to become an engineer but hard up, he was persuaded to go to university on a medical scholarship that required him to serve in the army after graduation, but Alexander von Humboldt negotiated his release from his duties so that he could teach physiology. He continued to do that in various universities until 1871, when he was called to the chair of physics in Berlin, the centre of the German empire. He invented the ophthalmoscope, allowing doctors to see into the eye. He became one of the great polymaths, doing important work on perception of colour and of music. Work on animal heat led him, as it had Mayer, towards conservation of energy, and his lecture of 1854 showed his capacity to generalize, and also to quantify. He spoke of:[29]

> a new conquest of very general interest [that] has been recently made by natural philosophy. . . It has reference to a new and universal natural law, which rules the action of natural forces in their mutual relation towards each other, and is as influential on our theoretic views of natural processes as it is important in their technical applications.

It followed that if electrical, magnetic, heat and light energies are equivalent to mechanical work, then they should all be expressed in the same dimensions of mass, length and time: that is, mass times length squared divided by time squared. The great task would be to unify and standardize this new territory, measuring and comparing and expressing units in terms of grams, centimetres and seconds (or older units in parts of the Anglophone world of science and engineering). 'Physics', a term from the Greek word for nature, had in 1800 meant a range of sciences, largely experimental but excluding mechanics; by mid-century, it came to mean the science of energy and its changes, including within its province electricity, magnetism, optics and other sciences that had been independent of or provinces of chemistry – now itself facing 'reduction' to physics, with the loss of fundamental status. Physics was becoming king.[30] With this advance, the physical sciences became more cohesive, the connectedness of things more apparent, and the authority of scientists greater.

THERMODYNAMICS

There remained the questions of whether there were further rules governing the conversion processes, and how to understand exactly what heat meant in gases. Carnot tried to answer the first of these in 1824.[31] The son of Napoleon's Chief of Staff, Sadi Carnot was a graduate of the École Polytechnique, but found himself out of favour and in the provinces under the restored monarchy. He pondered about the efficiency of steam engines (which he saw as the source of economic power that won the war for Britain): the improvements of Watt and then of Cornish engineers meant that more and more work was done for less and less coal. Adhering to Lavoisier's idea that heat was a weightless fluid, caloric, he compared (in a very abstract way) heat flowing through a steam engine from boiler to condenser with water flowing from source to sink over a wheel (something his father had analysed). Just as with waterwheels, the amount of work depends upon differences in heights and quantity of water, so in steam engines it depends upon the difference in temperature and amount of caloric. Neither water nor caloric is used up: only their capacity to do work is reduced. Neither can flow 'uphill' spontaneously. Greater efficiency can be achieved by raising the temperature in the boiler, in high-pressure engines, but there will be a limit. This was the first statement of what is nowadays called the second law of thermodynamics, a mantra for the scientifically literate, announced before the first, and (instructively) derived from a theory now seen as false – it would have been, and was to be, more difficult to show its truth if heat is motion of particles.

Carnot's work was sophisticated, in that he imagined another engine running backwards and working as a refrigerator to restore the state of things in a cyclical process, and as such it appealed to William Thomson. Carnot died young and unsung, but not before recognizing that the caloric theory was unsatisfactory. Thomson was a great admirer also of Fourier, whose analytical theory of heat (1822) comprised equations for its flow without commitment to what heat was. Thomson showed that these equations had a wider usefulness, notably in electrostatics[32] – later important for Maxwell in formalizing Faraday's ideas. Thomson found it hard to accept Joule's work, therefore, but could not avoid it: clearly, energy, while convertible, became 'dissipated' and unavailable. He, the engineer William Rankine (1820–72) and Rudolf Clausius (1822–88) evolved competing but similar explanations in the years around 1850. Clausius subsequently coined the term 'entropy' as a measure of the unavailability of energy. In a closed system, hot bodies will cool and cold ones warm up until all are at the same temperature: then no energy is available to do work. Extended, this idea pointed to steadily increasing entropy in the world, the 'heat death' of the solar system, and the gloomy thought that in a few million years everything on Earth would have come to a chilly end.[33]

In September 1859 the BAAS met in Aberdeen, and Maxwell presented his dynamical theory of gases, in which they are composed of elastic particles moving faster as the temperature rises, colliding with each other and the walls. This was not a new model, but Maxwell could not be ignored because of his respected work on the stability of Saturn's rings, and because he brought statistics to bear. Individual molecules will, like dodgem cars, move at a range of speeds, sometimes after a collision coming to rest; we cannot follow them, but can say that their average speed will be higher in a hot gas. He later inferred that a minute 'finite being' (Thomson nicknamed him a demon) able to see individual molecules, and placed by a frictionless door between two containers of gas at the same temperature, could let fast-moving ones through one way, and slow ones the other. One side would warm up, and the other cool down – contravening the second law, which must therefore have a statistical basis. The handling of chance was one of the great triumphs of nineteenth-century applied mathematics, but never-theless it was a shock to those who believed in strict determinism to find higgledy-piggledy at the root of physics.

EVOLUTION

At that very time, Charles Darwin, a man whose speculations could not be ignored because of his respected work on barnacles, was working on the proofs and index of his *Origin of Species*, and booked to go on 3 October to the hydropathic establishment at Ilkley in Yorkshire to recover from his exertions and the symptoms they had brought on. His principle of natural selection was, he believed, a *vera causa* as required by John Herschel, but it was an informal statistical law. On average, offspring that are better cam-ouflaged, quicker on their pins, sharper sighted and so on will survive and propagate, but amid the chances and changes of life, in individual biogra-phies nothing can be firmly predicted. Just as men both forgotten, like John Herapath (1790–1868), and eminent, like Joule and Clausius, had worked on gases, so evolution had many pioneers before Darwin. He disliked the word, because to him it evoked vague ideas of necessary and predictable progress. The last word in the *Origin* was 'evolved', but otherwise Darwin's term was 'development', at that time less teleological and theory-laden.

He was discouraged to hear that Herschel, to whom he had arranged for a copy of the *Origin* to be sent, had referred to natural selection as 'the law of higgledy-pigglety', evidently to Darwin 'contemptuous' and 'a great blow'.[34] He feared that his theory would fare no better than his predeces-sors' had done. His grandfather's evolutionary poem, *The Temple of Nature* (1803), had been published posthumously. It presented a progressive, optimistic vision, but did not flinch from nature's violence:[35]

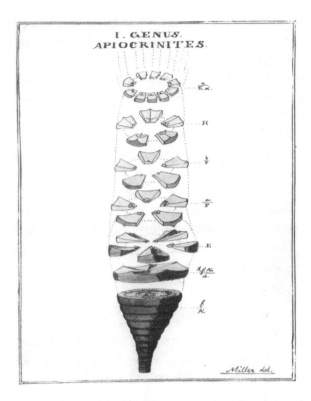

Figure 7 *An intricately designed fossil creature – exploded drawing of a 'stone lily':*
J.S. Miller, Crinoidea, *Bristol: Frost, 1830*

Air, earth, and ocean, to astonish'd day
One scene of blood, one mighty tomb display!
From Hunger's arm the shafts of Death are hurl'd,
And one great Slaughter-house the warring world!

This poem was much less successful than his earlier *Loves of the Plants*, because both the literary and political climates had become very different since 1789, as we saw. The left-leaning Erasmus had been mocked in the *Anti-Jacobin*[36] – wide-ranging optimism was no longer in fashion, and nature (though still seen as a source of morality) was looked at differently. Poetry was no longer a vehicle for conveying scientific information. Though men of science continued to write it, they moved towards light verse.[37]

Paley's contemporary comparison of the world and its creatures to watches was equally optimistic: 'The world was made with a benevolent design. Nor is the design abortive. It is a happy world after all.' [38] For him, each animal and plant was designed by God, and carnivores, by polishing off the ailing and elderly, actually reduced the pain in the world – if it came to the pinch, after all, we might prefer to be eaten up speedily by a crocodile rather than slowly by a cancer. Although his book was a great success, it did not satisfy those who sought a dynamical rather than a static pattern, particularly as people became aware of changes over time. Nor did it put an end to gloom such as that expressed by Hume, where blind nature pours 'forth from her Lap, without Discernment or parental Care, her maim'd and abortive Children', or the political economist Thomas Malthus who saw among humans the desperate 'struggle for existence'.[39] There was a long tradition of seeking moral authority in nature, as in William Wordsworth's early poem:[40]

> One impulse from the vernal wood
> May teach you more of man;
> Of moral evil and of good,
> Than all the sages can.

His next stanza mentions our meddling intellect, and how we murder to dissect; but while he and Coleridge (who recognized that we project our feelings on to nature) mocked Erasmus Darwin and Paley, they welcomed the dynamical science of Davy and the vitalistic physiology of John Hunter.

In France things went differently, but the evolutionary speculations of Lamarck (whose classification of invertebrates was generally admired) were derided just as Erasmus Darwin's were in Britain. Cuvier, his colleague and rival at the Museum of Natural History, was very contemptuous. Cuvier's great fame was based upon his reconstructions of extinct vertebrates from the Paris basin, and depended upon comparative anatomy and a neo-Aristotelian vision of nature doing nothing in vain. All the parts of an animal cohered to form a whole, and could not be the result of chance processes. To account for the astonishing succession of fossilized faunas he found, he invoked catastrophic events that had wiped out populations but he was anxious to avoid promulgating a 'theory' of geology in the eighteenth-century manner, though that was the title of the English translation of Cuvier's introductory essay.[41] Having outlived Lamarck, he took on other evolutionists, Étienne Geoffroy St-Hilaire (1742–1844) and his son Isidore (1805–61), pioneers in teratology (the study of monstrous births). Cuvier's prestige was sufficient to ensure that evolutionary ideas did not flourish in France as they did, however unrespectably, in Britain and Germany.

In Germany the writings of Oken, who saw polarity and analogies

Figure 8 *Buckland lecturing in Oxford, 1822: E.O. Gordon,* Life and Correspondence of William Buckland, *London: Murray, 1894, p. 32*

everywhere, and of Karl Reichenbach (1788–1869) who saw auras of odyle or od in a refinement of animal magnetism, discredited to some extent the romantic science in which unity and evolution had been prominent.[42] But despite the horror that such work aroused among the hard-headed, like Liebig, the close involvement of science and philosophy in Germany meant that great syntheses remained widely welcome there. In Britain, things were different but there were powerful radical undercurrents where phrenology, materialism and evolutionary ideas flourished – notably in medical circles, where there was huge resentment against the power of the elite, gentri-fied ranks of the Royal Colleges in London.[43] Direct contacts with France, where Britons flocked in the years after Waterloo to see a country more modern in its institutions than theirs, and then with Germany as its univer-sities developed, reinforced radical dissatisfaction.

William Buckland (1784–1856) at Oxford was very different. The son of a vicar, he was excited by geology, and became reader and then the first professor of the subject, turning up evidence it seemed for Noah's flood in the caves of North Yorkshire. One had been an antediluvian hyena's den: it was clear they had gnawed bones there, and not that their corpses happened to have been washed there from Africa by the flood waters. For these studies, which included feeding bones to living hyenas and comparing their faeces with those of the fossil hyenas, Buckland received the Copley Medal of the Royal Society from Davy as President.[44] Buckland was also excited by and involved in the discovery of dinosaurs. Among the students who chose to attend his lectures was Charles Lyell (1797–1875) who could not believe in a long string of catastrophes (the flood being the latest) with which God had visited the world. He proposed that geology should explain past changes in terms of forces acting at present (and at the same intensity), and in *Principles of Geology* (1830) set out to do that. Geologists had come to accept that the world was very old, but Lyell's 'actualism' entailed vast tracts, an abyss, of deep time.[45] One of his converts was Charles Darwin, just setting forth on HMS *Beagle* after being taken on a field trip by Buckland's Cambridge opposite number, Adam Sedgwick (1785–1873). Buckland himself went much of the way with Lyell his pupil, and in his *Bridgewater Treatise* (1836) the frontispiece that unfolds to over a metre shows the great age of the Earth and its succession of epochs. He saw progress leading up to the relatively recent appearance of mankind, on a planet made ready for us. But despite abandoning the notion of a world-wide flood, he adhered to catastrophes, recognizing with the Swiss Louis Agassiz (1807–73) evidence of an ice age in Britain.

Lyell, believing in a steady-state rather than progressive world, included a chapter intended to refute Lamarck. This was a dangerous thing to do because it calls attention to opponents, and in 1844 an anonymous evolutionary work, *Vestiges*, was published and caused a sensation.[46] The author, Robert Chambers (1802–71), was an Edinburgh publisher who realized that embracing evolution would harm his company's textbook business, and took great precautions to preserve anonymity, as he sketched a world in which from an initial fire mist the Sun and planets coalesced, as in William Herschel and Laplace's well-known nebular hypothesis.[47] Life then began with an electric spark acting upon a globule. Evolutionary progress under law followed; Australia's creatures were more primitive because it was younger. Mankind had been shown to be subject to statistical laws by the Belgian mathematician Lambert Quetelet (1796–1874), published in English by Chambers, and 'vital statistics', and 'the average man', were becoming facts of life.[48] The 'error curve' that Gauss had devised for astronomical averaging turned into the 'bell curve' of human characteristics. Chambers thus included humans in his inexorable evolutionary process, modelled partly upon the new railway system with its limited number of junctions. All embryos started at the same point as a single cell. On the main line of development, there were definite points at which they could

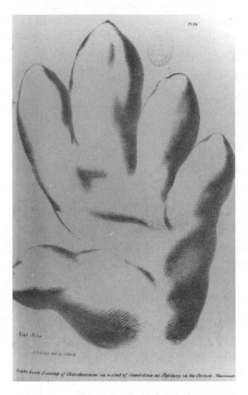

Figure 9 *Dinosaur footprint: W. Buckland,* Geology and Mineralogy Considered with Reference to Natural Theology (Bridgewater Treatise), *London: Pickering, 1837, pl. 26*

diverge, becoming fish, reptiles or different kinds of mammals as the side lines branched. When some ducks' eggs went further down the line, platy-puses emerged, and in due course from their eggs, rats – no doubt in both cases, to their parents' astonishment. Mankind had gone furthest but it was probable that some higher form would emerge from mankind, in this jerky evolutionary process.

Darwin was appalled by the book's conjectural family trees, and by the outrage it aroused among the respectable, but it sold very well, and impressed Alfred, Lord Tennyson, who echoed some of its ideas in his great poem *In Memoriam* (1850), with its famous stanza:[49]

> [Man] Who trusted God was love indeed
>> And love creation's final law –
>> Tho' Nature red in tooth and claw
> With ravine, shriek'd against his creed.

Chambers's world seemed frighteningly godless, though he was a deist rather than an atheist.

Meanwhile Darwin – who had sketched his theory of natural selection in 1842 and in more detail in 1844, in a fair copy to be published should he die – got down to work on barnacles as well as collecting evidence for variation, divergence and development. Barnacles were striking because a naval surgeon, J.V. Thomson (1779–1847), had very recently (1828–34) demonstrated that they hatch looking like little shrimps, and settle down on maturity to a mollusc-like life.[50] Unlike their crustacean cousins, shrimps, lobsters and crabs, they have gone down in the world – and have greatly increased and multiplied. Evolution was not a matter of 'upward' progress, but of divergence into niches, driven by natural selection. Darwin experimented with plants as well as pigeons, and was in constant correspondence with the botanists Joseph Hooker (1817–1911) at Kew and Asa Gray (1810–88) at Harvard, who came by 1859 to accept evolution to explain the distribution of plants, and with Lyell, who sat on the fence. When in 1858 Darwin received a letter from his admirer Wallace, who after a time collecting on the Amazon was now hunting for plants and animals in Malaysia, outlining his new theory of natural selection, he was appalled to find himself forestalled. Hooker and Lyell arranged for Wallace's letter and Darwin's earlier sketch to be read and published at the Linnean Society in London. Darwin was unwell and could not go, and the occasion was a non-event.[51]

Darwin then got down to compressing what he had intended to be a three-volume work on natural selection, with extensive annotations and bibliography, into an abstract which became the *Origin of Species*. As a result, this classic of science has no notes, looks like a popular work such as *Vestiges* and is very accessible. Like Paley's, it is a long argument, and a cumulative one; not a chain of reasoning, like a geometrical proof, but a rope in which weak fibres twisted together will bear a great weight – a legal kind of proof beyond reasonable doubt, rather than a logical proof beyond all doubt. The selection of domestic animals for breeding is the analogy for natural selection (though no Cosmic Breeder is required), and then the distribution, classification, structure and vestigial organs of animals and plants, and the (imperfect) fossil record are all accounted for in a book in which words like 'probably' abound, and where there is even a chapter on 'difficulties'.[52] Among the many critics of *Vestiges*, picking upon its errors of detail, was the young and fiery Thomas Henry Huxley (1825–95), anxious to get away from Paleyan 'Design' but missing the wood for the trees. Darwin was much more careful than Chambers, and speedily won Huxley over, although he was never easy with natural selection. But one consequence of Huxley's sharpening his beak and claws to promote Darwinian evolution was that it polarized those for and against.

Darwin's strategy had been to get people to go some way with him, to accept parts of the story and to toy with evolutionary ideas when facing puzzling facts in natural history. He believed they would find themselves on a slippery slope which sooner or later would bring them into full agreement. An affluent and respectable Cambridge man, he was not keen to upset apple-carts.[53] Huxley was proud to be plebeian, promoting a secular world and loving a fight.[54] He sought to make science a career open to the talents, and beyond that a major cultural force displacing the churches and the Oxford and Cambridge 'establishment'.[55] He was already engaged in a feud with Richard Owen, a patron who had patronized him: they quarrelled publicly at the BAAS in Oxford in 1860 about the brains of humans and gorillas (where Huxley was right), and Owen primed Bishop Samuel Wilberforce (1805–73) for the notorious and confused debate on evolution there – an event which has entered the realm of myth.[56] Huxley pained Darwin by describing evolution as a hypothesis for some years, his standards for 'theory' being rigorous.

RECEPTION OF EVOLUTIONARY IDEAS

They won some converts quickly, but scientists are conservative and as Darwin commented (and Lavoisier had found), it was the young looking for problems to solve who welcomed the theory, rather than their elders who had for years successfully classified organisms – often in museums where they saw perhaps only one example of any given species and were unaware of how variable creatures are. Evidence for the evolution of horses came from expeditions into the still Wild West of the USA, as did many dinosaur fossils, and the early bird – archaeopteryx – from a lithographic stone quarry in Germany, came to delight Huxley. Darwin himself studied the hard case of orchids, showing how their extraordinary flowers were the result of the plant's evolution in step with the particular insect that fertilized it, and upon which a dab of pollen was duly plonked. He published on variation, on climbing plants, on sexual selection and the descent of man, on the expression of emotions in men and animals, and fascinatingly on earthworms. All the time, he sought to reveal our close connections with the animals, and the unity of the human race: sexual selection was not only

Figure 10 *Photographic illustration:* C. Darwin, The Expression of the Emotions in Man and Animals, *London: Murray, 1872*

the key to the peacock's tail, but also to the way humans had diverged.[57] On the Amazon, Henry Walter Bates (1825–92), who after collecting with Wallace had stayed on, noted how butterflies from genera tasty to birds had evolved to look like distasteful species: evolution was happening now.[58]

It would be wrong to see clergy as united in opposition to Darwin: he had some immediate supporters, such as the parson-naturalists Charles Kingsley (1819–75), and William Samuel Symonds (1818–87), in whose lectures and book *Old Bones* (1860) he was compared to Newton.[59] They welcomed the idea of God governing through natural law, and Symonds pointed out that the lion just could not eat straw like the ox – poetic passages in the Bible should not be taken literally. God, he added, had been very kind to the shark. A few weeks after the *Origin* appeared, seven Oxford men published *Essays and Reviews*, introducing to the British public the German research on the text of the Bible, and also reflecting sceptically on miracles and the account of creation in Genesis 1 in the light of science.[60] This created a furore, and an upsurge of conservative reaction; but it is certainly fair to say that more people lost their faith through biblical criticism than through natural science.[61] One of the notorious authors was Frederick Temple (1821–1902), who in due course became Bishop of Exeter and then

London, and ultimately Archbishop of Canterbury. He gave a series of prestigious public lectures in Oxford in 1884, two years after Darwin had died and been buried in Westminster Abbey, concerned with religion and science: one was focused upon evolution. He (and later Anglicans) saw no problem – indeed, for him evolution added force to Paley's argument from design.[62] Roman Catholics had at first no serious problems over Darwinism either, but just as the papal states in Italy were being forcibly absorbed into the newly unified country, the first Vatican Council (1869–70) promulgated the infallibility of the Pope and other resolutions which led to the condemnation of 'modernism' and persecution of liberals. The consequences of this conservative 'Ultramontane' change of direction were profound. Christians in other churches were dismayed. In France the tone of the republic which followed the Prussian siege of Paris and the end of Napoleon III's empire in 1870–1 was anti-clerical, and in the new German empire under Prussian domination Bismarck's *Kulturkampf* was directed against Catholics, seen as a threat to the new political order: the word was coined by the pathologist Rudolf Virchow (1821–1902).[63] In Britain, Roman Catholic evolutionists like Huxley's pupil St George Jackson Mivart (1827–1900) found themselves in a difficult position, condemned by their church as liberal, and by Huxley's 'Church scientific' as 'jesuitical'.[64] Jews, used to a tradition of rabbinical commentary and argument, seem to have been less perturbed than Christians about the status of Genesis.[65]

In 1880s Britain evolutionary accounts of everything were fashionable.[66] Darwin had seen natural selection as too bloody to be the way of working of a loving God, and Huxley came to believe that morality meant opposition to natural selfishness.[67] But despite *fin-de-siècle* gloom about degeneration, Darwin's bleak vision was widely replaced by a more optimistic and teleological one: progress was everywhere. Basing morality in nature still seemed possible, and thus Henry Drummond (1851–97) wrote in his *Ascent of Man* (1894) that:[68]

> Evolution has ushered a new hope into the world. The supreme message of science to this age is that all Nature is on the side of the man who tries to rise. Evolution, development, progress are not only on her programme, these are her programme. For all things are rising, all worlds, all planets, all stars and suns.

In France, positivism was powerful, Lavoisier's and Cuvier's reputations high, and both atomic theory and evolution suspect. But in Germany

Romantic thinkers had made evolutionary thinking easy.[69] There, Ernst Haeckel (1834–1919) played a role akin to that of Huxley, affronting respectable opinion in the stuffy Wilhelmine empire while delighting in the beauty of nature and depicting it in superb illustrations of invertebrates. But whereas Darwin had envisioned a shrubby evolutionary tree with humans at the end of one branch, the best known of the various trees imagined by Haeckel had us at the very top like the angel on the Christmas tree.[70] In the same vein, Drummond's Darwinism was presented in his successful Lowell Lectures delivered in Boston. Indeed, progressive evolution was widely accepted in the USA, among clergy and lay people across denominations on the East coast.[71] But whereas the plebeian Huxley had succeeded in portraying opposition to Darwin in Britain as coming from an effete church and ruling class, and his *Man's Place in Nature* (1863) had underlined our relationship with the apes, in the USA evolution went with the winners in the Civil War, and in the defeated South was resisted along with other attacks on traditional beliefs and institutions by carpet-baggers from the North.[72] The results are still with us, and whatever the truth was in the nineteenth century or now, the picture still prevalent in conservative evangelical circles is that of Huxley and his fellow agnostics: religion and science, especially Darwinian, seem enemies.[73] This 'conflict' thesis of the Scots-American John William Draper (1811–82) remains potent in the hands of Richard Dawkins and other atheists also.[74]

CONCLUSION

These two grand syntheses, energy conservation and evolution, must rank high among the intellectual achievements of their century. They brought together a great number of apparently unrelated facts and observations, and reunited separated sciences under the umbrellas of classical physics and evolutionary biology, opening doors to new research. The first seemed to point towards a world running down, the second to a world moving up. They engaged, as we shall see later, when questions about time and the age of the Sun and Earth came to the fore, and thermodynamic calculations and the fossil record led to different conclusions.[75] Thomson and Huxley talked past each other in what was in part a debate about whether physics was fundamental, laying down the rules by which everyone doing science must play. Soon after Huxley's death, Antoine Henri Becquerel (1852–1908) discovered radioactivity. Confirmed by the elderly Thomson, the discovery was to transform this discussion because it revealed sources of energy undreamed of in the nineteenth century.

It is curious that as the twentieth century opened, both energy conservation and evolution were generating scientific controversy, and indeed both seemed open to doubt. Darwin had left unsolved the problem of the cause and limits of variation, to be worked on by his cousin Francis Galton (1822–1911), who pioneered fingerprints, and August Weismann (1834–1914) who observed chromosomes and assailed the Lamarckian idea (accepted by Darwin in deference to critics) that acquired characters could be inherited. But in 1900 the forty-year-old work of Mendel on peas, and the way characteristics such as being green or yellow, smooth or wrinkled, were passed on, was belatedly and posthumously recognized by scientists in different countries simultaneously, and genetics born. At first, this implied a jerky evolution rather than Darwin's accumulation of tiny variations, and in 1909 when the half-century since *Origin* was celebrated, the aged Hooker and Wallace were among the very few who adhered to a strictly Darwinian view. Meanwhile, Albert Einstein (1879–1955) in 1905 suggested that matter and energy were inter-convertible, and thus not separately conserved – something confirmed with the harnessing of nuclear energy in the twentieth century. Science is provisional, and the greater the circle of light the greater the circumference of darkness. We should not expect that even the most powerful generalizations will survive, unmodified by experience, indefinitely.

Intellectual stimulus thus was and remains an essential part of science, and vital to its cultural significance. But it should not be over-estimated, because science has to be useful too. And just as we get a very distorted view of the science of the seventeenth century if we concentrate upon the astronomy and forget the medicine in which everyone was necessarily interested, so with the nineteenth century the search for health was crucial. Especially in the first half of the century, medical schools were also (as for Mayer, Helmholtz, Darwin and Huxley) important for the training of men of science in many disciplines. It is to the search for healthy lives that we must now turn our attention.

5

HEALTHY LIVES

As the years passed, there were more and more people. Censuses and common perceptions show that populations were rapidly increasing in Europe and North America in the early nineteenth century. Nevertheless, death rates were high, especially among young children, and premature adult deaths from accident, war and disease (including infection in childbirth) ensured that orphanages, stepfamilies and children brought up by an aunt or grandparent were commonplace. In the late eighteenth century, marriages did not on average last significantly longer than they do now: death did us part. Although most marriages were fertile and big families the general rule, a considerable number were childless: Banks, Davy, Whewell, Faraday and Tyndall, for example, had no children. This was a source of sadness, as it had always been, but could (as could the single life) release the energy of the women involved for other purposes, maybe scientific. At the beginning of the nineteenth century, vaccination against smallpox was just coming in, replacing the much more dangerous inoculation from the pustules of a sufferer (which spread the disease), but 'consumption' (which included tuberculosis and other wasting diseases) was rampant, and so were 'fevers' like typhus and typhoid. People lived with the idea that children were loaned by God, who might reclaim them at any time from this sinful world, taking them to heaven, but we should not suppose that such consolation made deaths easy to bear. It is plausible to suggest that the Western world:[1]

has developed a professional and scientific model of medicine that has
probably contributed more to the spread of secular values than all the
insights and discoveries concerning the age of the earth and its place
in the universe, or the story of human evolution. Louis Pasteur and
Alexander Fleming may have altered attitudes to the role of religion far
more than Galileo and Darwin.

We saw how the method of inventing based upon common sense increasingly became 'applied science', technology, during the nineteenth century.
Doctors had always based their claims to know best upon not only their clinical experience but also their scientific knowledge, and in an age of science
much of that was novel. The collection and interpretation of 'vital statistics', beginning in the early eighteenth century for life insurance purposes,
was greatly advanced by Lambert Quetelet (1796–1874) in Belgium, and
governments knew far more about their populations by the end of the nineteenth century than they had at its outset. Statistics pointed to causal links
in disease, and lay behind much of the activity in public health that made a
huge difference to expectation of life, so that by 1900 Western countries at
last had clean water, drains and better food (the result of prosperity and of
legislation). First in Bismarck's Germany and then in Edwardian Britain,
the Welfare State began to take form. Statistics of this practical kind worried those running the BAAS, because they had political implications – as we
know from Benjamin Disraeli's remark about 'Lies, damn lies and statistics'
– and scientific societies were supposed to be neutral in attitude to politics
and religion.[2] Moves towards social sciences and value-judgements made
and make natural scientists uneasy. Moreover, inferences from such statistics seemed rather down-to-earth and commonplace. Whereas Maxwell's
dynamical theory of gases was high-flown and arcane, and Darwin's natural
selection much less commonsensical than it was made to appear by Huxley
in his lectures to working men, public health really was a matter of trained
and organized common sense.[3] The great life-saving medical advances of
the nineteenth century were dependent not on sophisticated theorizing, but
on views of dirt and disease now largely discredited but then perceived as
good Baconian science. Error is better than confusion, all science is provisional, and these ideas both worked in practice and formed an accessible and
comprehensible theory.

MEDICAL MEN AND WOMEN

By the later eighteenth century, there was, especially in England, a free market in medicine, where the physician trained in a university had to compete with midwives, wise women and cunning men, with clergy like John Wesley, and with quacks offering aphrodisiacs and electrical or magnetic treatments, as well as with surgeons and apothecaries who had served an apprenticeship (and were the general practitioners of the day).[4] The Germans had *Wunderaerzte*, and the French *officiers de santé*, who were similarly destined to work mostly with the poor and in the provinces. The status of surgeons gradually rose: in the Royal Navy they had by 1800 ceased to be warrant officers like the boatswain, and counted as commissioned officers, entitled to wear a sword.[5] In London, the [Royal] College of Surgeons was founded in 1800, and the Apothecaries Act of 1815 required apprentices to attend formal courses in chemistry in addition to their training on the job, often by then in a hospital rather than with a local practitioner. Medical schools had fostered academic chemistry for centuries, and in Scotland courses at Edinburgh University had been available to surgeons. Similarly, surgeons and doctors studied together in revolutionary Paris from 1793, and medical courses there set new standards for others to follow. During the eighteenth century there had been a drive to establish hospitals in self-respecting towns, in France and Germany through governments, and in Britain voluntarily.[6] In London, augmenting the provision of the medieval St Bartholomew's and St Thomas', Guy's Hospital was established to deal with desperate diseases,[7] and from 1807 courses of chemistry lectures were given there.[8] This was serious, up-to-date science, with a laboratory. Students could buy a printed 'syllabus' of lecture summaries, periodically revised and reprinted in new editions and interleaved with blank sheets on which to take notes, which sometimes survive and transport us back into the world of Keats.[9] By the early nineteenth century, surgeons in Paris were learning a great deal of anatomy and undertaking some bold operations, as can be seen in the magnificently illustrated *Treatise* (1831–53) of Jean Marc Bourgery (1797–1849) and his lithographer Nicholas Henri Jacob (1782–1871). In that great tome, as in the famous textbook *Anatomy* (1858) by Henry Gray (1827–66) and his illustrator Henry Vandyke Carter (1831–97), Renaissance artistic conventions were finally given up.[10]

The moving spirit behind surgical education in England, raising surgeons' status from artisan to gentleman, was John Hunter, a Scot who after army service opened a museum and medical school in London in 1785. He

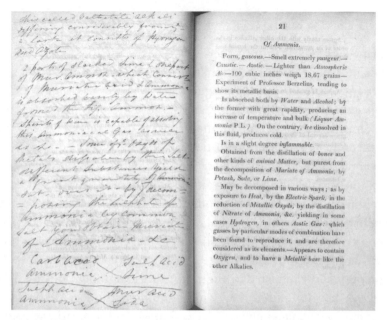

Figure 11 *Notes taken by Robert Pughe, a medical student, beside text:*
W. Babington, A. Marcet and W. Allen, Syllabus of a Course of Chemical
Lectures, *London: Guy's Hospital, 1816, p. 21*

became famous for his anatomical preparations and specimens, kept in the
museum, and for his researches in comparative anatomy and physiology,
and on venereal disease, where he conducted alarming self-experimenta-
tion with pus from a patient.[11] He was convinced that processes in living
bodies went differently from those in dead matter or in test-tubes. In particular, we maintain our identity and form through the continual flux of
our particles. As already noted in Chapter 1, Hunter was particularly struck
by the way that after death the juices in the stomach may begin to digest it.
In life, they break down any lumps of meat that we eat, but not their own
meaty container, which is somehow protected by the vital process. Living
and dead matter demonstrably obeyed different laws.[12] Another version
of this generally held way of thinking, loosely called 'vitalism', came from
his eminent contemporary Johann Friedrich Blumenbach (1752–1840) of
Göttingen, whose collection of human skulls was noteworthy and whose
lectures Coleridge attended on Thomas Beddoes's (1760–1808) recommendation. Blumenbach invoked a *nisus formativus* guiding the development,
or 'evolution', of the embryo in the womb. Medical men were beginning to
invade what had been the feminine realm of childbirth, and John Hunter's
brother William (1718–83) published a magnificently illustrated study of

the gravid uterus. But the results for mothers – especially the poor concentrated together in lying-in hospitals where infections spread fast – were for some time very mixed.

It was not until 1827 that Karl von Baer (1792–1876) at Königsberg first observed the mammalian egg. Improved microscopes then made embryology a major field of interest, with implications for evolution and classification in zoology (rather than for mothers), and also led to new understanding of living cells, associated with Matthias Schleiden (1804–81) and Theodor Schwann (1810–82). Meanwhile, in London, Hunter's disciples were involved in the Animal Chemistry Club, a subset of the Royal Society that included William Babington (1756–1833) of Guy's Hospital, and the future Presidents Davy and Benjamin Brodie (1783–1862).[13] When visiting France became possible again, tourists found different fashions not only in dress but in physiology. Much more materialistic explanations were in vogue – in Britain and the USA, Priestley's unorthodox Christian materialism had been discredited by his left-wing political views. The new star of Parisian medical science amid the fertile chaos of revolution and war was Xavier Bichat (1771–1802), who began lecturing in 1797. He had called the attention of medical men to tissues in his book on membranes, and had written a classic treatise on life and death, noting the flux of our component particles and distinguishing 'animal' life (to do with perception and motion) from merely 'organic' life (to do with assimilation and common to animals and plants). Lamarck coined the word 'biology', life-science, to separate the animal and vegetable kingdoms from the mineral, and, like Hunter, Bichat steered medical study away from anatomy towards physiology, investigating the living organism. For him, the 'vital powers' made the study of life a sphere distinct from physical science. But – especially when he deals with death – his book makes unpleasant reading because of his vivisection experiments, with dogs as the usual victims. Thus, investigating asphyxiation, he reported:[14] 'Open the carotid and precipitate the respiration of the animal by tormenting it (for pain will constantly have this effect) and the jet of blood will be visibly increased.' He also performed autopsies on human corpses, 'a prodigious number of dead bodies', remarking by the way that it was 'very fatiguing and unsatisfactory' to dissect the bodies of those who had been hanged or suffocated because of 'the fluidity and abundance of the blood'. We find a similar observation when the Glaswegian Andrew Ure (1778–1857) was later performing a public dissection of a felon, and the reporter noted that 'a profuse flow of liquid blood' inundated the floor.[15] But cadavers were hard to come by in Britain, so grave robbing, and even murdering to dissect, were a feature of medical schools, especially in Edinburgh where the anatomist Robert Knox (1791–1862) was known to ask no questions and pay on the nail when corpses were delivered.[16] From 1832, with the Anatomy Act, the unclaimed bodies of paupers were available and criminal activity no longer sustained British medical education. On the other hand, the poor dreaded that their bodies or those of their loved ones would be whisked from the 'workhouse' (established from 1834) to the dissecting room, as raw material for anatomical science.[17]

After Bichat's early death in 1802 from fever caught when attending patients in hospital (there were many such medical martyrs), his famous book *Physiological Researches on Life and Death* (1822) was edited by François Magendie (1783–1855), with new notes that often contradicted the text by taking it in a more materialistic and positivistic direction – exciting to the young and rebellious.[18] Magendie carried on the researches that had brought Bichat fame, and acquired in Britain his epithet 'murderous' because of his experimental vivisections elucidating the nervous system. Britons were proud that Charles Bell (1774–1842), who simultaneously demonstrated the separateness of motor and sensory nerve endings, had been able to do so without cruelty to animals.[19] But in 1819 William Lawrence (1783–1867), star pupil of the Hunterian John Abernethy (1764–1831), delivered a course of lectures at the citadel of vitalism, the Royal College of Surgeons (where Hunter's collection was preserved, and being augmented[20]), dedicating the published version to the elderly Blumenbach, but introducing new French ideas – suspect at best to a generation that had lived through the Revolution and long war.

Lawrence remarked on the need to protect science and society from 'odium theologicum'; then, after introducing the word 'biology' into English, he added:[21]

> Life . . . is merely the active state of the animal structure. It includes the notions of sensation, motion, and those ordinary attributes of living beings which are obvious to common observation. It denotes what is apparent to our senses; and cannot be applied to the offspring of metaphysical subtlety, or immaterial abstractions, without a complete departure from its original acceptation – without obscuring and confusing what is otherwise clear and intelligible.

He was denounced by Abernethy, and there was an enormous row – such things were a feature of nineteenth-century science, but this one became so nasty that Lawrence (who was taken to deny the existence of a soul) agreed to withdraw his book. It was extremely popular with medical students, however, and the enterprising publisher William Benbow brought out unauthorized 'pirated' editions.[22] Lawrence challenged his right to do so; but Lord Eldon's judgment in court was that the work was blasphemous, and there could be no copyright in blasphemy. The law was thus proved an ass, helping the wicked work to circulate effectively, and in due time Queen Victoria, appreciating competence above religious orthodoxy, appointed Lawrence one of her doctors. Radical and disrespectful thinking became a feature of London medical circles where Oxbridge and Royal College privilege was resented. Thomas Wakley's *Lancet* was their voice.[23]

MAINTAINING HEALTH

Despite advances in physiology such as Bell's and Magendie's, and in surgery, there was often very little that the doctor could actually do, except advise on lifestyle and hope that nature would, if not prevented, work a cure.[24] Neither the Hippocratic 'humours' nor wide-ranging Enlightenment mechanical or chemical notions seemed helpful. The physician John Ayrton Paris (1785–1856), first biographer of Davy and prominent figure in the Royal Institution, wrote on diet (an ever-popular if not often scientific topic), puzzling over ageing:[25] 'If every part of the machine be thus capable of immediate and constant repair, why should it ever wear out?' He was in a tradition going back to Hippocrates, in which a regimen to avoid disease was crucial, prevention being better than cure (a maxim that was also to guide those involved in public health). Paris wrote in the spirit of Hunter, and at a time when (like Parson Woodforde[26]) the well-to-do ate enormous meals. Before and after him, as today, publications to do with healthy diet were legion, and the advice they gave curiously varied.[27] Lettuces were for Paris 'narcotic', or soporific, as they famously were for Beatrix Potter's (1866–1943) Flopsy Bunnies,[28] and he recommended that cabbage should be boiled twice, changing the water in between. But mostly his advice was prudent, explaining that small changes in habits such as meal times could be a big help in tackling indigestion.

John Kidd (1775–1851), regius professor of medicine at Oxford, writing in 1833 a Bridgewater Treatise on the goodness and wisdom of God, concluded that of God's gifts in the realm of plants, there were really only three things that were much help in medicine: quinine, opium and alcohol. Quinine did wonders for sufferers of fever and ague, while the others were of more general application. The Guy's Hospital lecturers referred to 'the intoxicating effects' which nitrous oxide (laughing gas) produced.[29] When people wonder why it was not used in surgery for forty years after Davy demonstrated its properties (in a medical institution) in 1799, an answer must be that opium in reasonably pure form had become available, and was much easier to administer:[30]

> How often has not *opium* lulled the most excruciating agonies of pain? How often has it not restored the balm of sleep to the almost exhausted body? . . . And let me not omit the restorative virtue of that gift of Heaven [wine] . . . when rightly used, not only to revive the drooping energies, but to rekindle the almost expiring spark of life.

Indeed, opium was prescribed by Beddoes, Davy's boss, to Coleridge. In the early nineteenth century, it was the aspirin or paracetamol of the day, often dissolved in alcohol to yield laudanum.[31] With few resources, the doctor's bedside manner was crucial: Charles Darwin's father Robert (1766–1848) was reported to have a wonderful way of raising the morale of patients. Prognosis even when gloomy was also important: interpreting symptoms and foreseeing a fatal outcome, the good doctor gave time for the patient to settle affairs and collect the family at the deathbed.

Scurvy, no longer a serious threat in Kidd's time, had been a disease primarily of mariners, and Richard Watson (1737–1816), professor of chemistry at Cambridge and subsequently a bishop, was aware of an industrial disease, the mercury poisoning that afflicted those who silvered mirrors.[32] With the Industrial Revolution came other new illnesses, the result not only of poor housing in overcrowded and rapidly expanding towns and cities but also of working conditions. In 1831 Charles Turner Thackrah (1795–1833), a founder of the medical school in the industrial and commercial city of Leeds, published his study of the effects of trades on health, with a second expanded edition a year later shortly before his early death: like Keats, his fellow student at Guy's, he was consumptive. He estimated from records of births and deaths that at least 50,000 people died annually in Britain from the effects of manufacturing, and the 'intemperance' connected with it on which the Edinburgh medical graduate Thomas Trotter (1760–1832) had published a book in 1804.[33] As well as operatives, Thackrah looked at the health of those in traditional trades, at *bons vivants*, and at professional men. Faced with stuffy factories, dusty and dirty jobs, the pressures of keeping up with machinery for long hours, extremes of heat and cold, eye-strain among milliners and dressmakers, and general bustle, he made sensible recommendations for better light and ventilation, proper dinner hours, diet and lifestyle. His work became a model for those pressing for regulation of workplaces.[34]

He was not the only doctor promoting public health measures. James Kay, later Kay-Shuttleworth (1804–77), published in 1832 his little book on the condition of the working classes in booming Manchester, which was dismal indeed.[35] Irish immigration had forced down wages, working days were long, housing was overcrowded, diets were poor, pollution was everywhere, and the result was reckless turbulence, drunkenness, immorality, disease, suffering and premature death. Properly planned new streets and housing, public works in Ireland to provide employment there, sensibly run trade unions, and better social relations between masters and men were required to combat these evils. Friedrich Engels (1820–95) took up this and other reports in his *Condition of the Working Classes in England in 1844*, and Elizabeth Gaskell (1810–65) in *North and South* (1855) and other novels brought this state of affairs to a general readership, as part of a new genre of 'two nations' writing. Their object was not only to increase empathy: the problems were practical and urgent. As events would show, while the well-off enjoyed much better health, disease in the poor quarters of a town would spread. Sickness was no respecter of persons.

CHOLERA

The year 1832, the time of the great Reform Bill, was momentous also because by the autumn of 1831, after spreading across continental Europe, and despite quarantine restrictions, the 'Asiatic cholera' had reached Sunderland, from where it spread rapidly through Britain. Endemic diseases were one thing, but a plague was another. Within hours of first showing symptoms, including extreme diarrhoea, a high proportion of sufferers had turned a horrible blue colour and died an agonizing death. There had been something called cholera before, but this was 'cholera morbus'. Doctors were powerless: their usual nostrum, bleeding, was worse than doing nothing because the cholera victim dies of dehydration, loss of body fluid. There were two ways of reacting to the epidemic. One was to see it as sent from God, affronted by the wicked behaviour of the nation as he had been in biblical times; the other to see it as the result of filth. We should not be surprised at that. Illness has two aspects: the subjective (what is happening to me, and why, and how can I cope with it?) and the objective (what is the diagnosis, the cause and the prognosis?); and both have their place. It need not be a simple matter of either/or, though some chose and choose to see it that way. In the event, a two-pronged approach to deal with both possibilities or aspects was adopted. Days of national humiliation and prayer were called, with services in churches and chapels,[36] and local boards of health were set up.

Our ancestors had always lived in smelly circumstances, but overcrowded towns stank worse than small villages had, and the burning of coal led to smog. The old idea that disease was due to mysterious 'influenza' had given way to 'malaria' (bad air), or the generally preferred term 'miasma', being blamed – the result of damp and sickening effluvia from stagnant dirty water, middens and dung hills. Rotting matter could be expected to induce something like putrefaction in living human bodies. The clue was to follow one's nose: bad smells, dirt and disease were inextricably linked. Boards or committees set up in panicky towns, composed of local worthies like the mayor and aldermen, doctors and surgeons, clergy and lawyers, set to work seeking out filth and getting rid of it. Dung hills were removed, culverts unblocked, hovels whitewashed, earth closets emptied and streets cleaned. The epidemic ran its course, and ceased.

Everyone relaxed, and resumed the untidy habits that led again to the build-up of filth and rubbish. And after a few years, the cholera returned: in 1848–9, and 1853–4, in a cycle that went on and on. It became perhaps harder to see each time why God should go on and on punishing the citizens of Britain and other European countries (which suffered equally) for their sins, and another and more lasting clean-up rather than more prayer seemed to increasing numbers the right answer. People began to suggest that intercessory prayer did not work.[37] William Farr (1807–83), armed with data from the newly established Registrar General's office, produced statistical analyses based upon the available data, and became an important propagandist for public health measures generally. But the disease remained terrifying and unpredictable in its outbreaks, leaving governments worried about contagion and quarantine.[38] The answer came in the famous work of John Snow (1813–58), apprenticed in Newcastle where he met cholera on its first appearance, and then more formally trained in London.[39] Noting that cholera acted very rapidly on the digestive system, he believed that it must come from something eaten or drunk. His practice when he qualified was in London's Soho, where water was supplied through pumps in the streets and squares. In the area around the pump in Broad Street, cholera was widespread; around neighbouring pumps, rare. He perceived that the families (some working class, some middle class) who got cholera all drew their water from the first pump, a social centre as well as a necessary facility, and had its handle removed.[40]

Convinced that water was responsible, he used the statistics Farr and others gathered to establish it beyond reasonable doubt. In 1849 the Southwark and Vauxhall and the Lambeth water companies supplying south London all drew their water from the Thames in central London, and cholera had raged there. By 1853, the Lambeth company was drawing it from Thames Ditton, upstream, and in that year's epidemic customers using Lambeth water, by then available in houses in a large zone where mains from both companies were laid, proved significantly less likely to get cholera than those using Southwark and Vauxhall water from further down the river. Naked-eye inspection, and a sniff, had indicated that Broad Street water was contaminated with sewage. The same was true of Southwark and Vauxhall, and confirmed by microscopic examination magnifying the flocculent particles and animalcules. Having also made a reputation for himself in the new field of anaesthetics, giving chloroform to Queen Victoria in childbirth in 1853, Snow died in 1858. By then, the connection between foul water and cholera was gaining general acceptance. Although Snow insisted, against accepted opinion, that dirt and bad smells were not themselves the cause of disease, water in great cities was manifestly so impure that nobody could ignore it, and the connection with disease was easy to perceive.[41]

PUBLIC HEALTH

Meanwhile, in 1842, Edwin Chadwick (1801–90), a disciple of Jeremy Bentham (1748–1832), published on behalf of the Poor Law Commissioners an official report on the sanitary condition of the 'labouring population' of Britain – the first industrial nation, with about half its population already living in towns and cities.[42] Picking up the evidence of doctors such as Kay, Poor Law officials connected with the new workhouses and other informants, Chadwick quoted and summarized a vast amount of information about living and working conditions. He deplored the waste of life, and reminded parliament, to whom the report was addressed, that if the lives of working men were healthier, then trade union leaders would be older and wiser, less liable to preach revolution (a real possibility in the 'hungry forties') and enjoying lives that were happy rather than desperate. Public administration was weak, and there were urgent problems facing government in these years of poor harvests, agitation for the abolition of the corn laws that protected agriculture by maintaining high prices, and the potato blight bringing famine to Ireland. Nevertheless, as a result of Chadwick's report, in 1848 the Public Health Act was passed, marking the (rather small) beginnings of government involvement in health. Here, Britain was backward. The more paternalistic regimes in continental Europe (who had their Chadwicks too[43]) had started much earlier, with town physicians (like Mayer in Heilbronn) and legislation giving an example to follow. But these first laws in heavily urbanized Britain were merely permissive, allowing towns and cities to raise local taxes and make provisions to improve cleanliness and health.

Chadwick was seen as hectoring, and was unpopular; and his hopes of being in charge of national public health with substantial powers were never realized. Liverpool, and then the City of London (the square mile around St Paul's Cathedral, not the whole metropolis) took advantage of the Act to appoint medical officers. London's was John Simon (1816–1904), whose tactfully persuasive reports went down better than Chadwick's with those in power; and when in 1854 there was new legislation, Simon became in effect the government's chief medical officer, and worked to make public health provision compulsory.[44] Meanwhile, in 1847, as a result of further evidence on the health of towns, a new Metropolitan Commission of Sewers had been set up to replace seven separate previous bodies in London, in other respects still a congeries of distinct local authorities. This body was the precursor of the Metropolitan Board of Works, the first body to act for London (then the world's biggest city) as a whole.

Sewers had been designed to carry off rainwater, and with the introduction of the water closet it was, strictly speaking, illegal to connect the outflow to the sewer, but everyone did. This meant that a great improvement in hygiene had disastrous consequences. Great cities like London were built on tidal rivers. The sewers discharged at low tide, and when the tide turned so did the sewage, sloshing back upstream. In dry times, when there was not much fresh water surging down, the tidal reaches became fouler and fouler, and at low tide foetid beaches were exposed. In the summer of 1855 came the Great Stink in London, when things got so bad that the Houses of Parliament were disgusting and something had to be done. The elderly Faraday was consulted, and from a boat on the river he dropped white cards to see how long they took to sink from sight. In the turbid water they disappeared immediately. He had wondered whether chlorine might be used to disinfect the Thames, but clearly that wouldn't get to the root of the problem, and on 21 July 1855 *Punch* carried a cartoon of Faraday, top-hatted and holding his nose, 'giving his card to Father Thames'.[45]

The answer lay not in chemistry but in engineering. Joseph Bazalgette of the Metropolitan Board of Works oversaw the construction of great embankments to contain the river, with main sewers running beneath them and discharging well downstream from London. Branches made with glazed pipes brought sewage into them. This, coupled with provision for drinking water to be taken in well above the city, and filtered, meant that in 1874 the Registrar General could rejoice that in London the population had become self-sustaining for the first time in history, and that less than a third of children born there died before the age of three.[46] This investment in drains, copied elsewhere in Britain, in Paris (where the engineer in charge was Eugène Belgrand, 1810–78, and the sewers have become a tourist attraction) and in cities in the USA where things had been no better,[47] was the most important factor in reducing death rates in the nineteenth century, and it depended upon a common-sense equation of dirt and disease rather than any fuller understanding. On 15 November 1862 *Punch* had foreseen a glorious future:

> The passenger of Chelsea boat
> Unwonted salmon shall admire,
> Where dogs and cats he used [to] note,
> Defunct that on thy breast did float,
> Emitting exhalations dire.

And indeed, a century on, salmon returned to the Thames.

NURSING

One of Chadwick's allies was Florence Nightingale (1820–1910). Wealthy and well-connected, unmarried and uncertain what to do, she stayed in 1851 with deaconesses at Kaiserwerth on the Rhine, for whom nursing was a vocation rather than, as in England, usually a job for the unskilled, and certainly not to be undertaken by young ladies.[48] Knowing she had much to learn from examples on the continent, she then tried to train with the Sisters of Charity in Paris in 1853, but caught measles and had to give up. Nevertheless, she resolved to set up a hospital for gentlewomen in London, and did so. But then in 1854 war broke out between Russia and Turkey, and Britain joined in with the French on the side of the Turks. There was action in the Baltic but most fighting was around the Black Sea in what came to be called the Crimean War. This gave her the opportunity to use her social position and contacts, and formidable energy, in a much bigger cause. Inventing the role of war correspondent, William Russell (1821–1907) reported for *The Times* in London on the mismanagement of the war, and contrasted the French medical arrangements and field hospitals, where Sisters of Charity were nursing the sick and wounded, with the incompetent British ones.[49] In response, Florence Nightingale, taking a team of thirty-eight nurses, set off to Scutari, across the straits from Istanbul, and set about reforming the base hospital there. Impatient, tactless and bossy, cutting through red tape, armed with both official government backing and money given by well-wishers, facing and creating enemies among the medical and military high-ups, she showed a genius for organization and administration.[50] She realized that illness was a far greater problem for the army than losses of life in battle. Determined to catch up with the French, she began to work out how to reform the training of nurses, who would work in an authoritarian regime under a matron in a well-organized hospital.

Her efforts alone did not end the high death rate at Scutari: only when engineers discovered and drained cess pits beneath the hospital did that improve radically. But she was welcomed ecstatically when she toured the front, and on her return after the fall of Sebastopol and the ensuing peace, she was a heroine, 'the lady with the lamp' who had comforted and eased the wounded soldiers. Other nurses such as Mary Seacole (1805–81) had heroically looked after soldiers at the front, but Florence Nightingale was from a different social class, accustomed to hobnobbing with the mighty, and could begin to transform the system. In the Crimea and then at home, she took advantage of her status to bring about reforms. After testifying and publishing on health in the army and in India, she wrote *Notes on Nursing* (1859), which was very humane and indicated clearly that while she was a terror to doctors and generals, she had very much the right idea about patients and their needs.[51] Her school, founded in 1860 and run by disciples, turned out the newly professional nurses who became the matrons in hospitals around the country in a general raising of standards, and now (like some other eminent Victorians) an invalid, she campaigned and lobbied from her bed, living on until 1910.[52]

Like many contemporaries, she never had much time for the theory that hypothetical germs rather than dirt and miasma were the cause of diseases, taking practice rather than theory from the French. Down to her day, hospitals had been places for the sick poor; anyone with a comfortable house would expect to be nursed, undergo surgery if necessary, and give birth, at home, where cleanliness and care could be relied upon. That began to change, and the sawbones surgeon who had been a skilled craftsman until the mid-eighteenth century, 'Mr' rather than 'Dr', became in the course of the nineteenth an elite member of the medical establishment in consequence.

Anaesthetics were introduced in America, nitrous oxide (rather unsuccess-
fully) by Horace Wells (1815–48) in 1844, and ether by William Morton
(1819–68) in 1846. These were rapidly picked up in Europe, and in 1847
James Young Simpson (1811–70) in Edinburgh introduced chloroform,
with which Snow became expert. Surgery became much less desperate, and
surgeons could take their time, but the risk of infection was high.

POISONS

As the place of science in medicine, and the role of tests backed by laboratory and statistical evidence rather than clinical experience, was expanded,
so pathology became a field of great interest, determining more accurately
the causes of death.[53] Poisoning was rife among our ancestors. It could be
accidental. Substances were after all found to be toxic the hard way, by
trial and error. Doctors, ever since Paracelsus began prescribing mercury
for the new disease of syphilis, have used desperate remedies for desperate
diseases, and still do. Venomous snakes, spiders and plants like hemlock
and deadly nightshade were dreaded, and Erasmus Darwin related fright-
ful tales of the Upas Tree in the Indies, deadly to all who approached it.[54]
Mephitic vapours lurked in caverns. Arsenic was favoured as a pesticide
and readily available; it was also, like lead, used in paints. But poisoning
could also be the result of utter culpable negligence, or of murder. Food
and drink sold in towns and cities were all too often adulterated: milk was
watered, flour and tea bulked, confectionery sweetened with sugar of lead
and coloured with metallic dyes, beer impure.[55] And deliberate poisoning
haunted the imagination because it was necessarily premeditated, planned
in cold blood (often, it was supposed, by women), and was extremely difficult to detect. Gradually, chemists and doctors improved the methods of
analytical chemistry to help with these problems, making legislation on
food and drink possible (and thus opening professional opportunities for
the trained analyst), and murder trials less questionable.[56]

Because of the excitement that trials generated, the pathologists who
gave evidence became celebrated figures. They wished to be seen as expert
witnesses, called to give evidence on matters of indubitable fact rather than
opinions to be tested by cross-examination, but because tests for poisons
were unreliable, there were many years when the defence and the prosecu-
tion could find doctors who would take opposite views of the case.[57] The
greatest figure, whose authority extended internationally, was the Spaniard
Mateu Orfila (1787–1853), who after medical training in Barcelona went

(like others from all over occupied Europe) down the brain drain to Paris in
1807, where he worked with Louis-Jacques Thénard (1777–1857), an up-
and-coming pharmacist who lectured at the Collège de France, and other
prominent chemists. He remained in France for the rest of his life, writing
the standard work on toxicology, *Traité des poisons* (1814) and rising to the
top of the medical profession. In 1819 he was appointed a professor in the
Faculty of Medicine. He followed Thénard's example of using experiments
didactically in his lectures, and like many contemporaries was suspicious of
speculative and wide-ranging hypotheses in chemistry and medicine.[58]

A great advance came in 1836 when James Marsh (1789–1846) devised a
reliable test for arsenic, sensitive enough to detect it in tissues rather than
just in stomach contents. Orfila was enthusiastic, delighted that an objective test had replaced the garlicky whiff and the estimate of symptoms over
which experts could easily disagree. He saw that the test could be applied
if suspicions were aroused long after death and burial, so that corpses
could be exhumed and examined if necessary, and that the absorption of
arsenic in the body could also be followed. He shone as an expert witness,
with his Parisian position and status, his confidence, showmanship and
rhetoric. But there was a problem, not infrequent with new techniques and
instruments: the test was too sensitive. Orfila came to believe, because he
kept getting positive results, that there was always some 'normal arsenic'
in human bodies – which therefore meant that the test was inconclusive.
Clever lawyers argued that all the arsenic found in a corpse was 'normal'
rather than administered by their client. Standards of chemical purity
have, as we saw, changed over time, and probably the zinc orfila used in
the reaction contained minute quantities of arsenic. With refinements in
practice, the positive results disappeared. Orfila had made a mistake.

The episode dented his reputation, and that of chemistry, but forensic science was here to stay, and in Britain Robert Christison (1797–1882) – who
had also studied in Paris, but as it happened had no direct links with Orfila,
whom he admired – became the great authority on toxicology and what
came to be called medical jurisprudence. Gradually, as analytical techniques
were further improved, other poisons than arsenic could be unequivocally
detected, and public analysts became a profession with their own society.[59]
Theirs was a branch of science in which exact manipulation and agreed and
repeatable procedures were far more important than any theory behind the
tests that worked (that had to wait upon the rise of physical chemistry[60]).
Because medical schools included chemical instruction, and many eminent
chemists had begun their career with a medical education, doctors competed

with chemists for jobs in this field. Diagnosis had always been a bit like detection, and Arthur Conan Doyle's Sherlock Holmes was modelled upon an Edinburgh medical man, Joseph Bell (1837–1911), famous for the clues he picked up from patients' appearance and gait. While murder trials were the exciting end of the field, it was the steady and methodical monitoring of water, milk, beer and food, focused upon life rather than death, that held importance for everyone. The work was often done in a private laboratory or doctor's surgery, but by the end of the century there were government laboratories with salaried employees carrying out these tests.[61]

MEDICAL SCIENCE

Physicians had been respected as gentlemen of wide learning, wisdom and clinical experience, who knew their patients. By the middle of the century, formal training not only in comparative anatomy and pharmacy but also in chemistry and physiology was becoming the key. The laboratory in Guy's Hospital and Thénard's and Orfila's Parisian workplaces were a sign of this. Expertise in testing and the interpretation of tests were crucial, bringing authority. It was not clear for some time how far scientific knowledge actually improved clinical practice, but some instruments came closer to it than chemical analysis. The stethoscope, now the badge of the profession, had been introduced by René-Théophile Laënnec (1781–1826) in 1819, as a short tube like those still used by midwives; the present form dates from the 1850s. Bichat had used an injecting syringe in his animal experiments, but the hypodermic syringe only came into use in the 1850s, as did the ophthalmoscope, invented by Helmholtz and making it possible to see inside the eye; then in the 1870s came the clinical thermometer.[62] But perhaps most important was the microscope.[63] Invented in the seventeenth century, the compound microscope was difficult to use. Spherical lenses did not focus sharply unless only the middles were used, and then it was hard to get enough light, and coloured fringes formed around images, making them fuzzy. Just as Orfila's tests generated rather than detected arsenic, so astronomers saw such things as canals on Mars through telescopes, and microscopists perceived things like homunculi in sperm that weren't there. In the early nineteenth century, improvements in optical glass, and the adaptation to small lenses of the system of colour-correcting used for telescopes, led to reliable compound microscopes replacing the 'simple microscopes', magnifying glasses, that prudent men had trusted hitherto.[64] The world of cells opened up. By 1850, medical students were being taught to use achromatic microscopes, which, having been very expensive, were becoming everyday instruments. Tissues and cells, rather than gross anatomical structures, became the focus of medical scientists increasingly divorced from the care of actual patients.

An extreme example is Louis Pasteur, who was not medically trained. A chemist, his first research was on crystals of tartrates, which he carefully observed and found to be asymmetrical, affecting polarized light differently – an important discovery in the history of structural chemistry. Synthetic and natural products (from fermentation) were different. His training in scrupulous examination and his prepared mind led him to test bold hypotheses. Yeasts were crucial in the tartrate story, and minute organisms might also (as others had conjectured) be involved in the fatal diseases of silkworms he investigated, and then of chickens and farm animals, whose fatal anthrax could be passed to humans. He identified bacteria as responsible, and found that inoculation with dried bacilli protected rather than infected animals. He loved controversy and showdowns, and his public demonstration of the vaccine he had prepared, with invited newspaper reporters, on a farm in 1881, attracted huge attention. The unprotected sheep and goats died on cue, and the vaccinated survived. In 1885 he prepared a vaccine for the much-dreaded though mercifully rare disease, rabies. He could not himself treat patients but supervised its administration to a boy, Joseph Meister, from Alsace, who did not develop the disease after being bitten by a rabid dog. Pasteur was perceived as one of humanity's great benefactors, and in 1888 the Pasteur Institute was set up to continue his research and to prepare vaccines.[65]

GERMS

Tyndall at the Royal Institution took up the idea of our invisible friends and foes, extending and popularizing Pasteur's discoveries in the English-speaking world, while Joseph Lister (1827–1912) in Glasgow applied the germ theory in introducing antiseptic surgery. Finding that carbolic acid (phenol) destroyed bacteria, he used swabs impregnated with it and sprayed it about in the operating theatre, greatly reducing the risks of surgery. Lister became the first medical man to be made a Lord. Wounds doused in carbolic, however, did not heal as quickly as those kept clean, and stringent cleanliness, with rubber gloves, and surgical instruments sterilized as Pasteur recommended, by boiling rather than floods of antiseptic, became the norm by 1900. Pasteur's much younger rival and successor in making new vaccines was Robert Koch (1843–1910), when in the second half of the century Germany took over from France as the great centre of medical innovation. Koch identified the bacillus responsible for tuberculosis in 1882, and then in 1884 that responsible for cholera.

Though this was in a sense completing the discovery made over thirty years earlier by Snow, the germ theory did not easily catch on. Everybody was exposed to countless germs, so how was it that only some people became ill? 'Typhoid Mary' in New York infected people with the disease without displaying any symptoms herself. Doctors had been primarily interested in patients and their 'constitutions', making them liable to certain conditions, but now the focus was on diseases and their supposed causative agents. Koch drew up a list of criteria that had to be satisfied in claiming that a bacterium was the cause of some disease. But conservatives, steeped in clinical practice, would not readily abandon miasma. Nevertheless, vaccines against tetanus, a common result of infected wounds, and the diphtheria that, no respecter of persons, had killed the philosopher Bertrand Russell's eminent parents as well as numerous children, were available from 1890. Disciples of Pasteur ('Pastorians', as they were called) in Paris and beyond were prominent in preparing and administering vaccines.[66]

MEDICAL EDUCATION

The great figure in German experimental medicine had been Johannes Müller (1801–58) at the University of Berlin, where he had taught physiology. His eminent pupils included Karl Ludwig (1816–95), Emil Du Bois-Reymond (1818–96), Ernst Brücke (1819–92) and Helmholtz, who in 1847 published a joint manifesto calling for physiology to be based on chemistry and physics, in effect repudiating the master's vitalism. They all went on to become successful teachers and researchers, attracting students from abroad as Germany became the place to go rather than France for eye-opening experience, and as the German research-oriented university became the model for Britain, the USA, and indeed generally. Rudolf Virchow (1821–1902), another of Müller's pupils, was appointed to a chair in Berlin in 1856, and published his vastly influential *Cellular Pathology* in 1858. His pupils went far and wide, propagating his emphasis upon cells and functions like nutrition. French medical schools remained important, and the great physiologist Claude Bernard (1813–78) carried on the tradition of Bichat and Magendie, reflecting brilliantly upon method in his *Introduction*

to the Study of Experimental Medicine in 1865.[67] Highly suspicious of statistics
and premature quantification, and in the reigning French positivist tradition
uneasily equivocal about hypotheses (though in fact influentially reasoning
hypothetico-deductively), he called attention to 'internal secretions' from
glands, and to homeostasis, the way animals maintain their 'interior milieu'
in the face of changes in external circumstances.

In the second half of the century, there was a general tightening up
of medical education, with the German model perceived – sometimes, as
in France, reluctantly – as the gold standard. In England apprenticeship
gradually faded away and training in hospitals became the norm. From
the 1830s, as in London and Newcastle, this was in connection with a university so that students could proceed to degrees including the MD; such
formal courses to bachelor level were required by 1860. Unlike most on
the continent of Europe, British hospitals remained 'voluntary', supported
by charitable contributions, and patrons and supporters sometimes had
the right to nominate patients. Doctors usually worked there part-time in
an honorary capacity, which brought them status and experience: patients
who were poor had to obey doctors' orders. The emphasis was generally
clinical rather than experimental, though gradually information about what
remedies actually worked was accumulated. Oxford and Cambridge granted
medical degrees, and had professors, but did not offer genuine courses until
after the reforms of the mid-century, and even then expected students to get
clinical experience in London hospitals. Their graduates could then enter
the Royal Colleges of Physicians or Surgeons, and set up as consultants, at
the head of the profession. Huxley, who lacked such privileged access, in
his teaching in London transformed comparative anatomy into physiology
(otherwise neglected in Britain), and wrote a textbook on the crayfish, the
creature he used as his exemplar.[68] While himself avoiding vivisection of
vertebrates, he recognized that for scientific and medical progress it would
be necessary and played a part in resisting the pressure groups opposed to
it, in what he saw as a battle for true science. His disciples came to occupy
important posts in other universities, and in 1870 Michael Foster (1836–
1907) was appointed at Cambridge, where, promoted to a chair in 1883, he
managed to found a school of physiology that by the end of the century had
closed the gap between British and German standards. In Germany, Paul
Ehrlich (1854–1915) hit upon the 'magic bullet' remedy for syphilis, salvarsan, in 1909, and had the idea that there are definite receptors in cells where
particular molecules can fit – a clue to how drugs work.

Scottish universities offered academic and clinical training, and many

practitioners in England, and indeed in imperial outposts too, were educated at them. Eventually, in 1858, the General Medical Council was set up
as an official regulatory body for the profession, setting standards both educational and ethical, and playing a curious role, midway between a learned
body and a white-collar trade union. The tradition was liberal, and benevolence and wisdom were supposed to characterize the practitioner, as in the
writings of William Osler (1849–1919).[69] Osler, born in Canada, had gone
from Montreal to the new Johns Hopkins University, and ended his career
in Oxford, a successor to Kidd. In the USA professional regulation came
late: the moving spirit was Abraham Flexner (1866–1959), who in 1910
published a survey of medical education in North America, comparing most
of it unfavourably with Germany. There were many proprietary medical
schools, often very small – like Hunter's had been in Britain. But whereas
in Europe only recognized universities had been able to award degrees, in
the democratic America it was a free-for-all. Homeopaths and followers of
other alternative traditions set up colleges and issued diplomas, sometimes
(as had happened in some impoverished universities in eighteenth-century
Europe) with minimal requirements for residence or study. Gradually,
the German model came to prevail with Johns Hopkins as the exemplar,
followed by other private and state institutions. Dentistry was gradually
separated from general surgery, and amalgam was used in fillings from
1834, giving an alternative to extraction for decaying teeth, which with the
coming of false teeth meant that our nineteenth-century ancestors were
no longer toothless in old age. The famous explorer William Moorcroft
(1770–1825) was trained in medicine, but went into veterinary science and
travelled in India seeking horses suitable for cavalry regiments stationed
there.[70] Training for veterinary surgeons began in the early nineteenth
century, bringing professional recognition, and they gradually replaced far-
riers, who had learnt their trade on the job.

So far, except in nursing, the story has been one of medical men. Intrepid
women began to break in: in 1849 Elizabeth Blackwell qualified as a doctor
in the USA, in 1869 Sophia Jex-Blake matriculated in Edinburgh, and in
1874 (because universities prudishly closed loopholes that had not explicitly
excluded women) she founded the London School of Medicine for Women.
Women could thus become doctors. Midwives had always learnt on the job,
but in 1881 the Institute of Midwives brought professional status to these
women who had previously lacked it. Hospitals for women, where they
would be cared for by women, opened. But the 'Lady doctor' remained a
rare creature throughout our period, medical schools lagging even behind

other university institutes and departments, and gynaecology remained an almost entirely male preserve. In hospitals the matron was a formidable figure – perhaps like the Regimental Sergeant Major in the army, vital but emphatically not in command in what was definitely a man's world.

THE FRINGES

In-patients in hospital were confined in a total institution, subjected as Michel Foucault put it to a cool medical gaze, but expected in due course to emerge and resume their ordinary life.[71] But the nineteenth century also saw an enormous rise in the number of asylums for the insane.[72] Lunacy had by the eighteenth century come to be seen as a medical problem, not greatly to the advantage of sufferers, who were often chained up. In 1796 the Quaker William Tuke (1732–1822) founded the York Retreat, where a mild regime facilitated recovery, and this example was followed by John Conolly (1794–1866) in the Middlesex Asylum at Hanwell, who argued that 'restraint' meant neglect. Intended, as their name implied, to provide safe and compassionate separation from the busy world, asylums were located in the country, and many patients became institutionalized and stayed for life. Meanwhile, in Paris, Philippe Pinel (1745–1826) had been appointed in 1793 to the Bicêtre Hospital, where with Jean-Baptiste Pussin (1745–1811), a former inmate, he replaced fetters with the less-constraining straitjackets for otherwise uncontrollable patients. He subsequently moved to the Salpêtrière Hospital, which he also reformed, and where later the eminent neurologist Jean-Martin Charcot (1825–93) drew large numbers of students, including Sigmund Freud (1856–1939). Charcot's spectacular demonstrations with 'hysterical' patients attracted great attention, not all of it scientific.

The 'mad-doctors' or 'alienists', not always well-regarded professionally, who presided over asylums tried to classify mental illness, just as others were reclassifying diseases using germ theory in addition to exact clinical observation. They had little success, though important legal definitions of insanity were agreed, leading in Britain for many years to the curious possible verdict 'guilty but insane'. Just as expert witnesses could disagree over poisoning, so they could and can over the state of mind of people accused of crime – but in this case laboratory tests offered no hope. In the field of mental disability, the taxonomic urge led to various degrees of 'idiocy' being recognized and named, often in association with the fear of degeneration, as we shall see.[73]

Disliking that clinical gaze that reduced them to the status of examples of measles or neurasthenia, and the prospect of heroic but ineffective procedures, patients in the nineteenth century (as nowadays, but with more reason) turned to 'alternative' practitioners, gradually being excluded from the regulated medical profession. Thus Charles Darwin found relief from his debilitating symptoms in water cures, involving cold showers and other discomforts at home as well as periodic visits to spas. Many others took cures at spas, in Germany as in Britain, where social life might be as important as therapy,[74] and others consulted practitioners such as homeopaths who gave them the attention they did not get from orthodox practitioners, and relieved their symptoms, sometimes using medical traditions coming from other cultures. Those who did not choose, or could not afford, to consult any kind of doctor could buy much-advertised pills such as those that enriched Thomas Holloway (1800–83), who endowed a college for women, Royal Holloway, opened by Queen Victoria and now part of London University.[75]

CONCLUSION

Medicine was for long the only scientific profession, and the prestige of physicians depended upon their wide learning as well as their clinical experience. Nevertheless, it was only in the course of the nineteenth century that science, in the form of physiology, pathology and chemistry, came to be seen as essential for the training of all medical practitioners. For much of the century, more commonsensical science sufficed, and far more lives were saved by public health measures based on connecting disease with dirt than by advances in surgery and germ theory. In the first half of the nineteenth century, many scientists, like Helmholtz and Huxley, had had a medical training, or were, like Davy and Darwin, medical dropouts. Interest in anatomy and physiology might lead into palaeontology or biology, while pharmacy involving the isolation of active constituents from natural products and the testing of remedies led readily into chemistry. Because, as Huxley found, making a living by science was hard, many did their science in their spare time from medical practice. Later, as specialization increased and the expansion of universities and science-based industries led to more jobs, medicine as a scientific training became less important. But it had been, as with Orfila, important as a gateway to laboratory life, our next subject. Different traditions, educational regimes and practices had evolved in different countries, and a period of study abroad remained important throughout our period. In medicine as in other fields, the coming of rail-ways and steamships meant a proliferation of conferences, national and international. And expanding medical schools and subsequent specialization led, just as in other sciences, to the formation of new societies and journals, and also to book-publishing clubs like the New Sydenham Society. By the early twentieth century, medicine was based upon a congeries of sciences and its ethos had been transformed.

6

LABORATORIES

Newton's investigation of how a prism splits white light into its component colours was done in his rooms in Trinity College, Cambridge, admitting sunlight through a small hole in the closed window shutters (and when these were opened his mathematical work could be done at the desk). But for his chemical experiments, he needed, like alchemists before him, an elaboratory, with a furnace and other equipment. There are splendid pictures of alchemists at work, sometimes indeed conveying a sense of futility and failure but often showing an optimistic team at their labours.[1] Desperate remedies for desperate diseases were one fruit of such labours, in the tradition of Paracelsus. Much was learnt about minerals and metals, and techniques such as distillation were practised and improved. The experimental tradition in science came out of chemistry.[2] A chemical philosophy, a rival to the mechanical world picture, emerged[3] in which the Creation was perceived as a chemical process, and in which the world would be improved (and perhaps even life generated) through the application of 'chymical' knowledge.[4] These labours did not require a specially designated room or space; the kitchen with its stove and sink might be used. But as chemists came to acquire, and require, sensitive balances, delicate glassware and other expensive items, and to handle dangerous substances, so well-to-do men of science, like Lavoisier at the Arsenal in Paris, fitted up a laboratory. This was indeed part of the process in which those dedicated to the science separated themselves from amateurs, who lacked equipment and facilities – but who might be able to improvise and make do, for elaborate equipment can make the scientist its slave.

Visiting the city in 1818, Berzelius exclaimed that 'the amount of work in chemistry that is done in Paris is completely incredible. I believe that there are more than 100 laboratories devoted to research.'[5] Almost all these would be a room in a house, like the one James Smithson (1765–1829, founder of the Smithsonian) had in Paris at that time, devoted to experiment, and fitted up with more or less refined equipment. Right through the nineteenth century, such private laboratories remained extremely important. William Crookes (1832–1919), for instance, discoverer of thallium, investigator of cathode rays and President of the Royal Society at the outbreak of the Great War in 1914, did all his science in his private laboratory.[6] But during the course of the century, institutional laboratories and institutes, attached to universities or to industries, became increasingly important. Nevertheless, most work in the nineteenth century was attributed to one person: long lists of the whole team (including the director of the laboratory, who may have had only a remote connection with the particular topic) became a feature of twentieth-century research. When William Allen (1770–1843), the Quaker pharmacist, and William Haseldine Pepys (1775–1856) published an important joint paper on the composition of carbon dioxide and the nature of the diamond in 1807, the perplexed Royal Society could not award them the Copley Medal because 'it is an indivisible thing, and cannot be given to two'.[7] Faraday worshipped God by finding out about His works in the laboratory, alone except for the taciturn and obedient Sergeant Anderson, his assistant or acolyte. Nevertheless, laboratories were usually sociable places, dens where friends dropped in and made suggestions, joining in the work going on perhaps, and where a paid helpmate might be doing work for which his or her master would be credited. Berzelius's housekeeper Anna washed up the test-tubes along with the dishes, while Mme Lavoisier (née Marie Anne Paulze) had an important role assisting her husband in the laboratory, though not in the splendid clothes of their set-piece portrait by David,[8] and other wives also played important but largely anonymous roles in the laboratory.[9]

We are accustomed to the white coat as the uniform of the experimentalist, but that is a feature of the twentieth century. Indeed, half a century ago one could, preserving one's amateur status, do a chemistry degree without ever acquiring or choosing to wear one, though most did so. In the nineteenth century the protective clothing was an apron, similar to those worn by artisans and reminding us that practical chemistry or physiology was an art as well as a science. The girls in *Conversations on Chemistry* (1806) by Jane Marcet seem to have been doing experiments in their muslin dresses without pinafores, for Caroline's gets holed by acid.[10] Other safety precautions were usually taken in response to catastrophes, rather than anticipating them – and this carelessness of danger increased the macho appeal of chemistry. Thus at the Royal Institution, Faraday was first employed when Davy was disabled by an explosion, getting fragments of glass in his eyes. Subsequently in this research they wore goggles – but avoided such cumbersome precautions until the danger was proven. At the end of his life, Davy wrote:[11]

> Patience, industry, and neatness in manipulation, and accuracy and minuteness in observing and registering the phenomena which occur are essential. A steady hand and a quick eye are useful auxiliaries; but there have been very few great chemists who have preserved these advantages through life; for the business of the laboratory is often a service of danger, and the elements, like the refractory spirits of romance, though the obedient slave of the magician, yet sometimes escape the influence of his talisman and endanger his person. Both the hands and eyes of others however may sometimes advantageously be made use of.

And Davy's friend Joseph Cottle (1770–1853) was surprised that so enthusiastic and courageous an experimentalist should have survived 'to the vast age, for him, of fifty years'.[12] The French chemist Pierre Dulong (1785–1838) lost an eye and a finger in discovering the explosive properties of nitrogen trichloride in 1811, and Davy, who took up the work, was thereby warned to take precautions. Chemistry included not only stinks but bangs.

APPARATUS

Davy had at his disposal in London a large basement laboratory, in what had been the kitchens, with a furnace and other equipment on a large scale. It is illustrated in the textbook by Brande, his successor, and now preserved as a museum in the state it was in the time of Faraday, the next director.[13] Allen was among those who supplied reagents, but purification was a major feature of nineteenth-century practical chemistry, and only towards the end of the century could one buy chemicals of analytical purity (another factor making the repetition of classic experiments tricky). Again, experimental research often required the making of apparatus, but from the early days instrument makers were connected with the Royal Institution, and wealthy enthusiasts like Allen, Pepys and Smithson were among its backers.[14] Apparatus that had been contrived for the first time by someone like Davy, famous for his rapid and creative misuse of his equipment in research (sometimes performed in front of a select group of observers on banked seats in the laboratory), would be improved and then put on sale by Frederick Accum (1769–1838), Richard Knight (1768–1844) or other tradesmen. Right through the century, the research chemist had to be handy, but the gap between apparatus being invented and made commercially available was not long. The inventor had to describe it and its use carefully, with illustrations, and perhaps even like Newton with his prisms, distribute examples to colleagues. When it could be bought, and the classic work with it readily repeated, it could be taken for granted, or 'black-boxed' in today's jargon.

This happened with Davy's apparatus of 1807 for isolating potassium using an electric current. In line with his demonstration in 1806 that chemical affinity was electrical, he used electricity to melt and decompose potash in a spectacular experiment, after which he danced about the laboratory in ecstatic delight (a tradition that continued after successes in other laboratories too) and scrawled a description of this 'capital experiment' in his notebook. Pepys improved it, and it was adapted for other unknown metals by using a negative pole of mercury rather than platinum. Davy's battery was temperamental, indeed worn out, and he played the patriotic card (with moderate success) to raise money from Royal Institution supporters to build a bigger and better one, while in France Napoleon funded the building of a rival battery.[15] Davy's cost about £600, at a time when the Royal Institution's average annual expenditure on apparatus and reagents was £140. These huge and expensive toys did not lead to further amazing discoveries: indeed, in France, Gay-Lussac found that a different method, passing the potash down a red-hot gun-barrel, was more convenient. Laboratory science was not the expensive 'big science' of the day – that was exploration in the tradition of Captain Cook – but keeping up to date, and thus keeping in the forefront, was not cheap.

Faraday hoped to end this financial straitjacket. A blacksmith's son, trained as a bookbinder, neat and meticulous, he was proud of his manual skills and his capacity to make do and mend. *Chemical Manipulation* was his only real book, the others being collections of papers from journals.[16] In it he described the processes of chemistry, weighing, grinding or trituration, heating, dissolving, filtering, titrating, making one's own litmus papers, and even how to open bottles when their stopper is stuck.[17] He was justly proud of his 'tube chemistry', in which a tube bent into a zig-zag was used for fractional distillation: he had used one in 1825 when isolating

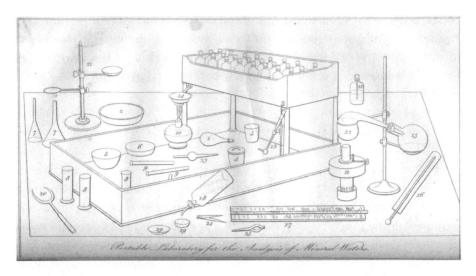

Portable Laboratory for the Analysis of Mineral Waters.

Figure 12 *Portable laboratory, by Faraday:* Quarterly Journal of Science,
Literature and the Arts *10 (1821): 216*

benzene from whale-oil. Different fractions collected at each bend in the
tube. Davy had noted how in his lifetime the apparatus chemists needed
for most of their experiments had become much less bulky and expensive:
a spirit lamp could be used in place of a furnace, and much smaller quantities of all reagents, as demonstrated by Wollaston and Smithson. He
remarked that everything necessary might be contained in a small trunk,
and had taken such a 'portable laboratory' when he toured the continent
with Faraday as his assistant cum valet in 1813–15, though in completing
their work on the nature of iodine they took advantage of Chevreuil's
laboratory.[18]

Faraday illustrated such a portable laboratory, for mineral analysis, in
the Royal Institution's *Quarterly Journal* as well as in Brande's book.[19]
Ships' surgeons took chests of equipment on their voyages, and doctors at
home had chests full of instruments and potions to take in the carriage on
their rounds. Such things may have been a model for the portable labora-
ries available from dealers such as the Henry family in Manchester, friends
of Dalton. They might be unpacked for use on travels (perhaps in the field
by mineralogists), in schools up to date enough to want science teaching
but having no dedicated laboratory, or at home, where users might divert
themselves like Jane Marcet's characters or perform original researches of
a not-too-ambitious character.[20] They came in wooden boxes like doctors'
medicine chests, and were ancestors of the chemistry sets of the first half

of the twentieth century, marketed for boys, but subsequently tamed by health and safety regulations, and laws governing the sale of chemicals. The young Oliver Sacks's love affair with practical chemistry, not unlike that of nineteenth-century boys of scientific bent, could not be repeated nowadays.[21]

CRAFT SKILLS

As Faraday made clear, laboratory skills in manipulation were not easy to acquire, and like learning musical instruments, required practice: his book has exercises at the back. His meticulous laboratory notebooks give evidence of his own skills, as well as his mind at work.[22] Glass vessels and tubes were of course much used, because one could see what was going on, but Josiah Wedgwood had supplied his friend Priestley with ceramic apparatus, and sold similar pieces. Ceramic crucibles, evaporating basins, mortars for grinding (better than marble, which might crumble slightly) and pneumatic troughs for collecting gases over water or mercury were essential in any laboratory. As Lavoisier had proved, the glass of the late eighteenth century was by no means inert chemically. He found that after water had been boiled in it, a flask became slightly lighter. Blowing and bending glass was not as easy as it looked. Making a gas- and water-tight T-joint in glass tubing was a required exercise for chemistry students down to the mid-twentieth century, and tricky to do – a tube can easily kink in bending and will crack if not annealed by gentle cooling. The great problems with glassware were to do with mechanical strength and with heat. To be robust, and stand up to being clamped in retort stands of iron, and joined together in arrangements for such processes as washing and drying gases by bubbling them through water and passing them over calcium chloride, the glass had to be thick. But to stand heating without cracking, it had to be thin. Patience, ingenuity and determination were required of the experimenter.

To spread heat around a flask and reduce the risk of cracking, a sand-bath on the furnace was often used, while for lower temperatures a water-bath, or bain-marie, was appropriate. Sometimes a glass tube required heating to decompose a solid or make it react with a stream of gas: here again, the experimenter would try to spread the heat rather than concentrate it. Spirit lamps were more manageable than furnaces and easier to control (though not so hot as a furnace with a good blast of air); they might, like the high-performance lamps of Aimé Argand (1755–1803), have a cylindrical wick with air coming up the middle. Joining apparatus together, necessary for distillations and other processes, was tricky. Connections must be gas-tight and firm, but not so rigid as to be fragile. Chemists had their favourite sticky mixtures, called lutes, with which to pack joints – hoping that they would be chemically inert during the reactions going on. In the course of the century lutes were replaced by corks flexible enough not to crack the necks of flasks, pierced using cork borers (hollow sharpened steel cylinders of various sizes) for the tube to go through. Corks were rolled (in paper) underfoot until soft enough for use. Where flexibility was needed, rubber tubing was introduced in Faraday's time. He wrote of 'caoutchouc connectors', tubes made as required by the experimenter out of sheets of India-rubber: these were thin and soft, but hard rubber pressure-tubing came later in the century. By the twentieth century, rubber bungs were also available in place of corks, and Pyrex glass, much more resistant to cracking though harder to work, was also becoming common.

Crucibles were made to stand intense heat in the furnace, which in well-equipped laboratories was a blast-furnace. These were hard to manage, but Wedgwood's pyrometer, designed as we saw for quality control in pottery kilns, provided a way of measuring the melting point of metals and minerals, in 'degrees Wedgwood', which could be tabulated along with other data, though these temperatures were arbitrary and purely empirical, and could not be connected to the everyday Fahrenheit or Celsius scales of glass thermometers.[23] Crucibles were themselves liable to crack, or might be attacked by whatever they contained when it got extremely hot. Here, Wollaston came to the rescue of experimentalists with metallic platinum, or platina as it was often called (in defiance of the usual rules of naming), a metal that could not then be melted.

Wollaston was a physician, but more interested in science than in patients. He devised and operated a practicable process for purifying the grey powder in which form platinum was known, and transforming it into pure, shiny and coherent metal – by 'swaging', hammering it at high temperature. He and his partner Smithson Tennant (1761–1815) isolated the new metals palladium, rhodium, osmium and iridium in the course of their labours, but platinum was the profitable one.[24] In 1801 Wollaston, benefiting from a legacy, bought a house in the West End and converted the back rooms into a laboratory, veiled in secrecy, where he ran the business as well as participating fully in the scientific life of London. Platinum was not thought of for jewellery, looking as it did like silver, but it found industrial uses in touch-holes of guns and in vessels for distilling corrosive materials, notably sulphuric acid. For the chemist in the laboratory, it was invaluable.

Platinum happened to expand with heat just as glass did, so wires could be sealed into glass tubes or vessels to carry an electric current; moreover, being inert, platinum poles did not react with substances like potassium. Platinum spatulas enabled chemists to fish things out from beneath liquids, crush small crystals and do all sorts of other manipulative tasks: they were, until the coming of stainless steel, an essential part of the chemist's equipment. Platinum crucibles resisted attack from caustic substances even in the furnace, and the electrical resistance of platinum varied with temperature so the metal could be used to bridge that gap between scales and replace Wedgwood's gadget. Platinum later proved valuable as what Berzelius called a catalyst, making reactions happen. Wollaston cornered the market, and was reputed to be very rich.

LEARNING HOW TO EXPERIMENT

Platinum apparatus was, like Davy's electrochemical cells, an example of innovation in equipment opening up a new field. In experimental science, this is what frequently happens. Thus Liebig devised ways of putting the analysis of organic compounds on to a new footing. He had worked in Paris with Gay-Lussac, and met Alexander von Humboldt, who became his patron, pressing for him to be given a chair at the little University of Giessen, his home town. There he introduced laboratory teaching, and a research laboratory for graduate students. He was not the first to do so, others in Germany and Thomas Thomson in Glasgow having begun to build up research schools, but Liebig was the most successful and his model prevailed. His students took PhDs, a degree previously not of much importance. The doctorates that mattered were in divinity, law and medicine, the traditional higher faculties ranking above Arts (which included philosophy, moral and political as well as natural).[25] Liebig published a journal, normally described as *Liebig's Annalen*, in which they could publish their work, and he also took over from Berzelius the publication of an annual report on the progress of chemistry. He thus became a central figure in the scientific community, and his close links with Britain, notably through the BAAS, gave him an international reputation. Indeed, time spent working in Liebig's laboratory, and a testimonial from him, became extremely important in getting academic posts in Britain as well as Germany.[26]

Liebig in effect invented the research student, something that had an enormous effect upon the development of science. Not all his doctoral students could expect to find academic posts, but new opportunities came from the rising pharmaceutical and dye industries, especially in Germany where an industrial revolution was getting under way, but also in Britain where textiles were of supreme importance and many Germans found jobs. As natural dyes were improved and then aniline dyes invented, chemical industry offered great opportunities to bright and well-trained men.[27] Liebig's laboratory specialized in the analysis of natural products. The isolation and purification of active components was an essential part of work with drugs and dyes, but Lavoisier's ideal of getting accurate elemental analyses, as with mineral chemistry where Berzelius, Wollaston and Thomas Thomson were famous names, proved very difficult (and not very helpful) for organic substances, composed of carbon, hydrogen, nitrogen and sometimes small quantities of a few other elements.[28] The classic first step was distillation, an ancient craft, and in this process (as in Faraday's book) the neck of the retort might be wrapped in damp cloth or paper to cool and condense the product. Liebig adopted but did not invent the 'condenser' that bears his name, in which the vapour passes through a tube surrounded by a glass jacket through which a stream of water is passed, entering at the bottom and leaving at the top, thus giving much more efficient cooling. For catching the carbon dioxide given off in analyses of compounds of carbon, he invented an intricate glass vessel, the 'kaliapparat', in which it was absorbed by potash. Devices like these led to reliable and repeatable analyses, in which his students could be trained, opening up organic chemistry so that it became more organized and popular than the inorganic branch, and just as useful. Distillation might begin with the dry compound, or a solution in water, alcohol or ether. From the distillate, if crystals formed (perhaps in combination with a standard reagent), their melting point would identify the starting substance as results became tabulated in handbooks.

FROM ANALYSES TO FORMULAE

Fitting formulae to results was much trickier. Dalton's atomic theory yielded atomic or equivalent weights for the elements, but it was not even clear whether water was HO or H_2O, and there was complete disagreement among chemists until after the famous International Congress at Karlsruhe in 1860. Skill and judgement were required of those graduate students and their supervisors, and delicate negotiation between theory and practice.[29] Thomson and Berzelius had had a major row over precision and accuracy in achieving repeatable results and deducing atomic weights and formulae from them.[30] When Faraday showed in 1825 that benzene, which he had just discovered, had the same components in the same proportions as acetylene, and Liebig's friend Friedrich Wöhler (1800–82) found in 1827 that ammonium cyanate transformed itself into urea (previously known only from urine), they revealed the importance of structure and arrangement in this sphere of chemistry. Wöhler did not demonstrate to himself or to Liebig that there was no vital force: after all, his synthesis required conditions very different from those in a living organism, and 'vitalism' was not the sort of thing destroyed by a single experiment but by a slow change of world-view.[31] These experiments showed the existence of isomorphs, different compounds with the same elemental composition: analysis was not enough. Auguste Laurent (1807–53) called for a deductive approach to chemistry, a plea taken up by August Kekulé (1829–96) who introduced into chemistry the idea that one should postulate structures, deduce consequences and test them experimentally.[32] The French chemist Marcellin Berthelot advocated synthesis as equally important with analysis in putting structures beyond doubt. Gradually, it became possible to work out just how reactions developed. On his return to London from Paris in 1849, Alexander Williamson (1824–1904) followed the stages in the formation of ethers from alcohols,[33] Cato Maximilian Guldberg (1836–1902) and Peter Waage (1833–1900) in Norway studied the rates of reactions and the effect of excess of one or other reagent, and Vernon Harcourt (1834–1919) in Oxford investigated a reversible reaction. It was possible by 1900 to work experimentally on mechanisms rather than the properties of particular substances: such physical chemistry was by then an important division of the science, with its journals and societies. At last chemists could begin to understand how the standard procedures, and the named reactions with specified reagents, that they had learned and practised actually worked. But the first steps in chemistry remained qualitative and quantitative analysis.

EQUIPMENT

In quantitative work, filtering, drying and weighing were essential steps, fully described by Faraday.[34] He bought sheets of unglazed 'bibulous paper', and cut strips to be impregnated with litmus to indicate acids and alkalis, and circles with which to filter. These filter papers were then folded to fit into a funnel, either flat or fluted like a fan. The ideal filter paper for quantitative work in mineral chemistry would burn without leaving ash (or, more realistically, leaving a consistent amount of it) and the residue could then be weighed or further treated.[35] For drying, there were glass desiccators, with calcium chloride or sulphuric acid at the bottom, a shelf above on which the damp substance would be put for a long time, and a close-fitting lid. Titrations were carried on using a pipette to suck up a definite quantity of alkaline solution (the tastes of things became familiar to careless students), and a burette to add acid. These were originally a calibrated glass jug, but then took the form of a long tube, with at the bottom a piece of rubber tubing and a pinch-cock. Glass taps were liable to stick, and were generally avoided. Weighing, the most accurate process in the laboratory, was an art. Balances, with pans resting on knife-edges made usually of agate, were kept in glass cases with a lifting front. The substance was placed on one pan, and weights on the other, with the pans being lifted from rest by turning a knob, and speedily returned to it as weights (the small ones handled with tongs) were added and taken away to get equilibrium. The fine adjustment was made by a 'rider' of twisted wire, pushed with a special hook along a scale on the beam from outside the case, the front being closed to exclude draughts. Where possible, balances were kept away from reagents and corrosive vapours in an anteroom, and the same applied to microscopes, required along with chemical apparatus in laboratories dealing with mineralogy, botany and zoology, pathology, food and drink, or textiles.

The stinks, so evocative after years away from laboratories, were a real problem, especially in rooms smaller than the Royal Institution's cavernous laboratories (with their bank of seats for Davy to research in public, rather as surgeons performed in the operating theatre for a select audience). This was especially true for hydrogen sulphide, used for qualitative analyses and copiously generated in Petrus Jacobus Kipp's (1808–64) apparatus of 1844 (familiar to generations of chemists), a three-decker erection of glass from which the evil-smelling and lethal gas issued through a tap. Indeed, poisoning, slow or quick, was as much a risk of laboratory life as injury through explosion. Experiments might be performed by an open window or beneath the chimney, but gradually fume cupboards, communicating with the outside and closed from the laboratory by a sash window, were introduced, most readily in purpose-built institutional laboratories which gradually developed to incorporate such novelties. One of the first such places had been in the basement of the Old Ashmolean Building in Oxford, now a museum, and at the end of the eighteenth century a splendid laboratory was built at the University of Coimbra in Portugal, as part of the reforms instituted by the Marquis of Pombal (1699–1782) after the suppression of the Jesuits who had previously been in charge of education there. A fine collection of apparatus was ordered, mostly from London through John Hyacinth

Magellan (1722–80), for demonstration lectures, not only in chemistry but in natural philosophy.[36]

LABORATORIES FOR PHYSICS TOO

Faraday was trained by Davy as a chemist, made his early reputation as a chemist, and was generally regarded as one during his lifetime. In the BAAS he held office in the Chemical Section, and never in the Physical (which went with mathematics, alien to him). A division of the Royal Society of Chemistry is named after him, Whewell saw him as inaugurating a new epoch in chemistry,[37] and John Frederick Daniell (1790–1845) (himself the inventor of a kind of electrical battery) at Kings College, London, wrote his chemistry textbook following Faraday's approach to the science[38] – which we might a little anachronistically call 'physical chemistry'. When, after Davy's death in 1829, Faraday returned to electrochemistry, the Royal Institution's laboratory changed into a place where (instead of analysis) electricity, and then magnetism and optics, became central – topics we would think of as physics, but which Faraday approached in the manner of a chemist, inductively. One problem was measurement. Cavendish had measured electricity with surprising accuracy by the effect the shock had on him.[39] Galvani and Volta used the legs of frogs and in this they were followed by Beddoes and Davy, whose requirements suggested to some in Bristol that they must be harbouring a French spy. Faraday had a 'froggery' at the Royal Institution,[40] but his work and others' on electromagnetism led to galvanometers where the needle pointed to a number on a scale. Further refinement, with the accurate measurement of very small currents using a beam of light reflected from a mirror, were a feature of electric telegraphy, and of electrical laboratories in the later years of the century, when precision became increasingly important.

Experimental physics had been a recognized science in the French Academy of Sciences since before the Revolution, and required laboratories – easier to keep clean and less smelly than those used by the chemists. Physics remained a sphere of demonstration experiments in the lecture theatre longer than chemistry, and only later adopted the laboratory teaching which had indeed been an 'extra' for undergraduate students of chemistry (to be paid for specially by those who opted for it) but which became an integral part of courses by the mid-century. Thus Thomas Thomson got a new laboratory in Glasgow in 1831, while in lagging England, University College, London, opened its chemistry laboratory in 1846, and a picture

Figure 13 *Teaching laboratory at University College, London, 1846,* Illustrated London News

of it shows a number of students at work, some of whom are wearing top hats.[41] The room is lofty and lit from skylights; there are long benches with drawers and cupboards beneath to work on, with racks of bottles down the centre so that students are accommodated on both sides; tall stools, as in a bar; and further racks of bottles along the walls. Students are distilling with a spirit lamp, rummaging for apparatus, sitting to observe a reaction, and making notes. Over a century later, teaching laboratories such as the one I began in had not changed all that much, except that gas and water supplies are not much evident in the 1846 plate.

In 1871, while the Royal Commission on Scientific Instruction was hearing the evidence it would begin publishing in 1872, its chairman, the Duke of Devonshire (after whom it was called the Devonshire Commission) founded the Cavendish Laboratory in Cambridge. The professor, whose first task would be to oversee the building, was Clerk Maxwell.[42] He had to begin his lectures where there were rooms, calling himself a cuckoo laying eggs in other birds' nests, but by 1874 directed what became one of the most famous science institutions in the world. Oxford already had good facilities in its museum, completed in 1860.[43] The Commission looked also at science in schools, such as there was, and its report of 1875 included plates and plans of the chemical laboratories at Rugby, Eton, University College School in London, and Manchester Grammar School, and at Harrow, Dulwich and Clifton, where there were also laboratories for physics. They look not unlike that at UCL, with various arrangements for seating while watching demonstrations as well as for performing experiments.[44] By this time, serious study of science as well as of classics was entering syllabuses. A decision to become a scientist, specializing at school and university, was becoming possible and lab work was a crucial part of it. The same was true for those following the alternative route into science, as part of the great army of qualified technicians trained by apprenticeship in a shop or works, backed increasingly by classes at Mechanics' Institutes, technical colleges like that set up at Finsbury Park in London or polytechnics on the German (and ultimately French) model, where there were laboratories.[45] Occasionally, as with Frankland, a pharmacist's apprentice might rise to great eminence in science.[46] He became President first of the 'learned' Chemical Society of London, and then of the new 'professional' body, the (later Royal) Institute of Chemistry, founded in 1877. This was followed in 1881 by the Society of Chemistry and Industry. These latter two bodies, along with the Society of Public Analysts, represented those making their living from the practice of chemistry and chemical engineering. And by this time, works laboratories for quality control and industrial research laboratories for product development were spreading from Germany to Britain, the USA and France. The role of the assistant thus became formalized. There was still a place for consultants with their private laboratories, called in usually when something went wrong (very instructive in science and technology), but their role was diminishing.

STERILE AND TIDY?

In the world of practical chemistry, the laboratory had evolved from busy places, the kitchen or the foundry, but now they were required to be sterile and tidy so that results could be depended upon. In 1809 Davy grumbled to his staff that the laboratory was 'constantly in a state of dirt and confusion'.[47] 'Cleanliness, Neatness and Regularity' were much wanted; standard solutions had not been made, glasses were unwashed, apparatus that had been ordered was not assembled, and (quill) pens, ink and paper were kept in a slovenly manner – 'I am now writing with a pen & ink such as never was used in any other place' – the writing is his usual scrawl. All these sound like classic instances of the servant problem, and indeed Faraday as his assistant from 1813 in due course collected and bound Davy's stray papers, and kept the exemplary record he called his *Diary*.[48] By the middle of the century greater cleanliness in the laboratory, as in the hospital, was becoming the rule – though in the early twentieth century, the Nobel Laureate Sir William Ramsay chain-smoked and dropped ash around the laboratory, which sometimes contaminated the chemicals, confusing the research.[49]

In general, though, laboratory conditions became increasingly standardized and different from those of ordinary life. Light levels and temperatures mattered, and were allowed to vary much less – there had been experiments suitable for the summer or the winter. In doing experiments, it was important to know that only one condition was being varied at a time: by the mid-century, the laboratory had become a place of artifice and artificiality, order and tidiness, where hypotheses could be tested and correlations confirmed or dismissed, very different from the messy world outside. As quantitative work in chemistry became even more important, the chemist had to be increasingly careful about impurities, and after 1859 the first 'physical' method of analysis entered his laboratory with a new means of heating. Flaring gas jets had been used for illuminating public spaces, including laboratories, for fifty years, but Bunsen (and his technician C. Desaga) invented the burner that bears his name, where the air is mixed with the gas in variable quantities so that when required a very hot blue flame is produced. This gave controllable intense heat, far superior to that from a spirit lamp: crucial in chemistry, where a rise in temperature of 10°C generally doubles the speed of a reaction.

In analyses, small samples of minerals had been heated on a charcoal block, using a candle or spirit-lamp with a blowpipe – hard to do, because the chemist must breathe in through the nose while steadily blowing.[50] By using different parts of a flame, oxidation or reduction could be achieved. A 'flux' such as borax to make melting easier might be helpful, and compressed air, and then an oxy-hydrogen blowpipe (protected from explosions by washbottles and wire gauzes) increased the scope. The presence of sodium was indicated by a yellow flame, potassium by a lilac one, and other metals had their characteristic colours. But sodium is abundant, and its brilliant yellow masks other signs. In their collaboration, unusual for that time, in 1860 Bunsen and his physicist colleague Kirchhoff used a spectroscope to analyse the light given off by minerals heated by a Bunsen burner. They found that each metal had a distinctive spectrum of bright lines, and identified two new alkali metals, caesium and rubidium, with their apparatus. Crookes was an avid disciple, and discovered the rare and poisonous metal thallium. But for the rest of the century, older chemists remained uneasy about analyses done without chemical manipulation. Nevertheless, instrumentation, which would in due course include self-recording spectrometers, would inexorably deskill chemists who had diligently learned how to do things from manuals like Faraday's.[51] Here again, innovation in apparatus and new techniques associated with it opened new territory for science, this time on the fuzzy frontier of chemistry and physics.

For the later nineteenth and early twentieth century, however, the chemistry laboratory was a place of test-tubes, retorts, condensers, bottles and Bunsen burners (requiring numerous gas-taps, to which they were connected by rubber tubes as necessary). In succession to Coimbra, and in line with the new flourishing of the research university, increasingly well-equipped laboratories were built, especially in Germany, often to attract scientific stars to migrate from one little state to another. Those at Bonn and Berlin were particularly splendid, palaces with teaching and research laboratories, balance rooms, lecture theatres and accommodation for the professor whose little empire it was, and who was expected to run it like the captain on a battleship. The plans and elevations were published not only in Germany but also in Britain, as examples of what was needed if there was to be any hope of keeping abreast in this new world of dyes, explosives and drugs. In the 1860s the laboratory in Zurich had cost £10,000, that in Bonn twice as much, and Berlin's would be even more expensive.[52] The Devonshire Commission duly heard evidence about laboratories abroad, with Germany hailed as exemplary by several witnesses. After the Franco–Prussian War of 1870–1, Strasbourg came under Prussian rule and became Straßburg, with a showpiece German university and excellent laboratory facilities. In Germany and France such facilities were built by the state, while in Britain and the USA there was a long history of suspicion of such patronage so private benefactors were sought, sometimes in vain. The BAAS, and then the Royal Society with a small government grant, had made sums available to individuals for research, but could not finance laboratories, and it was not until 1889–90 that the British government began to make annual grants to English universities, primarily for science and technology – some American states did better.

LABORATORIES IN OTHER INSTITUTIONS

Sometimes laboratories were built as part of a science site that would include facilities for various kinds of activity. Universities had long had

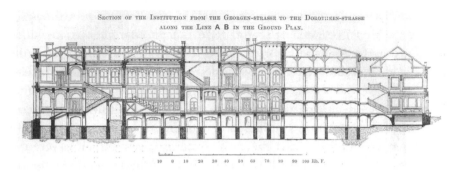

SECTION OF THE INSTITUTION FROM THE GEORGEN-STRASSE TO THE DOROTHEEN-STRASSE ALONG THE LINE **A B** IN THE GROUND PLAN.

Figure 14 The latest thing – a section through the new Berlin chemical laboratory, 1866: A.W. Hofmann, The Chemical Laboratories in Course of Erection in the Universities of Bonn and Berlin, *London: Clowes, 1866, p. 66*

botanic gardens, and more recently observatories.[53] In Oxford, some colleges had had small laboratories rather like Newton's, but used also for (optional) teaching. In 1860, just in time for the BAAS visit, a new museum in Venetian Gothic style (but incorporating cast-iron columns and a glass roof like railway stations) was opened to hold natural history specimens, and particularly the fossils that Buckland and his successors had collected. Attached to it were lecture theatres, and a chemistry laboratory, modelled upon a medieval abbot's kitchen – it is still in use. Equipped under the direction of the geologist John Phillips (1800–74), the whole marked an impressive commitment to science in a university at last modernizing itself.[54] It was financed by the University Press's sales of Bibles and Prayer Books, so it is curious that the first use of the building was at the BAAS, where after the main proceedings were over Wilberforce confronted Huxley and others in debate over the origin of species.[55] The bringing together of museum and laboratory was a significant step, and seen to be so by William Flower (1831–99), Director of London's Natural History Museum, when in 1893 he deplored the fact that in older museums, focused simply on display, there were 'no storerooms, no laboratories, no workrooms, connected with the building'.[56] Galleries and museums now think his way. Zoos and botanic gardens added laboratories as first dissection and microscopy (involving chemical staining), then by the end of the century chemical data, became important in taxonomy and in economic botany, and later added observatories also, as the spectroscope (coupled with photography) opened up the chemistry of the stars – a proverbially impossible subject.

The new chemistry laboratories in the German universities were the most commodious of their date, while the smallest was probably that on board the oceanographic survey ship, HMS *Challenger*, on her voyage around the world between 1872 and 1876.[57] A photograph shows a blow-pipe and crucible tongs ready on the table, with an array of flasks and bottles neatly stowed around the very confined space. There is a cane-bottomed chair to sit on amid the racks and drawers in this shipshape closet. Another photograph shows apparatus: distillation is in progress, using a bain-marie and a spirit lamp, with a condenser that seems to be arranged to reflux condensed vapour back into the flask. Twenty years earlier, a portable laboratory was recommended by Henry de la Beche (1796–1855) of the Geological Survey, for mineralogical analyses in connection with a scientific voyage, but more could now be expected.[58] The gap between the field, where things were observed, and the laboratory where experiments were done, was narrowing.

This was particularly marked in the USA. Americans, with a continent to explore and settle, expediently began with natural history.[59] The Coast Survey and the Topographical Engineers were funded by Congress, and on the expedition Charles Wilkes (1798–1877) led to Antarctica and the Pacific in 1839–40 the wider world became the object of their curiosity. Franklin had been celebrated for his electrical researches, and Joseph Henry, first Director of the Smithsonian (1846), was an important figure in the science, making a large electromagnet in 1830. But practical rather than theoretical science was the American forte, and the work of Josiah Willard Gibbs (1839–1903) on thermodynamics was unappreciated at home. In the later years of the century, Americans became well known for precision work with high-tech equipment like the gratings made by Rowland for spectroscopes. Physicists did not need to be handy in the way chemists did – Poisson and Wolfgang Pauli (1900–58) were notoriously ham-fisted – but it helped. Those who studied it will have fond memories of pulleys and weights, lenses and prisms, magnets, soldering irons, wires festooned in helices insulated with shellac or cotton thread, brass knobs, cathode ray tubes, and, like Cavendish, occasional electric shocks.

FUNDAMENTAL EXPERIMENTS

In Britain, after Maxwell's premature death the Cavendish Laboratory under Lord Rayleigh went in for very accurate determinations of units and constants, in competition with Berlin, as physics became the science of energy in its various manifestations. Crookes had meanwhile taken up Faraday's research on the passage of electricity through gases, achieving lower and lower pressures and discovering rays which streamed from the cathode in straight lines, casting sharp shadows, spinning a small propeller and deflected by a magnet.[60] These extraordinary effects made Crookes famous, but led others to startling experiments that transformed physics. In 1895 Wilhelm Röntgen (1845–1923), working with a 'Crookes tube', found that rays were emitted from it which fogged photographic plates. Following up this observation, he discovered X-rays, publishing the famous photograph of his wife's hand. And in 1897 Joseph John (J.J.) Thomson (1856–1940), at the Cavendish with its experienced staff and armed with a more powerful pump than ever, managed to deflect the cathode rays with an electric field. Cleverly designing apparatus so that the rays thus deflected were brought back to their original point by a magnetic field, he demonstrated that they were composed of extremely small, negatively charged particles, which were the same whatever gas there was in the tube and whatever metal the cathode was made from. He called them 'corpuscles', the term Boyle and Newton favoured for the ultimate particles of matter,[61] but they were soon identified with the hypothetical particles of electricity, electrons, inferred by George Johnstone Stoney (1826–1911) and Helmholtz.

Many, especially in Germany, had believed that like X-rays the cathode rays were waves rather than particles. J.J. Thomson's experiment seemed 'crucial', deciding the question in favour of particles once and for all, like those of Young, Dominique Françoise Jean Arago (1786–1853) and others that had put the wave theory of light beyond doubt by the mid-nineteenth century – despite Newton's preference for particles of light. In the opening years of the twentieth century, Einstein was to demonstrate rather shockingly that in different circumstances light behaved like particles, and electrons like waves. Experiments, however careful and refined, do not entail theoretical conclusions: they are never completely crucial.

AMERICAN STYLE

In the USA the federal government was generous with money for surveys of the West, bought from Napoleon in the Louisiana Purchase or seized from Mexico. These lands were settled before they were surveyed, with little houses in the big woods and on the prairie. Particularly important were the Pacific Railroad surveys, to find a route across unexplored territory, controlled beyond the Frontier by hostile 'Indians', and including very high mountains, to link California to the Mid-West and the Eastern seaboard. These surveys provided wonderful opportunities for natural historians and geographers, in the wake of Meriwether Lewis (1774–1809) and William Clark (1770–1838), and analysis of the minerals by portable laboratory or in a bigger one back home would confirm whether or not there was gold in them thar hills. In 1884 the BAAS held its annual meeting in Montreal, in an imperial gesture boosting science in Canada. After the meeting, William Thomson, doyen of classical physicists, went to Baltimore to the new Johns Hopkins University, founded on the German model. There he held a seminar for professors from American universities, whom he called his coordinates, on the nature of light, matter and ether – taken down by a stenographer and mimeographed, they were finally revised, expanded and published twenty years later.[62] The ether was required as a basis for light-waves, just as the sea is for water-waves; and Thomson toyed with ideas that atoms might be vortex-rings in it, or perhaps centres of force in a great field of energy. In the group was Albert Michelson (1852–1931), a former naval officer who after teaching science at Annapolis Naval Academy went to the National Almanac Office in Washington, DC, and then for two years' study in Europe, in Paris, Heidelberg and Berlin. In 1882 he had been appointed to the Case Institute in Cleveland, Ohio. He specialized in exact optical measurements, and invented the interferometer, an instrument in which a light beam was split into two, traversing different paths and producing interference fringes when recombined. Measuring these fringes enabled him to calculate the velocity of light in 1879 with new accuracy. Also there from Case was the chemist Edward Morley (1838–1923), eminent for his extreme accuracy, notably in measuring the densities of oxygen and hydrogen (to one part in ten thousand), and the ratio in which they combine.

In the light of the discussions at Baltimore and experiments he had done at Potsdam when in Europe in 1881, Michelson collaborated with Morley in 1887 in a famous experiment with the interferometer, sending beams of light along equal paths at right angles, and then mixing them. If the light took slightly longer on one round trip than the other, the beams would interfere and produce a pattern of bright and dark fringes in the apparatus, which floated on mercury so that it could be rotated. If the Earth is moving through the ether, then at some point one path would be expected to be in the direction of its travel, and the other at right angles to it. The beam going in the direction of the Earth's motion would be expected to take longer, and fringes would be seen. We can see why through using a simple analogy. Anyone who has rowed a boat, or makes a simple calculation, knows that it always takes longer to row upstream and back than it does to cover the same round-trip distance in still water. The apparatus was extremely sensitive, and any effect would be small. The experiment was therefore done at night so that all around the laboratory would be still. Within the limits of error, there was no effect. This was the first American experiment since Franklin's with the kite that astonished, and also puzzled, the wondering international community of scientists.

It did not overthrow the whole ether theory – single experiments don't do that, as we saw with vitalism – but it meant that it had to be modified in ways that seemed ad hoc to some, notably to Einstein, in whose theory of relativity the null result was to be expected – though it seems that he did not know about the experiment until after he had rethought the assumptions about space and time that had dominated science since Newton's day. Equally important in the new directions science took in the twentieth century were the very exact analyses of Theodore Richards (1868–1928) at Harvard. Like Wollaston a century before, he was regarded as infallible. He confirmed the view of Rutherford and Frederick Soddy (1877–1956), then in Montreal, that uranium decays spontaneously to lead while emitting radiation, by showing that lead from a uranium mine had a lower atomic weight than ordinary lead. Radioactive decay, transmutation, isotopes and the nuclear atom came firmly into the world-view of physicists and chemists through the refined techniques of chemical analysis.

CONCLUSION

Lavoisier had well-equipped laboratories at his disposal, and so did Priestley in his house in Birmingham. Such dedicated spaces for carrying on experiments, in institutions or private premises, were increasingly necessary for chemists, though for some scientists 'portable laboratories' sufficed. Laboratory science needed to be learned on the job, like a craft: and informal apprenticeship was the usual way in, until Liebig and others introduced teaching laboratories for research students from the 1820s, and soon undergraduates received practical instruction also. Biologists, too, required laboratories, primarily for dissection and microscopy, and for physicists they became essential as experimental and mathematical physics converged. All these activities entailed training, informal and then later formalized. From 1860, 'physical methods' entered the chemical laboratory with the spectroscope. The laboratory provided both equipment and apparatus, and also the controlled environment in which to use it. In the laboratory there was both inductive accumulation of knowledge, trying things out and seeing what happened, recording data; and also deductive testing of theories. This became increasingly important as theories of the atom, the nature of light, and of chemical structure, started to make testable predictions. The laboratory came to be seen as the place par excellence where scientific knowledge was generated.

For Einstein, science was a free creation of the human mind, and for Davy's favourite poet, Alexander Pope, the proper study of mankind was man.[63] Human bodies had been dissected, subject to various taboos, ever since ancient times, but during the nineteenth century the human body and mind, our relationship to the animal world, and the supposed distinctions between different races became a subject of scientific study as never before. It is to that vast field that we now turn.

7

BODIES, MINDS AND SPIRITS

Linnaeus in his *Systema Naturae* applied his double-barrelled system of naming to humans, labelling us *Homo sapiens*. For the brief Latin description that identifies each species, he wrote 'Nosce te ipsum', know thyself.[1] This was not as easy as it seems. He then described the various varieties of mankind, and named the other species in our genus *Homo sylvestris*, man of the woods, a translation of the Malay 'Orang-utan'. Linnaeus and his contemporaries used terms like 'family' and 'genus', which imply relationship, as metaphors, but by the end of the eighteenth century they were being taken literally in evolutionary speculations. The idea that the orang-utan was our cousin, if not perhaps a kissing cousin, was around long before Charles Darwin wrote the *Origin of Species*. But for most of Linnaeus's contemporaries, nature formed a great chain of being, from the lowliest forms of life up to humans, and then (in the realm of spirit) on through the orders of angels. It was important that there were no gaps in the chain, and naturalists on their voyages and travels should, as their research programme, look for 'missing links' that would make the chain complete. The orang-utan being very near us, with the various species of monkeys below him, meant that the chain just below us must be fairly well understood; in other parts of it, there were bigger gulfs. Linnaeus's own hierarchy was nested rather than linear, but it was just as important to fill it as far as possible and his disciples were despatched to distant countries to collect 'nondescript' animals and plants. The idea that plants and animals might change their characteristics was anathema because that would upset the Creation, and confuse the whole orderly business of science: this was a taxonomy for a stable society. We shall see how changing science changed views of what it meant to be human over the course of the nineteenth century.

In the later eighteenth century, exploring expeditions were sent from many different countries, notably Spain, France, Russia and Britain, with papers exempting them from capture in war.[2] But with the Battle of Trafalgar in 1805, British maritime supremacy was assured for the next hundred years and the Royal Navy particularly prominent. James Cook, defeating scurvy and armed with the *Nautical Almanac* and then with chronometers, had founded a tradition of survey and charting that was crucial to building up commerce and a seaborne empire.[3] Joseph Banks, accompanying him on his first voyage, never forgot the experience of meeting what Charles Darwin later called the naked savage in his native land. In Terra del Fuego, Tahiti, New Zealand and Australia, Cook's party encountered a range of cultures that they sought according to their lights to understand and respect.[4] For Tahiti and New Zealand, classical Greece and Rome provided models that any educated gentleman could evoke, and they gave classical nicknames to Tahitian chiefs. They also expected a hierarchy of king and chiefs, and indeed helped to bring it about. They brought classical and Christian practice to bear in trying to understand religious observances. As people had done in other sciences, they imposed a pattern upon what they observed, drawn from their reading and discussion of history and political philosophy, and from their enlightenment assumptions, or what they thought common sense. But they found the Fuegians and Australians much more difficult. European taboos on nakedness and dirtiness, and the imaginative leap needed to comprehend a life of hunting and gathering, got in the way of empathy. Theirs was after all the age of improvement.

Jean-Jacques Rousseau (1712–78), philosopher and botanist,[5] had written of the noble savage, and some travellers hoped to find people uncorrupted by commerce, wage-labour, and the complexities and compromises of civilized life. The instructions drawn up for Cook, and for the ill-fated French expedition under the Comte de la Pérouse (1741–88) that called at Botany Bay in 1788 and was never seen again, were very liberal, forbidding the use of excessive force and urging the brotherhood of man. Unfortunately, intercultural misunderstanding proved all too easy and lives were lost on both sides. It was a sobering moment to find on Cook's second voyage that a boat's crew from one of the ships had been killed and eaten in New Zealand.[6] Trading had to be carefully controlled, across a line drawn in the sand, because different ideas of property were involved, and without constant vigilance, pilfering of attractive toys like navigational instruments happened. When it did, the line between firmness (which often involved detaining, or kidnapping, a prominent native until the object was returned) and violence was hard to draw. Indeed, Cook was to die in such circumstances, in Hawaii, in 1779.

LANGUAGE: THE DEFINING CHARACTER OF HUMANS

For Banks, as well as botany, the proper study of mankind was man. Speech distinguished us from the orang-utan. He and others dallied with Tahitian women, learned the language, and were later astonished to find themselves understood over 2,000 miles away in New Zealand. In other Pacific islands the language was quite distinct, and vocabularies of different languages were drawn up – like modern phrasebooks, they can have been of only moderate use, but they did indicate extraordinary differences and curious similarities, where a word seemed to mean much the same in widely different parts of the world. But as we saw, this kind of comparison was becoming obsolete, notably as scholars like Jones, Herder and Wilhelm von Humboldt studied languages in the context of other cultures.[7] Jones had seized the opportunity to learn Sanskrit from a pundit, finding to his surprise that it was like European languages. British nabobs who went to India either returned rich or more often died out there – and Jones was unfortunately in the latter category. Humboldt, diplomat, political theorist and patron of the great classical architect Karl Friedrich Schinkel (1781–1841), as Prussian Minister of Education in 1810 founded the University of Berlin, which became the model for the research universities of the nineteenth century and thus crucial in our story for the development of all the sciences.[8] Loving grammar and syntax, he studied American languages, then Basque, and went on to write the essay that was eventually published posthumously in 1836, *On Language*, tracing the development of languages, their structure and their families as an introduction to a huge work on the language of Java.[9] For him, language was creative power, *energeia*, the organ of thought and key to diverse mentalities. Just as Cuvier replaced the great chain or ladder, and the Linnean nests, by a branching shrub, so Humboldt saw large distinct families of languages, each member of which had an individuality like a person. The best, most expressive, in his (unsurprising) view belonged to our group stemming from Sanskrit; he saw linguistic change, but not a linear progressive evolution. Each branch remained separate. Philology became a science, one of the most prestigious disciplines in German universities especially, with alarming implications for Bible study in the form of 'higher criticism' as scholars analysed the text.

The biblical myth for the diversity of tongues was that after the flood when humans all spoke the same language, they resolved to build a great city at Babel (Babylon) with a tower of bricks reaching up to heaven. When God saw this he perceived that nothing would be impossible for them. He therefore came down, scattering and confusing them so they no longer understood each other's language.[10] This wonderful story about hubris, about what we might do if we worked together, and for Christians a fore-shadowing of the 'speaking with tongues' at Pentecost, was taken literally in the eighteenth century. This pedestrian reading of it as history became a casualty of science, of geographical and linguistic discovery: the world was so big, and languages so diverse, that without whole series of miracles it was inconceivable that the changes could have happened within the biblical timescale. Jones's studies revealed that Hindus accepted vast cycles of past time, and German studies of Semitic languages began to indicate how the writers of the Bible had incorporated and reworked material from different cultures and different dates into the books that have come down to us. The most serious challenges to organized religion in the West came from such studies rather than from geology and biology.

For the orthodox, the capacity for speech that separated even the dirtiest 'savages' firmly from orang-utans was a sign of an immortal soul.[11] Thus for Descartes, bodies and minds were wholly distinct: animals were automata, but through the pineal gland in our brain, the seat of the soul, we were able to choose what to do and move our muscles accordingly. Humans, dwelling amphibiously in realms of matter and spirit, were god-like in their separation from the rest of nature. For Newton, famously, 'brute matter' was inert; even gravity could not be innate within it.[12] Though the law was exact, how things attracted each other across space, acting at a distance, was a mystery, and so was the relation between our souls and bodies. Medieval Christians had envisaged the dead sleeping until woken by the angel's Last Trump, bodily resurrected and sent to Heaven or Hell. By the eighteenth century, the alternative view prevailed, that our immortal soul, like a butter-fly emerging from a chrysalis, left its inert matter behind and went straight to join other members of the family in bliss (or maybe, for some families, pain). Priestley challenged this dualism with his Christian materialism, based upon his dynamic understanding of matter.[13] Roger Boscovich (1711–87) had suggested that the problem of action at a distance could be solved if atoms were point centres of forces, alternately attractive and repulsive, rather than billiard balls.[14] Priestley seized upon this idea, seeing the implication that if matter were active rather than inert then why shouldn't suitable combinations of it live and think? Matter was wonderful. He had abandoned the Calvinism of his upbringing and become a Unitarian minister, denying what he perceived as corruptions of true Christianity coming from Platonism: the immortal soul, and the doctrine of the Trinity. For him and other 'rational dissenters', Jesus was unique but not God. There was no need to invoke 'soul' or 'spirit' or any kind of immaterial substance. And at the Resurrection we would all by a miracle be restored to life and judged. Like earlier Protestant dissenters, Priestley became a fiery political figure, supporting American and French democracy.[15]

PHYSIOGNOMIES

In the reaction to the French Revolution, Priestley's version of materialism did not catch on, but in continental Europe another form of it arose. The Swiss pastor Johann Kaspar Lavater (1741–1801) sought to give a scientific basis to the old idea that faces indicate character. He illustrated his *Physiognomy* (1775–8) with silhouettes and engravings of human and animal heads, and details such as noses and lips that betrayed our propensities. Poor Lavater was wounded when a French army occupied Zurich in 1799, and he never fully recovered; he had misjudged the violent character of the soldiers, perhaps because he had no time to scrutinize their faces. His book was translated into English by Thomas Holcroft (1745–1809), the famous radical,[16] and his work was soon caught up and elaborated in a medical context, in craniology or, as it was later best known, phrenology, in which the whole head was significant. The idea behind it was that the brain was the organ of the mind, that the various faculties were located in distinct parts of it and that the shape of the skull indicated that of the brain beneath.

The first exponent was Franz Gall (1758–1828), who settled in Vienna in 1785 and practised medicine there. He began his project of mapping the skull, finding correlations between definite characteristics, such as benevolence, musicality or destructiveness, and the 'bumps' or prominent features of people's heads.[17] Intellectual and moral capacities were found to be placed above animal passions, as well as superior to them: highbrows are better-endowed with them. Big heads, like Cuvier's, will indicate big brains, and if cerebral faculties have definite locations of variable size, then because babies' skulls are soft they will bulge where the brain is well developed. Patients and prisoners brought Gall the opportunity to test and refine his

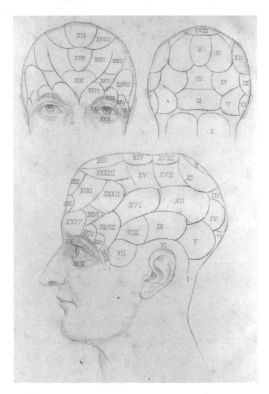

Figure 15 *Phrenological heads: J.G. Spurzheim,* The Physiognomical System of Drs Gall and Spurzheim, *London: Baldwin, Craddock and Joy, 2nd edn, 1815, frontispiece*

theory, adding faculties to the list as the investigation proceeded. His public lectures in Vienna attracted enthusiastic audiences until they were banned by the government in 1802, amid suspicions from orthodox medical men that this was quackery, and from orthodox churchmen that it was materialistic and undermined free choice. If character was set at birth by the shape of the brain, then someone who lacked the bump of benevolence could never feel the impulse to do good, and could hardly be blamed for that.

Phrenology was not crushed.[18] Indeed, rather like that other Viennese science of a century later, Freudian psychoanalysis, it spread through disciples and became an important aspect of culture for a hundred years, especially in the English-speaking world. Johann Spurzheim (1776–1832) came to Britain, and it was through his lectures and publications that the 'physiognomical system' of phrenology became well known.[19] His book set out its objectives on the title page: anatomical and physiological examination of the nervous system, and particularly the brain, would indicate the dispositions and manifestations of the mind. The frontispiece showed a head from the back, front and side with the various locations mapped upon it. Soon ceramic heads would be made on this pattern, and they can still be found in antiques shops. It was crucial that the brain was perceived as 'an aggregation of organs': the various faculties were separate from each other in a kind of federation. Disposition, propensities and character were thus the sum of parts. There were thirty-three faculties by 1815 as careful and close studies refined the classification: they included inhabitiveness (love for one's dwelling-place) and philoprogenetiveness (love of one's children), as well as combativeness, conscientiousness, secretiveness, self-esteem, hope, tune, comparison and wit. Our faculties were thus innate, and the phrenologist reading a head would, like today's aptitude and psychological testers, be able to predict behaviour and advise about educational and career choices.

Someone whose brain was combative, destructive, filled with self-esteem and covetous (but not benevolent, orderly or amative) would be dangerous to know, but the task of the educator was to work on other propensities to produce good citizens. Thus love of approbation, and perhaps cautiousness and causality (indicating that crime does not pay), might if present be encouraged, to counterbalance socially unwelcome traits, and trades might be found like butchering where even unpromising characteristics could be socially useful. One's better self might thus conquer baser instincts. The apparent determinism in phrenology (not so different from what many believe about genes in our day) was deeply shocking to many, and the evidence seemed shaky to others – a notorious murderer had the bump of benevolence unusually well-developed. But especially in Scotland, with its medical schools and Calvinist tradition of determinism or predestination, it caught on well, as we shall note in discussing scientific heresies. It is important in this story because it was a materialistic theory of mind that had wide appeal, especially among political radicals, and became widely if never fully respectable, a foundation for Victorian scientific naturalism. It was taken up by eminent educationalists, like the Scotsman George Combe (1788–1858) whose writings became best-sellers in Britain and America.[20] Painters and writers took advantage of phrenological findings, and race theorists noted that European males had larger heads than anyone else. Cranial capacity was taken to indicate mental capacity, despite Immanuel Kant's small head. Phrenology was perceived as more scientific than evolution, though it might well go with it, as in the work of the acerbic naturalist Hewett Cottrell Watson (1804–81, a pioneer of the study of plant distribution) and Chambers, the notorious author of *Vestiges* (1844).[21]

Few phrenologists managed to travel to exotic places, but Blumenbach had collected skulls from around the world, and Alexander von Humboldt brought some back from America – clearly, some were rounder and some longer than others. Human groups thus differed in the shape of their heads, and therefore in their characteristics and culture: their bumps were different. They also differed in the colour of their skin and in the form and proportions of their bodies. Explorers sketched Native Americans, Fuegians and Australians in characteristic attitudes, as best they could, but back in Europe the professional engravers who worked up their drawings for publication had studied classical statuary and knew how the nude should look. Wiry, scrawny, vigilant humans were transformed into Greek gods. Just as it took time for Dürer's magnificent rhinoceros to be superseded by more accurate representations, so European artists found it hard to depict the Antipodes and their inhabitants accurately.[22] Gradually, however, the squalid savage came to replace the noble savage in European consciousness. But even he or she had a soul to save, and as part of the evangelical revival from the late eighteenth century the flow of missionaries began, premised on the belief that we were all brothers and sisters under the skin.

ARE WE ALL ONE?

Not everybody believed that. Although wherever Europeans sailed, they found women to impregnate, and the offspring of such unions were themselves fertile (the normal test for membership of the same species), and although all were capable more or less of learning each other's language, it was possible and convenient to believe that we were not all descendants of Adam and Eve. After all, their sons had found wives, and it was possible though unorthodox to speculate about pre-Adamites, separately created.[23] All languages have disparaging terms for non-members of the group, indicating sub humanity. Perhaps non-Europeans were distinct species.

It was evident from experience in America that on contact with Europeans, natives died in great numbers, not simply because they had no muskets or horses but through disease (and little resistance to alcohol).[24] It was easy for readers of Malthus to see them coldly as less well-equipped for the struggle for existence, and for even the attractively sympathetic (like Catlin among the Mandans, and P.E. de Strzelecki (1797–1873) in Australia) to see the process as inexorable and inevitable.[25] Cook and his contemporaries were appalled to find that since the arrival of Europeans, venereal disease had broken out among the Tahitians. The French, the British and the Spaniards all blamed each other for thus infecting an earthly paradise. Nevertheless, there were people very ready to hunt and destroy the aboriginal inhabitants of the places they colonized and had perceived as unclaimed, treating them like noxious beasts rather than fellow-humans. Further, if such people really were distinct species, then enslaving them was not very different from domesticating animals. Farmers corral and own those, breed them, buy and sell them, and make them work. Slavery could be justified in biblical terms, going back to the story of Noah, his various sons and their pecking-order, but it was even easier if one denied 'monogenesis' (the idea that we all sprang from a single pair) and accepted 'polygenesis' (separate creation of races) instead. The idea was popular amongst supporters of the Confederacy in Britain and abroad during the US Civil War of the early 1860s. These surprisingly included Richard Burton (1821–90), who was well known as an explorer of Africa, and an admirer of Islam who had made a pilgrimage to Mecca in disguise, got on easily in exotic cultures, and might have been expected to see all mankind as one.[26]

Slavery had begun to horrify Britons belatedly, in the second half of the eighteenth century. The evangelicals Thomas Clarkson (1760–1846), Granville Sharp (1735–1813) and William Wilberforce (1759–1833) created a furore, leading to abolition of the slave trade (then mostly in British hands) in 1807, and a generation later abolition of slavery in the British colonies.[27] The Royal Navy was to enforce these measures. Meanwhile, Napoleon had suppressed the free black government of Haiti whose leader Toussaint L'Ouverture died a captive in France in 1803. Of him, Wordsworth wrote:[28]

> There's not a breathing of the common wind
> That will forget thee; thou hast great allies;
> Thy friends are exultations, agonies,
> And love, and Man's unconquerable mind.

Many boycotted sugar, the produce of the slave plantations, and Banks and others were active in the African Association, founded to open up the continent and suppress slavery. As part of a programme of exploration, in 1795 they despatched the surgeon Mungo Park (1771–1806), who got to the fabled city of Timbuktu, though he died (like many of their emissaries) on a subsequent expedition.[29] Meanwhile, the Sierra Leone settlement had been founded in the 1780s as a home for the freed slaves who had fought with the British against their American masters. In 1808 it formally became a British colony, where slaves found by the Royal Navy on ships intercepted (whatever flag they flew) were put ashore, irrespective of where they had come from originally. These settlers were not well prepared or equipped for their new life, but Thomas Winterbottom (1765–1859) was sent there as a surgeon to look after the health of the colony.[30] African women had saved Park's life, but while many (black and white) were prepared to see mankind as one great family and rescue their fellow-humans from a horrible fate, race inexorably became an important factor in thinking about people, and was to remain so throughout the century.

OUR COUSINS THE APES?

The Scottish judge James Burnett, Lord Monboddo (1714–99), was a major figure in the Scottish Enlightenment. He had in his voluminous works, *The Origin and Progress of Language* (1773–92) and *Ancient Metaphysics* (1779–99), emphasized our affinity with the apes, seeing the orang-utan as a variety of human, accidentally speechless. His, and Erasmus Darwin's, writings meant that by about 1800 evolutionary ideas were familiar if not taken seriously. Thus the young Romantic, Thomas de Quincey (1785–1859, the 'Opium-eater'), playing games of world domination with his big brother, found not only that most of his territory was captured, but that in the little island he was left with the inhabitants had evolved less than other humans and still had tails. Elsewhere, these had (on Monboddo's authority) worn away through being sat on over generations.[31] Then in 1817 Shelley's friend Thomas Love Peacock (1785–1866) published *Melincourt*, a comic and rather rambling novel in which the hero is an orang-utan, caught very young in the forests of Angola. He is of immense strength and even nobler than a noble savage in his courtesy, and is bought a baronetcy.[32] As Sir Oran Haut-ton, the natural and original man, he is then bought a seat in parliament, representing a rotten borough, where being able to vote but not speak is advantageous. As a nursery joke and a vehicle for entertaining satire, the apes (their different species not yet distinguished) were a good topic, but not yet an occasion for serious reflection on what it meant to be human.

This might involve a closer look at human bodies and what made them special, as well as minds and their expression in language. The eminent doctor Charles Bell, famous both for his work on the nerves and as a talented artist for his medical illustrations, wrote a Bridgewater Treatise on how the human hand displayed divine forethought and design.[33] He believed there was more to communication than speech, and that our bodies

AGITATED TAILOR (to foreign-looking gentleman), "Y-you're rather l-long in the arm, S-sir, b-b-but I'll d-d-do my b-b-best to fit you!"

Figure 16 *Our cousin the gorilla being fitted for a suit,* Punch, *28 December 1861*

were contrived to make communication easy.[34] Body language, especially facial expression, was crucial in conveying emotion: we frown, we smile, we sneer and for all these things we have appropriate muscles. His book was taken up by artists, along with phrenological texts, as part of their anatomical education, and went on appearing right through the century. But whereas phrenology with its organs for animal passions, and physiognomy with its beastly faces, emphasized in some respects our relationships with animals and our place in nature, Bell's was not just a natural religion such as Combe privately espoused,[35] or Deism like Chambers's, but much more orthodox.

So were the writings of James Cowles Prichard (1786–1848) on mankind. Anthropology, or more strictly ethnology, as a separate science had begun in Paris, that great centre of intellectual activity, with Joseph Marié Degérando (1772–1842), who wrote in 1800 on how to observe 'savage peoples' for the expedition sailing under Nicolas Baudin (1754–1803) to Australia.[36] The voyage was not a happy one because the citizen savants were a quarrelsome lot, who united only in turning against Baudin because they wanted more time ashore to gather data. But they did produce some magnificent artwork, including portraits of Tasmanian and Australian aborigines, and scenes of them dancing and lighting fires by rubbing sticks. The expedition called at the recently founded and at last modestly flourishing British convict settlement at Sydney Cove, near Botany Bay, where they were welcome because of their scientific credentials, even though Britain and France were at war. That settlement also had its share of artists and natural historians, both among the naval and military officers and men, and also among the convicts, some of whom were artists who had been found out using their skills in forgery. By the end of the Napoleonic Wars, Europeans had a good idea of what other races looked like.

Prichard brought together in a conservative synthesis all the various strands that we have been examining. Brought up a Quaker, he studied medicine in Edinburgh, before joining the Church of England and attending both Cambridge and Oxford.[37] He embarked on a medical career in Bristol, specializing as an alienist (or 'mad-doctor'). Staunchly conservative, he embraced humoral theory rather than the innovations of William Cullen (1710–90) and John Brown (c.1735–1788), his teachers in Scotland. He devoted himself to proving the unity and common descent of the whole human race in vindication of the Bible. In 1813 he published *Researches into the Physical History of Man*, in which men were demonstrated to be one species, distinct from apes so that black men were not some 'missing link'.[38] Early men (possibly even Adam and Eve) had been black. Differences within racial groups were as significant as any between them. Prichard was averse to evolutionary speculations, and along with them the idea that characteristics acquired (or lost, like those tails in de Quincey's world) in response to environments might be inherited, so for him whiteness was acquired with civilization rather than in response to climate. In later editions he became more sympathetic to environmentalism. He was well-read in French and German works, and by 1843 when he published a more popular version, *The Natural History of Man*, fully and handsomely illustrated in black and white and colour, he was aware of the Danish idea that we had come through stone, bronze and iron ages.[39] Clearly, although some of the quotations are unpleasantly condescending, Prichard and the artists whose work he used were aware that black was beautiful too.

That book was dedicated to Christian Karl Josias Bunsen (1791–1860), the Prussian ambassador and an eminent philologist and biblical scholar. Prichard was very interested in the development of languages as evidence of the unity of mankind, though this required expansion of the biblical time-scale by a good many missing generations. Missionaries in the wake of the evangelical revival had by then translated the Bible into numerous languages, fixing them as Luther's Bible had fixed German, and provided grammars and dictionaries. Prichard argued that not only were all humans anatomically and physiologically alike, and speaking related languages, but that they all originated from one place, in Mesopotamia.

Prichard's caring conservatism extended to the insane, where it could take practical form in humane treatment of those who, whatever their symptoms, remain our brothers and sisters. His work in psychiatry, where he urged that there could be 'moral insanity' even without the delusions then required for legal pleas of madness, led him to oppose phrenology. He did not believe that the mental and the moral could all be reduced to brain states, and saw phrenology as grossly materialistic and thus liable to undermine and corrode society and its values. Not just the brain, but the viscera ('gut reactions'), nervous system, and indeed the whole body played their parts in people's emotional lives.

HUMAN ANCESTRY

At the mid-century, Prichard was, along with Darwin, a contributor to John Herschel's *Admiralty Manual of Scientific Enquiry*. But his armchair ethnology and his reliance upon the Bible were by the end of his life old-fashioned, in a more travelled, sceptical and racist world.[40] From his days as a medical student in Edinburgh, Darwin had been interested in psychology and materialistic accounts of the mind, and also in a wide range of animals, including dogs and beetles. He published the *Origin of Species* in 1859, but not until thirteen years later another classic work, the *Expression of the Emotions in Man and Animals*, where he used the observations he had made on his children, as well as pets and other animals.[41] Illustrated with engravings and photographs (unusual at that date), the book argues that animals display in their cries and body language emotions similar to ours – something familiar to dog-owners, but smacking of anthropomorphism to the stiff-necked. Darwin disliked the language of 'higher' and 'lower' applied to animals as a vestige of the 'great chain' idea; all organisms were the outcome of adaptation over time to their environment, by natural selection, not of some progressive thrust upwards and onwards.

Darwin shared Prichard's detestation of slavery, and so did his ally Huxley, who in 1863 had published *Man's Place in Nature*, filling out the

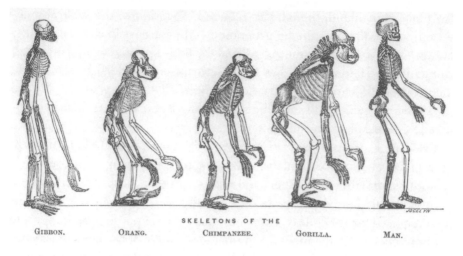

SKELETONS OF THE

GIBBON.   ORANG.   CHIMPANZEE.   GORILLA.   MAN.

Figure 17 *Skeletons of the primate family: T.H. Huxley*, Evidence as to Man's Place in Nature, *London: Macmillan, 1863*

case for human relationship with the apes (now including the chimpanzee, the gorilla and the gibbon as well as the orang-utan) that he had sketched at the famous Oxford meeting of the BAAS in 1860.[42] The frontispiece of this book, a procession of skeletons, has become an evolutionary icon: but it was not meant that way, because the skeletons are of modern apes not of our hypothetical ancestors. Huxley was doing taxonomy. His point was that an unprejudiced visitor arriving from Saturn would see families of creatures, like horses, donkeys and zebras, or bullfinches, chaffinches and greenfinches – and like the apes, including us. As well as anatomical and physiological material, this accessible book also contained descriptive natural history, including anecdotes about human cannibalism. There was no longer any serious hope of claiming that humans were a distinct zoological family, set apart from the rest of the animal kingdom. The orang-utan really was our cousin.

The great geologist Charles Lyell, Darwin's long-standing confidant who had persuaded John Murray to publish the *Origin*, had for many years sat on the fence over evolutionary theory, finding great difficulty in combining progress with uniformity. But in 1863, cautiously favouring Darwin's theory, he published his *Antiquity of Man*, summarizing recent archaeological research which established that the ages described by Prichard took humans back into the realm of mammoths and other extinct creatures. Huxley wrote part of the last chapter, on the similarities between our brains and those of apes. There are not many predictions possible about evolution, but Lyell suggested presciently that if we have common ancestors with the apes, then the places to look for early human fossils are where the gorilla, the chimpanzee and the orang-utan live today. In his concluding paragraph he faced squarely the accusation of materialism, a serious charge because it was associated with immorality and revolution:[43]

> It may be said that, so far from having a materialistic tendency, the supposed introduction into the earth at successive geological periods of life, – sensation, – instinct, – the intelligence of the higher mammalian bordering on reason, – and lastly the improvable reason of man himself, presents us with a picture of the ever-increasing dominion of mind over matter.

Mankind might uniquely be made in the image of God, but it was becoming harder to think so straightforwardly.

In his *Descent of Man* that came out in 1871, Darwin had no truck with polygenesis, and sketched out an evolutionary history in which human characteristics had gradually developed.[44] He devoted much space to his new idea of sexual selection. Among birds such as peafowl, the hen is inconspicuous but the cock has developed splendid plumage – which in fleeing from predators would be a great hindrance to rapid take-off. It was no longer good enough to say that God, delighting in beauty, had made them so. Darwin's conclusion was that peahens were cool creatures, only turned on by a magnificent show, and had consistently chosen the most extravagantly beautiful mates. In similar vein, lions' manes and the splendid antlers of the extinct Irish elk could be accounted for in the struggle not for existence directly but for a mate. Human racial characteristics, for this scion of an anti-slavery family, resulted not from environment but from sexual preferences. Mainstream Victorian science was less buttoned-up than emancipated Bloomsbury writers might have supposed.

Darwin's *Origin* had been written because Alfred Russel Wallace, living with Dyaks among the orang-utans in Malaysia, had in 1858 sent him an essay setting out a theory of evolution by natural selection essentially identical with what Darwin had been refining over nearly twenty years.[45] Earlier, Wallace (a natural history collector) had been on the Amazon with Henry Walter Bates (1825–92), whose studies of butterflies that 'mimicked' other species was an early confirmation that natural selection worked.[46] Nevertheless, Wallace could not accept Darwin's thoroughly naturalistic account of human origins: our capacities for pure mathematics and for music seemed to have no connection with survival. Wallace could not believe that we were just modified animals. He (like a number of other prominent men of science) became a convert to spiritualism, believing that our spirits survived 'bodily death' and could (through mediums) communicate with us from the 'other side'.[47] Without going so far, many contemporaries felt there must be more to humans than just the matter of which we are composed.

ANIMAL MAGNETISM AND THE SPIRITUAL REALM

This belief, and interest in strange phenomena, take us back to another Viennese doctor, Franz-Anton Mesmer (1734–1815), who in the 1770s had begun treating his patients with magnets. He believed that magnetism pervaded the world and that animal magnetism was a feature of living creatures, demonstrated by passing magnets over patients and inducing a trance, or perhaps convulsions, that might issue in a helpful crisis and cure.[48] In 1778 he went to Paris, where his cures created a sensation. But when a committee of the Academy (including Lavoisier, and Franklin, in Paris as American Ambassador) investigated his claims in 1784, they found that magnets were unnecessary (sticks would do the trick), and the magnetic fluid undetectable. He was denounced as a charlatan, his cures the effects of 'imagination'. Various eminent patients nevertheless continued as patrons and patients, but amid the turmoil of the Revolution he left France for a time. Eventually, he retired to Switzerland and died in relative obscurity.

Magnetic 'tractors' to draw and conduct the fluid became part of the armoury of alternative medicine, and patients might be mesmerized, their will mysteriously controlled. The nervous diseases with which Mesmer had enjoyed success continued to perplex the medical profession, and it was clear that bodies and minds were involved. In England, Coleridge is credited by the *Oxford English Dictionary* with introducing the term 'psychosomatic'. He made a profound study of medicine, particularly with Beddoes and then with James Gillman in whose house in Highgate he spent his later years, and understood that both mental and physical elements underlay his own conditions.[49] He was critical of Lawrence, as indeed were many in the medical profession: the mental could not just be reduced to the bodily, and should not be left to mountebanks. He was interested in electricity, and in the altered states of consciousness produced by Davy's laughing gas, and by drugs such as the opium prescribed to him and to which he became addicted. He also loved the mysterious, or the spooky, as his great poems 'Christabel', 'The Ancient Mariner' and 'Kubla Khan' show, and his reading indicates.[50] The younger poet Shelley also took great interest in the debate between Lawrence and Abernethy, and in vitality – but came down on the materialist side. And his wife, Mary, in *Frankenstein* (1818), showed where such ideas might lead the thoughtlessly ambitious physiologist.[51] The science of the Romantic period in Britain might celebrate with Priestley the wonders of active matter, or with Blake see Newton looking down, away from the spiritual realm that really mattered; and in Germany, the centre of Romantic activity, it was much the same.[52] Meanwhile, in Scandinavia, the mystical writings of Emanuel Swedenborg (1688–1772) had made him many disciples, and later Ørsted's anti-materialistic miscellaneous writings were published in Denmark, then in England, as *The Soul in Nature*.[53]

Mesmer's patients thrown into a trance or convulsions were like those overcome during the evangelistic meetings of the early Methodists and, in a long line going way back into prehistory, of shamans and mediums who communicated with spirits. None of the academicians investigating Mesmer had found his performance mesmerizing: it seemed that only the uneducated were affected, and particularly (it was claimed) women, who were susceptible – making it further disreputable. The industrial chemist Karl von Reichnbach (1788–1869) saw auras of 'odyle', the ethereal substance responsible for numerous phenomena, around the heads of sensitive souls, notably pregnant women.[54] But it was in the USA that the great flowering of activity in contacting the spirit world took place, in the 'burned-over district' of upper New York state, home of numerous religious revivals and camp meetings.[55] In 1848, two girls in the Fox family, respectable Methodists, heard mysterious rappings in response to questions, and a craze for communicating with the spirits of the dead spread rapidly through New England and in due course across the Atlantic to Europe.[56] The bereaved could suddenly be reassured that their beloved friends and relations were not lost for ever. They would be reunited in another world, from which messages (usually curiously banal) could come through a medium. And spiritualism empowered women, denied prominence in mainstream churches.

PSYCHICAL RESEARCH

Spiritualism caught on in Europe, especially in Britain where religious doubt was increasing. Some of those who lost their faith felt liberated, but for others it was painful, marking the distance from happy childhood on sunlit vicarage lawns to the world of drizzling rain, bald streets and blank days in dirty cities. Nostalgia, and a hope that lost faith might be restored through empirical evidence that the dead had not perished, along with the many bereavements our ancestors endured, propelled men of science like Wallace and Crookes into spiritualism. Seances took place in darkened rooms, and the happenings grew more spectacular as 'physical' manifestations of the spirit world were evoked by mediums. Usually they were female, and often held down by members of the party to demonstrate that they were not wandering around the room and producing the effects. (The opportunity to grasp young women in the dark did not improve the reputation of the investigations.) Some mediums were male, notably the Revd Stainton Moses (1839–92): 'Where was Moses when the light went out?' was a classic question, Another, Daniel Dunglas Home (1833–86), became the model for 'Mr Sludge' in Robert Browning's sardonic poem (written after the death of his wife Elizabeth Barrett Browning, who had been a devotee): Home was seen on one occasion to drift out of one upstairs window and back in again at another.[57] Reliable if puzzled witnesses attested to extraordinary manifestations, and received consolation. Crookes inferred a psychic force at work and submitted reports to the Royal Society, of which he was a Fellow (and would in due course become President), but they were rejected and he had to publish them in his own journal in 1870.[58] Henry Sidgwick (1838–1900) at Cambridge, a high-minded and well-connected agnostic, agreed in 1882 to chair a Society for Psychical Research to examine the phenomena.[59]

In Britain, Edmund Gurney (1847–88), Frederic Myers (1843–1901) and Frank Podmore (1856–1910) were set to work by the SPR to investigate phantasms of the living.[60] When people far from home were in mortal danger, they seemed quite often to appear at the bedsides of their loved ones, and a census of such happenings (among the respectable classes), interviews, and careful comparisons of times, indicated that this was a real phenomenon (and for us an extraordinary route into Victorian life and death). So they investigated telepathy, which might be the cause, with playing cards held up and with shapes drawn by one participant out of sight of the other. Sometimes it worked and sometimes not – the great problem with parapsychology.

Meanwhile, Sidgwick and his formidable wife Nora (1845–1936), with others, investigated scances, sometimes detecting fraud or conjuring tricks and sometimes not, but unable to satisfy themselves that any phenomena produced by mediums were incontestably genuine.[61] Myers took forward the 'phantasms' story, looking at cases where the apparition was of somebody already dead, and in 1903 his book was published posthumously, dedicated appropriately enough to Sidgwick and Gurney who had also died.[62]

Myers was fascinated by the way those hypnotized and woken seemed unaware of what they had done in the other state, though they carried out commands given then. This seemed rather like the phenomenon of multiple personality, dramatized by Robert Louis Stevenson in *The Strange Case of Dr. Jekyll and Mr. Hyde* (1886), where the bad persona loses the honorific title of doctor. Such cases were being investigated by the famous Parisian doctor Jean-Martin Charcot (1825–93) and others, and aroused wide interest. Myers believed that all of us are multiple to a degree, and also that in our minds the conscious part sits on top of the subliminal. Across that threshold, mediums dredge their messages, perhaps transmitted telepathically. And while most of us are uneasy about what is down there, geniuses are on good terms with it. It is no surprise that Myers introduced Freud to readers of English.

Meanwhile, on the continent of Europe things had been developing otherwise. In France, the physiological tradition of Bichat and Magendie was carried on with the laboratory studies of Claude Bernard, and then the clinical work of Charcot, physician at the Salpêtrière Hospital for nervous diseases in Paris.[63] He attracted students from all over the world, among them Freud, and became especially famous for his work with hysterics, sometimes in public sessions where his patients would, like Mesmer's, pass into a hypnotic state or become convulsed. Despite the secular foundations of the Third Republic after 1871, and the perceived conflict between science and the Roman Catholic Church, Darwinian theory did not catch on in the positivistic climate of Paris where Cuvier was still revered, but in triumphant Imperial Germany things were different. *Naturphilosophie* had prepared the ground for evolutionary ideas, medicine and physiology had made great advances, and in Jena Haeckel propounded evolutionary monism, often but unfairly taken to be materialistic, a naturalistic version of Darwinism in which humans were usually portrayed occupying the apex of a tree rather than just a branch of a bush. He had a bust made of our hypothetical 'missing link' ape-man ancestor and made amazing pictures of the creatures that compose plankton. In the rather suffocatingly respectable Wilhelmine world, he stuck up boldly for his monism, which he saw as a religion rather than a version of agnosticism.[64] His house is now a museum. Not far away in Liepzig, Wilhelm Wundt (1832–1920) was from 1875 doing the experimental psychology that promised definite answers to clear questions, and proved a model for psychology in the USA for the twentieth century.

CONCLUSION

Running through this story has been concern about materialism, sometimes welcomed but generally seen as reductive and as undermining morality. Humans were perceived as the talking animals, separated by our languages (evidence of mind and soul) from apes that were in other ways alarmingly like us. The phrenologists' suggestion that the shape of our brains, revealed by the bumps on our skulls, determined our propensities was for some liberating and radical; for others, it undermined responsibility and free choice. Either way, it created a stir and seemed like a way into understanding the mind. Campaigns against slavery, and voyages to the South Seas and Australasia, raised in acute form the question of whether all humans are brothers and sisters under the skin, and pioneers of anthropology were divided into polygenists, for whom the different races were distinct species, and monogenists who opposed such racism. For evolutionists, we were not merely very like apes but also shared a common ancestry, and they were our cousins. The gulf between humans and the rest of the animal world seemed to be disappearing. The spectre of materialism loomed larger. In the English-speaking world, many turned to spiritualism in the search for evidence that we had immortal souls, and here as elsewhere, science (in the form of psychical research) seemed to promise an answer. In the event, nothing firm came from it, and on the continent of Europe, psychology – experimental with Wundt and theoretical with Freud – prevailed as the science of mind.

Myers and his associates had hoped to establish that our immortal souls live on and can communicate with us, and mediums duly claimed to have received messages from him. But by the end of his life he saw humans as fragmented, with much of our personality submerged: it is not clear what he hoped would survive. And his friends found no certainty in their investigations. William James (1842–1910), a supporter of the SPR, looked at religion from the outside, writing his fascinating *Varieties of Religious Experience* (1902), while another supporter, Nora Sidgwick's brother Arthur Balfour, Prime Minister, President of the BAAS, and later in effect Minister of Science in the 1920s, retained his conviction that science rests upon faith, its field is limited and its explanations partial. For him, Darwin was right as far as he went, but goodness and beauty made life significant, and could not be reduced to survival values. There had been a few materialists at the beginning of the century and there were far more at the end, but *Homo sapiens* seemed as puzzling, and 'know thyself' as difficult, in 1900 as in 1800. Moreover, the century closed not only with interest in the spirits, but with *fin-de-siècle* gloom and even nihilism, while magic and irrationality played an important part in the 'modern' movement.[65] Nevertheless, it had been full of scientific triumphs, and to these and the growth of a Church Scientific, scientific 'religion' playing a cultural role like that of the Church in past centuries, we shall now turn.

8

THE TIME OF TRIUMPH

All over Europe, the 1840s had been a grim decade, the 'hungry Forties', culminating in the year of revolutions, 1848, when infuriated populations turned against their rulers. In Ireland there was real famine with the failure of the potato crop, and mass emigration to Britain and America – often to lives of squalor in the slums. Even in stable Britain, the wealthy who had grown up in the shadow of the French wars lived in dread of revolution from an ungovernable population. The Reform Bill of 1832 had given votes to the middle classes in industrial cities, but democracy (an alarming word in the nineteenth century) was still far off.[1] A big demonstration on London's Kennington Common, demanding the People's Charter with votes for all men, was contained by the powers that be. As the dust settled, all over continental Europe authoritarian regimes soon re-established themselves, little ameliorated, while exiled revolutionaries went to Britain (and its empire) and to America, where they formed an educated and energetic diaspora in otherwise philistine cultures. The rapid growth of cities was still maintained by immigration as the terrible death rates, especially of children, did not fall in London until the 1870s.[2] Other cities around the world were no better. Davy's visions of applied science conquering disease and bringing plenty were not apparently being realized in the 1840s, despite the coming of a maturing economy of railways, telegraphs and big industries.[3] The spirit of the 1850s was different: economic circumstances improved, and hunger was relieved by imports of grain from North America and the introduction of chemical fertilizers devised by Liebig and others.

PUTTING SCIENCE ON SHOW

Rumford's plan in 1799 for the Royal Institution had been that technical innovations would be exhibited as part of a project to improve the education of 'mechanics'. It had not happened, partly because running an institution in London's West End required the subscriptions of the wealthy rather than the small contributions of mechanics, and partly because manufacturers were unwilling to give away trade secrets. Industrialists like Wedgwood, Boulton and Watt lived in an atmosphere of secrecy, with anxiety about industrial espionage, because patents were problematic, being tricky and expensive to obtain and enforce. Wollaston's secretiveness about his platinum manufacture was notorious, but prudent and not untypical, and he was criticized only for getting scientific glory as well as money from his process.[4] By 1850, things were rather different. In France, where eminent scientists advised government-owned industries like the Gobelins tapestries, the Sèvres pottery and the le Creuset ironworks, there had been a series of exhibitions ever since 1797, culminating in a much-admired Paris exposition of 1849 in a temporary 'palace' on the Champs-Élysées. Also in Paris the Conservatoire des Arts et Métiers had a permanent exhibition of machinery. In Germany, exhibitions promoting crafts and industries were held in 'winter gardens'. Compulsory education on the continent, and Mechanics' Institutes and the March of Mind in Britain, were bringing closer Rumford's world of educated artisans, while tighter patent legislation, national and even international, was safeguarding innovations. The Society of Arts, which had long given 'premiums' for and publicity to new inventions, in 1817 appointed the distinguished chemist and mineralogist Arthur Aikin (1773–1854) as its salaried Secretary. Under his regime (1817–40), it turned towards the more modern activities of reading papers and advising on patents. In Germany, different states had come together in the Zollverein, a customs union, to promote trade and industry. There was a new confidence and the time was ripe for showing off about progress and improvement.

Though in France new inventions and machinery had been exhibited to international acclaim, Britain was the centre of the Industrial Revolution, and its industrial cities mounted exhibitions in the 1840s to promote their wares. But these were on a small scale, of regional interest. Queen Victoria's husband Albert had come to Britain filled with hopes of modernizing its institutions. In 1844 he was instrumental in bringing Liebig's favourite pupil Hofmann to London to run a new Royal College of Chemistry. Then, in

Figure 18 The Queen and Prince Albert inspecting machinery: The Illustrated
Exhibitor . . . in the Great Exhibition, *London: Cassell, 1851, p. 117*

1850, Prince Albert helped to persuade the Society of Arts, of which he was
President, to decide upon a great exhibition of the works of all nations, to
be held in London in the following year, with a Royal Commission being
duly appointed to arrange and organize it. A committee tried to design a
suitable structure, but could not reach final agreement. One member was
Joseph Paxton (1803–65), the former gardener's boy who had become the
personal friend of the Duke of Devonshire, for whom he designed enormous
conservatories. He sketched a design for a glasshouse a third of a mile (over
500 metres) long, lofty enough to contain the mature trees growing on the
chosen site in the wide-open spaces of Hyde Park.[5] Refreshments (taking
advantage of the shade of the trees), offices for tickets and for administration,
fire hydrants and public lavatories all had to be included. His idea received
public acclaim, and on 16 July 1850 his tender was verbally accepted, the
date for the royal opening being fixed for 1 May 1851. When we think of
the long run-up to projects like the Olympic Games in our day, we should
be duly astonished that everything was ready on time. And in the humorous
magazine *Punch* the building was christened the Crystal Palace.

The original intention of the Commissioners was that objects exhibited there should be arranged according to their nature and function, rather than their country of origin, but in the event this proved impractical and there were areas set aside for the products of different nations. The lion's share went to Britain and her empire, but the exhibition was truly international.[6] Juries were appointed to judge the exhibits in their various categories and award medals: they consisted of six or eight experts, half from abroad. The juries published their reports, in a thick volume densely packed with information about commodities. Guides to the exhibition, official and unofficial, were also published. One, *The Illustrated Exhibitor*, came out in weekly parts priced at 2d, which could eventually be bought as a volume. It contains, in eye-straining type set in double columns, a great deal of matter about the Crystal Palace and what was on show in it, including a full history of the construction, the girder and gutter arrangements, and how it was held up by snugs and bolts. Also included were plates, some folding out, of exhibits and events, letters from eminent foreigners, descriptions of exhibits such as printing presses and the model coalmine, lists of committees, jurors and medal-winners, and essays about the various countries represented. Everybody found the building astonishing. It was compared to a cathedral, and one awestruck journalist admitted early in the morning, before it opened to the public, declared that 'The GREAT GOD is in the midst!' [7]

The impetus for the exhibition was undoubtedly industrial, but what may surprise us is the emphasis upon the aesthetic. Looking through the *Illustrated Exhibitor*, the number of sculptures of chilly-looking nudes takes one aback, and the 'medieval court' with its Gothic furnishings was a very popular part of the exhibition. Thus the *Art Journal* published its catalogue of the exhibition, featuring beautiful rather than merely useful objects, such as manufactures or raw materials like gums or minerals, for which the jurors were awarding most of the medals.[8] Windows had previously been taxed, and the excise tax on glass had only recently been lifted, in 1845. That and new techniques made glass cheaper, better and more available – a modern material par excellence, and source for wonder and metaphor. Glassware displayed in this glasshouse was particularly striking to contemporaries.[9] Historians and cultural commentators have investigated the various messages that the millions of visitors to the exhibition picked up.[10] An important part of its significance was that the huge crowds, many visiting London for the first time (making use of the new-fangled railways), were so peaceable and interested: social harmony was being achieved, while on a larger scale confidence, and the promotion of international trade, peace and understanding was also evident. This can be contrasted with the exactly contemporary anti-Catholic hysteria about 'Papal aggression'.[11]

The Crystal Palace was demolished when the exhibition ended in October, and was re-erected in a modified form at Sydenham, where it stood until 1936. It had been damaged by a fire in the 1860s,[12] and it happened that another destroyed it on the night I was born. By then, it seemed musty, and the epithet 'Victorian' was a term of abuse. The dinosaurs reconstructed there remain, but the exhibition's legacy was much wider. From the profit it made (unlike all subsequent world fairs), the Kensington Gore estate (then on the outskirts of London) was bought, with the intention of making it a great cultural centre, Albertopolis,[13] like the museum-island begun in 1830 in Berlin. There, because the British needed to catch up with superior French taste and industrial design in particular, a museum (later named after Victoria and Albert) and Royal College of Art were founded. These were joined in due course by the Normal School (on the French model) for training teachers, the Albert Hall and Royal College of Music (for the British were well aware that they were behind the Germans musically), and, specifically for science, the Natural History Museum, the Science Museum, and the institutions (including the School of Mines and Royal College of Chemistry) that became Imperial College in 1907.

One lesson that might have been learned by earnest visitors from what was on show was that:[14]

There's nothing of honour, or wisdom, or worth,
But hard-fisted labour has been at its birth.

Victorians indeed admired hard work and self-help, but they were also encouraged to see science as the basis for all the improvements so evident in the Exhibition. Thus, appended to the *Art Journal Catalogue* there were essays, one by Robert Hunt (1807–87) on science, and others on manufactures and vegetable substances, as well as on colours and good taste (areas in which the British exhibits were felt to be deficient). Hunt was professor of experimental physics and a mineralogist at the Museum of Practical Geology, later to be amalgamated with the Royal College of Chemistry and moved to South Kensington. In line with Davy, Liebig and others, he claimed credit for science as the basis for technology:[15] 'The inventive genius, being closely allied to imaginative power, must be restrained by a philosophical [scientific] education to become of value to its possessor, or available for the benefit of his race.'

The inventor relying upon 'blind guesses' will, like the sorcerer's apprentice, be destroyed by spirits he has evoked but cannot control. Electric light and motors, useless toys in 1851, would (Hunt expected) become valuable as the primary laws of electricity were understood: mere empiricism will be good for nothing. Hunt derived satisfaction from observing that no novel application of science previously unknown in England had been disclosed at the Exhibition, and ended in a patriotic and theological glow, echoing Prince Albert:

> Believing that the entire world, from China in the east to Chili and California in the west will feel the exciting tremors of vitality which spring from the industrial heart of the world in 1851, we content ourselves with that excellent maxim . . .: 'Say not the discoveries we make are our own; the germs of every art are implanted within us, and God, our instructor, from hidden sources, develops the faculties of invention.'

Hofmann's laboratory, where synthetic 'aniline' dyes were first developed in the 1860s, would fit well with Hunt's picture of scientific knowledge as a basis for technology.

LEARNING FROM EXPERIENCE

While in the wake of the Exhibition a certain smugness about the home team having done well was characteristic of Briton, who could indeed take a proper pride in it, members of the scientific community were less happy and feared that the country might rest on its laurels. As we saw, Whitworth realized that Britons had much to learn from the United States about mass-production and standardization. Others were struck by the superiority of French design, and of German higher education, especially in science (but not by German industry, where few were far-seeing until the shock of Prussian victory over France in 1870). At Prince Albert's suggestion, after the Exhibition closed in October a course of weekly lectures about it was arranged by the Society of Arts through a committee including Brunel, Dickens, Faraday and the ubiquitous civil servant Henry Cole (1808–82). Held between November and March 1852, the lectures represent triumphant but critical thinking about the Exhibition's significance, especially for the future. We should remember that, despite those statues, the word 'art' meant craft and technique, making useful things, as often as decorative and aesthetic skill in fine art.

The inaugural speaker was Whewell, who emphasized the importance of science as the basis of arts, coyly drawing attention to his word 'scientist':[16]

> I have as yet said nothing of the effect which must be produced upon art and science by this gathering of the artists and *scientists* (if I may be permitted the word) of the world together; by their joint study of the productions of art from every land, by their endeavours to appreciate and estimate the merits of productions, and instruments of production; of works of thought, skill and beauty.

Other lecturers spoke plainly about British backwardness, especially in the science that underlay and could improve production across a range of industries, and find fresh uses for raw materials. Even British warships of Nelson's time had been inferior to French ones. British industry was backward and dependent upon foreign talent, and as Lyon Playfair (1819–98), chemist and politician, put it:[17]

> Our great danger is, that, in our national vanity, we should exult in our conquests, forgetting our defeats . . . A competition in Industry must, at this advanced stage of civilization, be a competition of intellect. The influence of capital may purchase you for a time foreign talent. . . but is not all this a suicidal policy, which must have a termination, not for the individual manufacturer, who wisely buys the talent wherever he can get it, but for the nation, which, careless of the education of her sons, sends our capital abroad as a premium to that intellectual progress which, in our present apathy, is our greatest danger?

Playfair singled out calico-printing, glass and ceramics, silversmithing, and diamond-setting as industries where foreign expertise and scientific knowledge were essential. The notion of British industrial decline has had a long history.[18]

Outside the charmed circle of those planning and commenting upon the Exhibition was the irascible mathematician Charles Babbage, scourge of the political and scientific establishment and of organ-grinders, and inventor of clockwork ancestors of our computers. Having in 1830 published on the decline of science in England (ironically just when Faraday was beginning his electrical researches, and Darwin setting forth on HMS *Beagle*), Babbage published a book on the Exhibition, largely full of how he would have done everything differently and better but also developing his theme that science should be a better-rewarded vocation. No great admirer of Whewell, he began his chapter on the position of science carefully avoiding Whewell's neologism:[19]

> Science in England is not a *profession:* its cultivators are scarcely recognized even as a class. Our language itself contains no *single* term by which their occupation can be expressed. We borrow a foreign word [*savant*] from another country whose high ambition it is to advance science.

He was sixty years old, a solitary and embittered man, and out of date in 1851 when the Exhibition, though it might seem to promote empiricism and utility, was used by those we can now call scientists to promote the cultural and national value of science. What had been the age of revolutions became the age of science.

THE NEW CONFIDENCE

One reason for confidence in science in the 1850s and 1860s was the success of powerful unifying theories, as we know happened with conservation of energy and evolution by natural selection. Physicists, biologists and geologists could feel confident that they had got the right end of the stick, and had built up formidable edifices of real knowledge. Some might be cocky, like Tyndall and Haeckel, but even for their more peaceable colleagues there was good reason for assurance in pressing forward and in training disciples. Even the more conservative, resisting Darwinian evolution as hypothetical or speculative, had reason for confidence. Their programme of classification was flourishing and, while the limits of species were recognized to be blurred, to dissolve the boundaries was unnecessary and metaphysically alarming. In the first part of the century, books had abounded in 'synonyms' for plants and animals, where in the author's opinion two people had assigned different names to the same organism. Proposals drafted by Strickland from 1840 for the British Association gradually became a basis for international agreement, taking Linnaeus as the baseline, and giving priority to the first formal name subsequently given to a new species, with authors' names in brackets according to agreed conventions. Strickland was also important in getting his contemporaries to reject bold overall patterns – such as the 'Quinary System' popularized by William Swainson (1789–1855) and tempting to Darwin, Wallace and Huxley – in favour of careful and open-minded assessment of multiple criteria. There were still differences between 'lumpers' who accepted a wide range of variety within species, and 'splitters' who loved to divide up such groups into separate species, but in general there was sufficient agreement to keep museum staff happily busy. Countless animals and plants were sent to the great metropolitan museums and gardens for classifying. Natural history was booming, highly popular, and useful in providing raw materials.

In chemistry, too, (following the agreement about atomic weights after the Karlsruhe Conference of 1860) atomic theory and consequent hypothetical models of structures at last proved genuinely helpful in understanding, and were even testable. Hofmann's croquet-ball and wire models of molecules, demonstrated at the Royal Institution in 1865, were constructional toys that supported the chemical imagination more effectively than diagrams on paper.[20] Moving from analysis to synthesis, chemists began constructing molecules. In Hofmann's laboratory, Perkin serendipitously made the first synthetic dye, mauve, from what had been a waste-product, coal tar.[21] A new epoch opened as chemists understanding structures prepared others (rapidly made thereafter on an industrial scale, in Britain, Alsace and Germany). Synthetic chemistry then spread, with the work of Kekulé, into other compounds containing benzene rings, and with Emil Fischer (1852–1919) and others on sugars, in the great laboratories of German universities.[22] Chemists worked out sequences of reactions that would determine structures in analysis, and others that would re-create them: they had become molecular architects. With the aid of models, elementary chemistry was easier to teach and, as the basic science, it was very widely taught and examined.

The revived atomic theory was not without its critics, who thought it still speculative and hypothetical, and flinched at the simplistic and dogmatic form in which it was taught. But it worked, and scientists are a pragmatic lot.[23] Another consequence of its revival was that, armed with new agreed atomic weights based on water being H_2O, several chemists – most notably Mendeleev – published 'Periodic Tables' of the elements, showing how they formed 'families' very like the groupings of botanists and zoologists.[24] The properties of a new or unfamiliar element (even an undiscovered one) could be forecast from its position in the table (where Mendeleev had left some gaps) and the properties of its neighbours. Chemists were also beginning to understand how reactions went in stages, and could be reversible as different conditions affected chemical equilibria: the role of heat and pressure could be predicted. Chemistry was ceasing to be a craft and an 'erudition', a mass of techniques and curious facts, and becoming a real science where empirical knowledge was ordered and explained through theory, outcomes could be anticipated, predictions made, and utility would follow. Hofmann, speaking about mauve and magenta, greeted the future, referring to Faraday's 'pure delight' in the elaboration of truth in his early investigation of benzene:[25]

> It was in that same spirit that his successors continued the work. Patiently they elicited fact after fact; observation was recorded after observation; it was the labour of love performed for the sake of truth; ultimately, by the united efforts of so many ardent inquirers, exerted year after year in the same direction, the chemical history of benzol and its derivatives had been traced. The scientific foundation having been laid, the time of application had arrived, and by one bound, as it were, these substances, hitherto exclusively the property of the philosopher, appear in the market place of life.

In the same lecture room, Davy had seen a bright dawn: now the light had come.

Even the weather began to yield some of its mysteries to science.[26] Clouds had been classified by the Quaker Luke Howard (1772–1864), as well as being taken up by Constable, and his French disciples, who made careful cloud studies as an important part of landscape painting. Captain Robert FitzRoy (1805–65) had made use of the barometer on the voyage of HMS *Beagle*, aware that it could predict the squalls and williwaws around Cape Horn. Back in England, and, after spells as an MP and a colonial governor, he was in 1854 put in charge of the new Meteorological Department of the Board of Trade in London. In the USA, Joseph Henry had used reports sent in from around the country by electric telegraphs to predict weather, while Arago at the Paris Observatory collected data and published essays and papers. Faced with very changeable British weather and frequent shipwrecks, FitzRoy began issuing public storm warnings in 1861, and devised a system of cones and drums hoisted at coastguard stations to pass the message to shipping. In his *Weather Book* (1863) he went beyond the weather reports collected since the seventeenth century by enthusiasts, including Charles Darwin's friend, the parson-naturalist Leonard Jenyns (1800–93), who had been the first choice for the berth on HMS *Beagle*. FitzRoy was by then an Admiral, but criticisms of his forecasts in parliament and elsewhere led to his suicide in 1865.[27] The published forecasts were discontinued, while the more cautious James Glaisher (1809–1903) and his

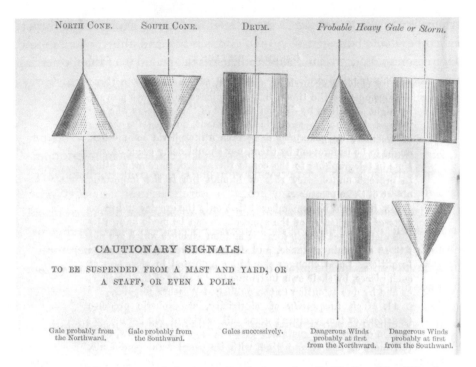

Figure 19 *Warning signals for impending bad weather: R. FitzRoy,* The Weather Book, *London: Longman, 1863, p. 350*

associates at the British Meteorological Society sought to build up a greater stock of facts from which to generalize. Glaisher had from 1846 prepared quarterly meteorological reports for the Registrar General, and had made many balloon ascents to make weather observations. But public demand, always mixed with mockery, meant that forecasts were resumed not long afterwards.

Meanwhile, Faraday's experiments on electromagnetism were being developed, on the one hand by a new profession of electrical engineers, bringing electric light and power. On the other hand, his insights into its relations to light and other radiation were being put into mathematical form, and extended experimentally, by Maxwell, Heinrich Hertz (1857–94), Crookes, J.J. Thomson and Guglielmo Marconi (1874–1937), yielding fresh understanding and giving the twentieth century its cathode-ray tubes, electrons and radio. Electricity no longer meant exciting demonstration lectures or quack medicine, but was perceived as one of the fundamental forces of the universe. At last, Newtonian dreams were being realized across the board, and scientists had good reason for proper pride. They pressed for a much greater role for science in education and the wide diffusion of scientific knowledge. Though they remained uneasy about specialization and the potential emergence of two or more cultures, they no longer had any reason to defer to classical learning, still the mainstay of the most prestigious educational systems.

SCIENCE AND DOUBT

The term 'science' was still in the 1850s, like 'art', used in a wider sense than it is today, but its derivation from 'scientia' implied organized, formal knowledge and certainty. Nevertheless, it was apparent that natural science was progressive and provisional. Even Newton was then perceived to have been wrong in concluding that light was corpuscular rather than a wave motion. Scepticism towards authority, and a readiness to suspend judgement and change one's mind in the light of fresh evidence, was a necessary part of science: doubt was an aspect of scientific confidence. At that time, academic philology and history, both rapidly progressing, were generally seen as scientific. Such science was already contributing, along with evidence from geology, to promoting the religious doubt which was a feature of the mid-nineteenth century in the English-speaking world, provoking a whole new genre of novels and memoirs, which still continues.[28] The most famous was Mrs Humphry Ward's *Robert Elsmere* (1888), which was a huge publishing success in Britain and the USA, arousing furious debate. Mary Ward, whose sister married Huxley's son Leonard, was a granddaughter of Thomas Arnold, a great promoter of women's education and an important figure in the literary world.[29] Loss of faith, complete or partial, was (and may still be) a matter of relief, seen as a part of growing up, like abandoning belief in Santa Claus and no longer having to go uneasily through rituals that have become meaningless. But often it was (and is) regretful. A nostalgia for lost content, for happy childhoods in peaceful countryside, was commonly a feature of those condemned to hard work in smoky, smelly towns. They had indeed tasted the sweet fruit of the tree of knowledge, but it had meant expulsion from Paradise. It would be wrong to see science, in our modern narrow sense, as uniquely responsible for this. Doubts were more likely to be raised by close examination of the texts that make up the Bible, with moral revulsion both at the behaviour of heroes of the Jewish scripture, and at the extreme humility and mildness recommended in Christianity, as well as the miracles and the inconsistencies in the various texts, the contrast between the opulent and worldly lifestyle of some clergy and churchgoers, and the simplicity of primitive Christianity, or a quarrel with a particular priest or minister. These books are, however, one indication that science was beginning to rival religion on the high ground of culture and morality in Britain, as it had in France – and in European countries conquered by the Grande Armée – ever since the Revolution. There, atheism often went with science, as was the case with Laplace and his successors, but in the English-speaking world it continued to be associated strongly with immorality, so that even admitting to religious doubts in public could be very damaging.

At the centre of a suspect circle of doubters was the publisher John Chapman (1821–94), among whose authors were two who suffered from excessive religiosity in their families: Francis Newman (1805–97), whose brother John Henry (1801–90) became a Catholic and a Cardinal; and James Anthony Froude (1818–94), whose bullying brother Hurrell (1803–36) was an ardent disciple of J.H. Newman and his fellow High Churchman John Keble (1792–1866).[30] Indeed, J.H. Newman's radical emphasis upon development, authority and tradition could as readily lead to scepticism as to fervour, as Frank Turner has shown, and physical, geological or biological science needed to play little part in it.[31] Froude's semi-autobiographical *Nemesis of Faith*, published in 1848, owed much to Goethe's novel of human and chemical reactions, *Elective Affinities*, which he translated. *Nemesis* lost him his fellowship at Exeter College, Oxford, where the book was publicly burned, but he had doubts about science as well as religion, writing:[32] 'What is man the wiser or the happier for knowing how the air-plants feed, or how many centuries the flint-stone was in forming, unless the knowledge of them can be linked on to humanity, and elucidate for us some of our hard moral mysteries?'

Francis Newman's contemporary confessional book, *Phases of Faith* (1853), shows a gritty liberal mind at work, and a hope of progress in religion like that in science with its 'strict adherence to impartial principle'. It will, he declared in conclusion,[33] 'Have movement, namely, a steady *upward* one, as the schools of science have had, since they left off to dogmatize, and approached God's world as learners.' Science should thus be an example to ill-tempered, hypocritical and frightened religious thinkers, whose claim that theology was queen of the sciences was absurd, and who were prisoners of Church dogma. The Roman Catholic Church that J.H. Newman joined in 1845 was, as we saw, at that time easy with science, but subsequently the Vatican Council of 1869–70 asserted Papal Infallibility and the Assumption of the Virgin Mary, centralized decision-making and took a hard line against 'modernism'. Liberals in Britain, who saw all this as a threat to science and

intellectual liberty, were outraged, and in Germany Bismarck launched the
Kulturkampf, perceiving the Roman Catholic Church as a threat to newly
achieved national unity under the Kaiser.[34]

Francis Newman's writings lacked the charm of his brother's, and it was
from the haunting rhythms of a great poem, Tennyson's *In Memoriam*,
that early Victorians (alerted by the controversy over *Vestiges*) came to
appreciate 'honest doubt' and to take science seriously as a contributor
to what seemed to them this new phenomenon.[35] The phrase 'Nature red
in tooth and claw' is his, and so are reflections upon geological time, the
irrelevance of science to our wrestlings with Death, and the hope that
future progress will bring into being people 'no longer half-akin to brute'.
Such a one had been his dead friend Hallam, 'a noble type/ Appearing
ere the times were ripe'. In the poem, science appears both as a leaden
echo, indicating an interminable and purposeless world, and also as a support for a belief in progress and improvement. There was no easy way of
reconciling the empirical and evolving ways of science with older cultural
authority. Science, dogmatic in its elementary parts, flourished in an
atmosphere of doubt at its frontiers. As it became more and more important culturally, so it brought doubt with it into the mainstream. This was
graphically displayed not only in secularist publications, but also in the
theological *Bampton Lectures* that the immensely learned logician Henry
Mansel (1820–71), who went on to become Dean of St Paul's in London,
was invited to deliver in Oxford in 1858. Published as *Limits of Religious
Thought*, the lectures took up the long-standing 'via negativa', stressing the
otherness and unintelligibility of God, which led Mansel to a position not
very far from the agnosticism of Huxley.[36] For him, God's existence could
not be demonstrated, and religious humility was in order.

Where atheism had been alarming, threatening the social order because
it seemed to deny any reason to be good, Huxley's agnosticism became
respectable. He and his associates lived exemplary lives, though Tyndall and
the mathematician William Kingdon Clifford (1845–79) might go scandal-
ously too far in mocking what they saw as ignorance and superstition, but the
more respectable perceived as pious practices.[37] Similarly, serious-minded
Positivists like Frederic Harrison (1831–1923), following Comte's philosophy and in some cases also his Religion of Humanity, could not be dismissed
as amoral hedonists. Respectable doubters collected on Sunday afternoons
at four o'clock in the London season (October to March) at St George's
Hall, Langham Place (in the West End) for the Sunday Lecture Society's
programme:[38]

To provide for the delivery on Sundays in the Metropolis, and to encour-
age the delivery elsewhere, of Lectures on Science, – physical, intellectual
and moral, – History, Literature, and Art; especially on their bearing
upon the improvement and social well-being of mankind.

Agnostics were thus not to be deprived of the advantages of churchgoing,
getting good sermons and sociability. They could pay a subscription of £1
for the season, or a shilling, sixpence or a penny for better or worse seats
on the day; the lectures were printed and sold at threepence. Here, as in
churches with their pew-rents, the class-system was as evident as on the
railways with their first-, second- and third-class carriages. The committee,
whose names were printed sometimes at the back of the published lectures,
included Huxley, Herbert Spencer (1820–1903), Tyndall and Frankland
(prominent members of Huxley's shadowy X-club), as well as Darwin.[39]
Although the topics were not exclusively what we would call science,
science and doubt were clearly bedfellows.

PUBLIC FACES OF SCIENCE

The success of the Great Exhibition ensured that international expositions
and world fairs would become a feature of the second half of the nineteenth
century. New York, Paris, Chicago and other cities promoted wonderful
displays of new technology, and it became increasingly plausible to believe
that this was the fruit of scientific research; that science and progress must
go together. Somehow, these great shows, blending entertainment, advertisement and instruction, never quite caught the sparkle of 1851 and the
Crystal Palace: it seems to have been an impossible act to follow. None ever
made a profit either. All this is not surprising. The Great Exhibition had got
its message across so well that others could not recover the charm of nov-
elty. The dynamic of modernity was there, and there were new generations
to be impressed by and won for progress, but in effect these shows were an
updating and amplification of an already familiar theme.

This was particularly true of the follow-up in London. It was intended
to have a sequel to the Great Exhibition ten years on, but Prince Albert was
taken ill in 1861 and in December died from typhoid. Moreover, a savage
Civil War had broken out in the USA, in which modern technology was to
result in great slaughter: Prince Albert's last dispatch to the Prime Minister,
Palmerston, had successfully urged him to tone down a bellicose note to

Abraham Lincoln's government. The exhibition went ahead in 1862, but opened under a cloud with the Queen in prolonged mourning, and great sadness for a lost leader among those who had promoted the Crystal Palace. This one was in a new building in South Kensington that completely lacked the magical quality and technical innovation of its predecessor, and was inevitably seen as a shadowy evocation rather than a worthy successor. Its highlight was perhaps the exhibit of crystals of synthetic dyes, mauve and magenta, hailed by Hofmann as a sign that Britain would triumphantly conquer this new realm of industrial chemistry, transforming coal tar into a new rainbow of colours.[40] But by then other nations were rapidly industrializing. As the first industrial nation, the British had come well out of 1851, but could not (as those far-sighted lecturers had seen then) expect to remain always in the lead. England was still thoroughly backward in higher education: Oxford and Cambridge (where Prince Albert as Chancellor had promoted reforms) had begun to offer a wider range of degrees, but many industries remained largely uninterested in graduates, and those based on chemistry and electricity went on recruiting Scots and Germans in the way Playfair had deplored. Hofmann himself was tempted back to Germany by invitations to both Bonn and Berlin in 1865 – and chose Berlin. In 1867 Britain did humiliatingly badly in the international exposition in Paris.

The power of applied science was vividly brought home not just in exhibitions and labour-saving inventions, but in warfare. The American war, bloodily fought with Napoleonic tactics but modern rifles in place of muskets, ground to a halt as the more industrialized North won. 'Reconstruction' in the South began, and a gilded age opened as easy access to the West began with the railroad across the continent, reducing the importance of the Mississippi for commerce. Industry of all kinds flourished (with alternating booms and slumps, furious debates about tariffs, and formidable strikes), and inventions were boosted, with Thomas Alva Edison (1847–1931) becoming a national hero. Designated the 'Century of Progress' following the Declaration of Independence, this new world was celebrated with aplomb in an exhibition in Chicago, booming again after its disastrous fire of 1871. State universities based upon land grants promoted agriculture and local industry. My American grandmother remembered hearing as a girl in 1890 the news that the Frontier had been declared closed: 'manifest destiny' was fulfilled, Indians were confined to reservations, and there was no more spare land. In reaction, the movement for National Parks took off.

Meanwhile, in Europe, too, the Kantian dream of perpetual peace and prosperity had been shattered by the outbreak of wars of unification in

Germany and Italy, and then in 1870 the Franco–Prussian War in which Paris was besieged. In the starving city, the animals in the zoo were eaten, as were the rats, and the Communards seized control, eventually to be overthrown by the French army amid slaughter and reprisals. The mighty empire of Napoleon III, under whom the boulevards of Paris had been driven through the old city to create a magnificent modern capital, fell; and for the third time the French had a republic. The lesson drawn all over Europe was that education had been the key to victory, and the Third Republic was also known as the République des Instituteurs, the Teachers' Republic. Firmly secular, looking back to the ideals of 1789, the state promoted science and sought to reduce the power of the Roman Catholic Church, whose basilica of the Sacré Coeur in Montmartre was built as a counterblast. For the Paris Exhibition of 1889, celebrating the centenary of the Revolution, the engineer Gustave Eiffel (1832–1923) designed the tower that bears his name, for forty years the world's tallest building and an enduring icon as powerful in its way as the Crystal Palace.

In more peaceful Britain, elementary education became compulsory and, following the Royal (Devonshire) Commission's deliberations and report, science began to find its way into school syllabuses, while new 'redbrick' universities emphasizing science and technology were founded. To ensure that standards were maintained, some became university colleges, taking London University examinations, while others followed Durham's example of appointing external examiners from established institutions. Tutoring, lecturing and examining became ways for scientists to support themselves, along with consultancy when available. There had been universities in the Spanish colonies since the sixteenth century, and by the early 1850s British colonies were following their example. Sydney and Toronto had the first such universities in 1850, followed by Melbourne's in 1854. With students coming from remote places, they incorporated or became collegiate bodies, in which different Christian denominations built and ran the different colleges, but teaching was more centralized than in Oxford and Cambridge. The young Huxley and Tyndall applied for new colonial posts, but (being outsiders) did not get them. Some of the profit from the 1851 Exhibition was used to fund scholarships to bring colonial graduates for further study in Britain: Rutherford is probably the most celebrated recipient. In Imperial Berlin, now the capital of a mighty European power, the museum-island quarter with its grand buildings flourished, providing the model for London's slowly developing Albertopolis.

CATHEDRALS OF SCIENCE

In 1876 the South Kensington (now Victoria and Albert) Museum held a special international loan collection of scientific apparatus. It still seemed rather remote from the centre of things, although the underground Metropolitan railway had reached it at the end of 1868. In connection with the exhibition, daytime lectures and conferences, and free evening lectures aimed at working people, were held, addressed by eminent scientists, technologists and industrialists. These were duly published in three volumes.[41] Some lecturers demonstrated historic apparatus, such as that used by Dalton, Davy, Faraday, Wheatstone and Thomas Graham (1805–69), and the geological maps of William Smith (1769–1839), but most were concerned with the latest innovations and what could be done with them. Science, especially physics, was coming to depend on instruments of high precision, with contrivances such as making use of sealing wax and string becoming less important. It would be expensive to keep up with the new level of accuracy. But at the exhibition, these gleaming pieces of mahogany and brasswork could be assessed by experts and contemplated by the general public. Unclaimed or donated items from this collection eventually became a nucleus for the new Science Museum established nearby in a building put up for another exhibition in 1885.[42] Adjoining it was the Imperial Institute (with a splendid campanile), showing colonial products, and the magnificent Natural History Museum with its flamboyant terracotta walls and its monster gargoyles. From 1885, a tunnel connected the museums with the underground station.

The British Museum had been established in the mid-eighteenth century by parliament in accordance with the will and bequest of Sir Hans Sloane (1660–1753), physician and President of the Royal Society.[43] It held and acquired, like an enormous cabinet of curiosities, maps, antiquities, ethnographical material and natural history, with a great library. All knowledge was there, and one can get from its present 'Enlightenment' gallery a vision of what was hoped for by the founders, and then particularly by Banks as an important trustee. When Banks died, he bequeathed to the museum his great collection of specimens and his splendid library (his house in Soho Square had been a wonderful resource, open to serious natural historians from around the world).[44] Along with it went Robert Brown (1773–1858), his librarian, who had sailed round Australia with Captain Matthew Flinders (1774–1814) in 1801–3 and had written an essay on its flora, based upon the large collections he and others had made.[45] This is a classic in the study of plant distribution, in which (with Humboldt) he was a pioneer. Brown's presence in the British Museum made it and its herbaria a centre for botany. But the most powerful person in the administration of the museum was the librarian. Anthony Panizzi (1797–1879) from Modena was librarian from 1856 to 1866, having been assistant from 1831. He was a good administrator, building up a national library worthy of the name, but plants, animals and minerals came far down his list of priorities.

With the newfound confidence of the mid-century, in 1856 scientists persuaded the British Museum to appoint a Superintendent of the Natural History Departments, who could speak on behalf of the various curators and would carry weight with the trustees. Richard Owen was the man.[46] At the Royal College of Surgeons he had turned John Hunter's museum into a magnificent collection of comparative anatomy and fossils, making himself the 'English Cuvier' with his monographs.[47] He had described the fossils Darwin sent back from HMS *Beagle*. He resolved to separate the museum's natural history, installing it in a new building, and began working on the trustees and lobbying in parliament, starting his campaign in 1858 (the lobbying began then). Owen was persuasive, but awkward. There was considerable opposition, notably from Huxley (who had a long-standing feud with Owen) and other Darwinians, who thought South Kensington too remote from Bloomsbury where the British Museum and the university were, and who also supported the efforts of Hooker to build up Kew as the national centre for botany, where they believed that Banks's, Brown's and other herbaria should go. It was more than two decades before the new building, worthy of an imperial power, opened in 1881. The separation from the British Museum was not without pain: the librarian refused to allow any books to go, and without a library the collections would be useless. Further money had to be voted to build up a magnificent collection, much of it duplicating what was held by the parent institution, where almost all of Banks's volumes continued to be held.

Museums like Cuvier's and Hunter's had been teaching collections, and the British Museum in its early days was, like great country houses, not exactly open to the general public but to ladies and gentlemen who made an appointment. By 1880, this was no longer so, but the new Natural History Museum was, from the start, for everyone, and unlike those run by showmen, it was free.[48] Owen was impressed by homologies, likenesses in structure, in the animal kingdom. He did not accept evolution by natural selection, seeing instead jerky progress as archetypes were developed. The museum, a cathedral of science indeed, displayed the types and homologies in its hall and galleries; it was only with his successor, W.H. Flower (1811–99), that its structuring became evolutionary.[49] As a symbol of the importance and sublimity of science, a great temple of nature, the museum presented natural history as worthy of contemplation, and was the site of serious work behind the scenes where taxonomists laboured over the collections brought back from the ends of the Earth. London's position at the hub of a seaborne empire gave it a particularly important position, but edifices similar in scale and intention were established in New York, Berlin and other metropolitan centres, with smaller versions in self-respecting provincial cities. One of the most notable was the Smithsonian Institution in Washington, DC, the 'nation's attic', founded in accordance with Smithson's will, holding remarkable scientific collections and publishing important works.[50]

It was not only in museums that nature could be worshipped in the light of science. The Crystal Palace was a giant greenhouse, and the Duke of Devonshire was not the first grandee in Britain or elsewhere in Europe to delight in having exotic flowers and fruits to show. At Kew the hothouses were impressive structures, and the gardens a tourist attraction. For a time, until the site was built over, the Royal Horticultural Society had gardens on the South Kensington site. Meanwhile, for rainy winter days, the nineteenth century was a great age of sumptuous flower and bird books, printed first from copper (like those of Audubon) and then from lithographs, like those of Gould, Lear and Joseph Wolf (1820–99), and hand-coloured.[51] These were extremely expensive and normally published in parts, being aimed at the leisured and opulent, rather than at individual scientists who needed to see them in libraries. Darwin was fortunate to obtain the substantial government grant of £1,000 to get the beautifully illustrated *Zoology of the Voyage of HMS Beagle* published, and later official voyagers were similarly assisted by governments in Europe and the USA.[52] For more ordinary purposes, uncoloured lithographs replaced engravings of fossils and plants, and wood-engravings became the standard medium for the booming industry of natural history publishing.[53]

The message of religions, of escape or release from illusion, of redemption, salvation and solidarity, was basically simple, though it could issue in bewildering sets of rules and in paradoxical creeds. Science was much more complex, demanding not just common sense but close study, careful observation and often mathematical ability. Beautiful books and popular expositions, programmes of lectures and exhibits in museums could get across some of the intellectual delight and the hope of usefulness that propelled the scientist, but while the evangelist could hope to make everyone a Christian or a Muslim, there was no reasonable expectation that everyone would become a scientist. To the alarm of serious scientists, it was often the most controversial or speculative parts of science that attracted both journalists and the general public. Thus there had been excitement over *Vestiges*, and the debate over whether there was intelligent life elsewhere in the universe ('plurality of worlds') was delightful in that, without a shred of evidence either way, men of the eminence of Whewell and Sir David Brewster (1781–1868) could get hot under the collar.[54] Nevertheless, science required specialization, and to make its teaching compatible with a wide-ranging and liberal education was a problem. As the sociologist L.T. Hobhouse wrote in the *Manchester Guardian* on 1 January 1901:

> To specialisation carried to this extreme we owe the efficiency and accuracy of modern science. To it we also owe a loss of freshness and interest, a weakening of the scientific imagination, and a great impairment of science as an instrument of education.

A world of 'two cultures' was coming into being.

RESEARCH, TEACHING AND PROGRESS

Franklin had retorted to someone asking him the use of some discovery, 'What is the use of a baby?' When Robert Peel, the Prime Minister, asked Faraday what use electromagnetic induction was, Faraday replied that he was sure one day the government would tax it. 'Pure' and 'applied' science began to be distinguished about the middle of the nineteenth century, and it became an axiom before the end of it that research in pure science must lead sooner or later to useful inventions, as Hofmann in his lecture on mauve and magenta had declared: the pursuit of truth, interpretation of nature, for its own sake was the foundation of knowledge and subsequent prosperity. In the hierarchy, the applied scientist ranked lower, but might well earn more. If science was indeed the key to progress, as those behind the Crystal Palace had claimed, then research must be critical to the future of nations, and something to boast about.

In France, the revolutionary École Polytechnique had begun with academicians educating the students in mathematics, physics and chemistry, but under Napoleon I it became more narrowly concerned with training engineers, particularly for the army. It later provided a model for polytechnics in

Europe and North America, with varying emphasis upon scientific research, but usually outside France ranking below universities in public esteem. In France, the highly centralized university system, augmented by the Parisian Hautes Écoles, was under-funded and bedevilled by *le cumul*, the practice in which eminent scientists occupying the limited number of lifelong places in the Academy held multiple posts, paying juniors a pittance for undertaking their duties.[55] In relative terms at least, the nineteenth century marked the decline of French science from its position of pre-eminence – just as it did for British industrial supremacy. By the later nineteenth century, it was one country among others in Europe with a potent scientific tradition and culture. And the leading one, though with nothing like the predominance of France (Paris) in 1800, was Germany.

As we saw, the ideal of the research university went back to Humboldt's University of Berlin, founded with ideals of *Wissenschaft* and *Bildung* rather than industrial utility. It was this ideal that became a model for the world during the nineteenth century, and in modified form still is. The German states competed, the will was there and money was found to support higher education. In the English-speaking world, adoption of the German model followed slowly upon the conviction that scientific research really was useful, in the long run at least, and that academically trained scientists and engineers were essential for the electrical and chemical industries where the second industrial revolution was happening. The shock of the Prussian victory of 1870 was salutary, but British higher education received no state support and the redbrick universities teetered on the verge of bankruptcy until in 1889 the government was persuaded to award grants to them. In America, state universities were supported by land grants, but with useful knowledge in mind; the idea of the research university made little headway until in 1873 the railway financier Johns Hopkins bequeathed his fortune to found a university in Baltimore.[56] This example was followed in other new foundations like Chicago and Cornell, and taken up by the venerable institutions of the Ivy League. Then, and ever since, there has been an uneasy cohabitation of those who with Humboldt see research as gloriously developing human potential and adding to knowledge, and those with narrowly utilitarian objectives. But one thing that was clearly emerging was that progress in, and support for, science was a source of national prestige.

CONCLUSION

After the 'hungry Forties', the 1850s opened with the Great Exhibition of 1851, where its magical building and international flavour put it in a different class from its predecessors in France, Germany and Britain. Its vision of industrial society based upon applied science was infectious, and it was clear to the perceptive that although Britain as the first industrial nation was indeed the workshop of the world, the USA and countries in continental Europe were catching up, and ahead in some ways. A museum and educational quarter was established in London, on the Berlin model. There was new confidence in science and progress, and scientists began to take their place alongside other sages. The scientific ideals of resistance to authority and readiness to adapt ideas to new facts spread into other fields, notably reinforcing religious doubt. Great museums, cathedrals of science, were built. However, science became visible not only in exhibitions but also in modern warfare, especially in the USA and France. National power, influence and progress were observed, especially in the Prussian victories culminating in 1870–1, to depend upon science and technology. Polytechnics were established, while newly expanded universities picked up the research ethos from Germany, and began to teach science and engineering formally to more and more students.

9

SCIENCE AND NATIONAL IDENTITIES

Francis Bacon had declared that knowledge was power, and if politics has always been power politics, then we should expect that as scientific knowledge expanded, statesmen as well as the public would be interested in it. Understanding and theory-making, technology and medicine form a spectrum within the umbrella of science, and were not much distinguished until late in the century when 'pure' science and mathematics came to be seen as more gentlemanly. Since the nineteenth century was the age of nation-building and nationalism, science also became a source of national pride and prestige, and a sign of modernity. The international competitiveness we associate with the modern Olympic Games began to be seen to some extent in nineteenth-century science, though it was exacerbated only after 1914 when international collaborative ventures collapsed as the Great War broke out and our story ends. Before that, modern Western science had been planted in the Islamic world, India, China and Japan, very different cultures with their own traditions of science, techniques and industries, and medicine: there, separating it from any particular basis in European history or natural theology was important. Technology was always easier to transfer, but the self-confident, independent science of the mid-century was more welcome than something saturated with European Christian assumptions and values. Equally, it was something in which European nations and the USA took pride, a way of facing down and mastering older civilizations, and a token of restless progressiveness to be contrasted with what many were happy to see as the unchanging cycles of Cathay.

THE GREAT POWERS

It seems natural to think about centralized nation-states, surrounded by stable frontiers, having national legal and educational systems, and national anthems – though nowadays challenged by multiculturalism. It was not always so. Britain is surrounded by the sea, and from the early eighteenth century Queen Anne had reigned over what Daniel Defoe (1660–1731) called the Empire of Great Britain, the kingdoms of England and Scotland and the Principality of Wales, which not only had a common sovereign but had lost their national parliaments. At the beginning of the nineteenth century, Ireland was incorporated in the same way in a shotgun marriage. That was never happy, but for a time at the acme of Victorian power Scotland was called (even by Scots) North Britain, and abroad the words 'England' and 'English' were often used to cover Great Britain and its inhabitants (the 'great' referred properly to size rather than grandeur). The borders or marches between England, Wales and Scotland had been lawless, but after union became fuzzy areas of mixed populations where ancient loyalties ceased to matter much, especially after the failure of 'Bonnie Prince Charlie's' revolt of 1745 and then the death of the last Stuart claimant to the throne. Nevertheless, the epic poetry of 'Ossian, son of Fingal' (constructed by James Macpherson (1736–96), who claimed to have heard and translated it from the Gaelic) had excited Romantic sensibility all over Europe in the 1760s. Highlanders had been perceived in enlightened Anglo-Saxon Edinburgh in rather the same way Bostonians might have contemplated Red Indians, but in *Ossian* they had their *Hiawatha*, and Highland heroes became a focus for a romantic Scottish nationalism. A generation after Macpherson's 'forgeries', Walter Scott (1771–1832), whose Waverley novels were the runaway publishing success of the time after the Battle of Waterloo, created a romantic Scottish history of clans and tartans and derring-do – very different from the realities of Enlightenment Scotland with its doctors, engineers and writers.[1] Romantic sensibility also created a Wales of bards and heroes, and an interest in the previously denigrated language. The English had King Arthur and his Round Table, King Alfred burning his cakes, Robin Hood and his Merry Men, Good Queen Bess, and No Popery – again utterly remote from the reality of industrial life. Folklore was studied in curious ambivalence.[2] There was no place for science in this nostalgia for never-never lands.[3] But despite all this, in and after 1789 those living in Britain were thinking of themselves sometimes, perhaps usually, as Britons, an identity that went with trade and power, steam engines and turnpikes, and the island was (despite its fringe of Welsh and Gaelic speakers) clearly something like a nation-state.[4]

When the Revolution broke out in Paris in 1789, France was not a nation in the modern sense. In the distant provinces, a patois or a different language (German, Breton or Catalan) might be spoken, and French hardly understood, while allegiances everywhere were local or regional. The revolutionaries sought to change that, and bring uniformity to the hexagon. Instead of historic boundaries between dukedoms they drew up a grid of departments, each with its *chef-lieu* and its administration, and imposed uniform laws, weights and measures, and schooling. With the Congress of Vienna after the allied victory in the revolutionary and Napoleonic Wars, a Europe of definite frontiers and clear jurisdictions went some way towards replacing the old jigsaw-puzzle of kingdoms, city-states, grand duchies and prince bishoprics. Revolutionary France, set up to replace a ramshackle *ancien régime*, had to be secular and modern, and although the revolutionaries had abolished the Academy of Sciences as elitist, they had very soon resurrected it in the Institut, a source of advice and prestige for the new state. The military engineer Lazare Carnot (1753–1823) was the associate of Robespierre and Saint-Just in the Committee of Public Safety, and was a great survivor, holding high office in the Directory and under Napoleon, by whom he was appointed to his first independent command in Italy. There were other men of science in the revolutionary governments, who also played a part in arming, equipping and mobilizing the armies that upheld the Revolution. The Jardin des Plantes and the museum were never closed, and their role in public instruction was increased, while educational reforms gave prominence to science – where the French were leaders. The execution of Lavoisier was almost at once uneasily seen as a martyrdom, but as he was followed shortly to the guillotine by Robespierre and most of his close associates, retribution could be said to have followed, and images and lives of this rather unattractive scientific saint could proliferate.[5]

Napoleon delighted in science, was gratified to be elected to the Institut, and promoted men of science to posts in his administration and to his Senate (which was in effect a sinecure). Subsequent French governments took science less seriously, and during the middle years of the nineteenth century it was starved of funds – hence *le cumul*, and the lack of laboratories and opportunities.[6] Nevertheless, France had been transformed into a nation-state. Britons might take pride in Newton as well as Shakespeare; the French had a longer list of men of science to put beside Molière and their other literary heroes, and had their kilometres, litres and kilograms to remind them of the power of science to change lives (though it was many years before these became universal). Germans lived in a number of states, fewer after 1815 when Prussia (whose army had clinched the Battle of Waterloo) emerged as the strongest. Napoleon had in 1806 abolished the Holy Roman Empire, the relic of a loosely federal union under an elective monarchy. Defeat by the French armies, revolutionary and Napoleonic, had shattered confidence in the previous status quo, and its effect was to promote German self-consciousness so that speakers of the language felt they belonged to one nation. They too had their folk tales, collected by Jacob (1785–1863) and Wilhelm (1786–1859) Grimm, and Richard Wagner (1813–83) used national myths and legends in his operas to powerful effect. In 1848, the revolutionaries hoped for a nation-state. They were unsuccessful, but by 1871 Bismarck had, through diplomacy and war, brought about a German empire under the Prussian Kaiser.[7] By then, Germany was no longer bucolic, but powerful and industrial. Universities were flourishing, and with them science and science-based industries. Men of science had been meeting in different parts of the country, moved between universities, and felt themselves Germans as much as Saxons or Bavarians (as the anthem 'Deutschland über alles' proclaimed); and in addition to Luther, Kant and Goethe, Germans had not only Kepler but also Gauss, Hegel, Liebig, Hofmann, Helmholtz and many more. Intellectual activity, notably science, and an industrial revolution based particularly on chemicals and electricity, brought the new nation prestige, power and influence as well as prosperity.

Austria, Protestant Prussia's great Roman Catholic rival for German allegiance after Napoleon's defeat, had a creaky empire (which later became the Austro-Hungarian Dual Monarchy), embracing speakers of German, Magyar, Czech, Polish, Serbo-Croat, Italian and other languages. Despite losing most of its Italian possessions, it had been gaining ground at the expense of the Ottoman empire, largely expelled from Europe by the end of our story, and it had imposed (as empires do) peace and some prosperity upon a wide area inhabited by otherwise quarrelling peoples, divided by religion (Protestant, Catholic and Orthodox Christianity, and Islam), by language, and by stirrings of nationalism. Vienna remained a great centre not only of music but of science, especially medicine, with the university keeping up with those in the German empire. The modernity associated with the Dual Monarchy was important, even if less so than in Germany, Britain or France. But her armies were less well-equipped and mobile than those of Germany which had to come to her rescue in the cataclysm of 1914–18, set off as that war was by the assassination of the Austrian

Archduke Franz Ferdinand by a Serbian nationalist. And as history worked out, the empire could not hold together in the face of national aspirations supported by the victors, and Austria dwindled into a small nation-state. The Russian empire, with its Polish, Finnish and Turkish subjects, was also brought down by the war, having previously been shaken by defeat in the Crimea and then half a century later by the Japanese. Its revolutionary successor-state was duly appreciative of science and technology, being founded upon aspirations for Soviets and electric power. Due honour was then given to the scientists of nineteenth-century Russia, who had been particularly important in the development of chemistry, but seemed alternately attractive and alarming in Tsarist Holy Russia.[8] Nevertheless, the Imperial government had supported a distinguished Academy of Sciences, and in 1839 had built and equipped one of the finest observatories in the world at Pulkovo, near St Petersburg.

ON THE EDGES

Britain and Germany thus joined France at the very centre of things in nineteenth-century science, and science was in different ways an important part of their self-image. The Netherlands, with a long tradition of important science but impoverished by the long wars, emerged from the Congress of Vienna as a single kingdom, but in 1830 Belgium seceded and became a small state, divided by language between Flemings and the more prosperous French-speaking Walloons, but united in a drive to modernity that made them a model for others. Belgium was exemplary in its collecting of statistics, duly analysed by Quetelet, its efficient administration, its communications and its industry. In science, like the Spaniards and Portuguese, Belgians looked to France, while the Dutch and the Danes looked to Germany. Thus Ørsted had picked up *Naturphilosophie* in Germany, and came back to his native Denmark full of dynamic enthusiasm. To the amazement of contemporaries, in 1820 he demonstrated the magnetic effects of a changing electric current, and thus inaugurated electromagnetism.[9] He did little more of any significance about it, and the discovery was taken up by Ampère in Paris and Faraday in London – a clear demonstration of how centres of excellence work. But Ørsted was a scientific star, and at home he became a hero who dominated science in Denmark – a big fish in a small pond – where he emphasized fundamental research in this quite poor country dependent upon agriculture, giving rather windy addresses collected under the title *The Soul in Nature* (transl. 1851). With

Tycho Brahe (1546–1601) the astronomer, he could be put alongside Hans Christian Andersen (1805–75) and Søren Kierkegaard (1813–55) as great Danes. Across the straits, Sweden with its strong tradition in metallurgy, mineralogy and chemistry (of great economic importance) as well as natural history, kept close links with Germany. Norway had been assigned to it at the Congress of Vienna, but in 1905 gained its independence; its scientific traditions were weaker, but important early work in physical chemistry was carried out there. Comparing the emergence of professional chemistry across the different countries of Europe at this time shows how very diverse the different countries' experiences were, but the perception was general that science was a crucial part of modern culture and of a modern industrial economy even where that was a distant aspiration.[10]

Science of the Western kind was less rooted in the European colonies and states situated further from these centres, geographically and scientifically on the periphery.[11] India had slowly come under British influence and rule through the East India Company, whose objectives were commercial and whose reputation was somewhat unsavoury. But by 1800, Banks was promoting the Calcutta botanical garden, and under his aegis and with Company sponsorship enormous and handsome volumes on the *Plants of the Coast of Coromandel* were published, partly for public relations at an uneasy time.[12] The Company's Governor-General maintained a menagerie at Barrackpore; Macaulay during his time in India inaugurated Western-style education there.[13] Soldiers and administrators from Britain, with essential but less-publicized Indian informants and colleagues, explored and surveyed the subcontinent.[14] They collected and sent home specimens and much information about Indian customs, history, topography, fauna, flora and fossils, with some splendid illustrations often done by local artists trained in the tradition of Mughal miniatures, to be exhibited in London and studied by experts there.[15] Company officials in Canton and Hong Kong made collections, including natural history specimens and pictures, while those from the Dutch East India Company did the same from their base in Nagasaki, and their annual tribute-bearing journey to the Shogun in Tokyo.[16] Stamford Raffles (1781–1826) governed Java from 1811 to 1816, during the French wars, when the Netherlands were occupied by the French army and Dutch colonies overseas were seized by Britain. He wrote about its history, got the American Thomas Horsfield (1778–1859)to describe its zoology, and made large collections – most of which were lost in a fire and shipwreck on his way home.[17] When after Waterloo the return of Java to the Dutch became inevitable, he settled upon Singapore as a wonderful site for a trading city. When he eventually returned to London, he founded the London Zoo with Davy.

Banks never forgot his botanizing at Botany Bay, and promoted science in Australia on survey voyages and expeditions, but also in the infant colony where he encouraged the 'acclimatization' of crops and animals. He was instrumental in the exporting of merino sheep that he had been supervising and breeding at Kew, thus launching the 'botany wool' trade vital to the colony's prosperity. Voyages of exploration under Admiralty auspices were undertaken by Flinders in 1801–3, Phillip Parker King (1793–1856) in 1818–22, J. Lort Stokes (1812–85) with HMS *Beagle* in 1837–43, and Owen Stanley (1811–50) in HMS *Rattlesnake* in 1846–50. Huxley was the assistant surgeon on this last voyage. On the *Beagle's* previous voyage, with Darwin on board, Tierra del Fuego and Cape Horn had been charted – vital for ships returning with the westerly winds from Australia.[18] Navigators sought in vain for large rivers, like those that had enabled the opening up of the Americas, but difficult overland travel was necessary in this huge and faraway land.[19] Rapid development came with the Gold Rush, at which time the geologist Roderick Murchison (1792–1871), who had taken up Banks's mantle as a patron of scientific exploration, claimed to have predicted gold in Australia and glowed with pride and satisfaction. Colonials were seen as primary producers, sending home wool to be made into textiles in Bradford, and specimens to be classified in London; if industrious and deferential to their metropolitan patrons, they could expect recognition, including election to the Royal Society. After 1848, Germans, from a well-educated country without colonies of its own, were also important in Australian science. In due course, with universities founded from the 1850s on, indigenous science became significant, with Melbourne and Sydney established as important cities, but the expectation remained that senior posts in the state, in churches and in academe would be filled by Britons or other Europeans rather than by the native-born.[20] Thus Bragg went from Cambridge to Adelaide, where he married, and in due course and in consultation with his son Lawrence (1890–1971) found that X-rays would indicate the position of atoms in crystals. This earned him a Nobel Prize and a call to a post back in Britain, rejoining his son, who had shared the prize and gone 'back home' earlier to study.[21] Cultural dependence, or 'colonial cringe', ended only in the second half of the twentieth century.[22]

Ireland was not a colony, being represented at Westminster, but its experience was not very different. There was a long tradition of cartography there (often associated with division of spoils), and like India it was used as a laboratory for social engineering. The first institution of higher education to be supported by the British government was the Roman Catholic seminary at Maynooth (in order to promote the education of Irish clergy at home rather than abroad, where they might acquire dangerous notions). The Ordnance Survey pioneered large-scale mapping in Ireland before undertaking it in England, and the Geological Survey was particularly active there.[23] Charles Molland in a hefty new collective biography seeks to demonstrate that science was and should be an important part of Irishness.[24] Unsurprisingly, many of his subjects left Ireland to work in Scotland, England or further afield where there were more opportunities: Tyndall and George Gabriel Stokes (1819–1903) are prime examples. But there was traffic the other way. Edmund Davy (1785–1857) had worked in London as his cousin Humphry's assistant, and in 1813 on the strength of this connection was appointed to the Royal Cork Institution as professor of chemistry, and then in 1826 to a post at the Royal Dublin Society, having just been elected a Fellow of the Royal Society. In Northern Ireland, where many Protestant Scots had settled after the campaigns of Oliver Cromwell and William of Orange in the seventeenth century, there was much movement to and fro across the narrow straits: William Thomson, Lord Kelvin, was born in Ireland though he spent his working life in Glasgow; his father James was teaching in Belfast, but moved back to Glasgow, where he had been a student, on being appointed to a chair in mathematics there when William was eight.

Molland includes some surprising 'Irishmen': Marconi because his mother was Irish, and Erwin Schrödinger (1887–1961), who took Irish citizenship in 1948 – small countries will always have some like that. But many spent their lives in Ireland, making his point that science is an important strand in Irish culture, to be ranked along with the writings of Wilde, Shaw and Yeats – after all, Ireland now sees itself as both Celtic and an economic tiger. Such Irish scientists include the chemists Sir Robert Kane (1809–90) and Thomas Andrews (1813–85), as well as the astronomer William Parsons, Lord Rosse (1800–67), President of the Royal Society from 1849 to 1854, and father of Charles (1854–1931), the engineer who pioneered steam turbines. In 1841, when he succeeded to the title and estate at Parsonstown, he set about building the biggest telescope in the world – a seventy-two-inch (1.8 m) reflector – overcoming huge technical problems casting, polishing and grinding the speculum metal, and erecting the Leviathan, as it was called, in which the observer sat fifty feet (15 m) above ground.[25] During the famine years of 1845–8, he was generous and energetic in relieving distress, and became popular as well as eminent. Ireland, with its damp and

Figure 20 *Lord Rosse's six-foot telescope,* Philosophical Transactions *151 (1861): pl. 24*

changeable weather, was not the ideal location for a great telescope (unsurpassed for seventy years), but it produced astonishing images of nebulae, dramatically illustrated for a wider public by John Pringle Nichol (1804–50) in Glasgow, who depicted them dramatically on a black ground.[26] The depths of space became fascinating rather than terrific. We should not be surprised that more Irish scientists came in the past from the ascendant and wealthier Protestant community, but Kane was one of a number of a Roman Catholics devoting themselves to science and industry.

Poles and Germans have argued bitterly over which nation Copernicus belonged to, a question that would have been meaningless to him, while biographical dictionaries often blithely ascribe nationalities like 'Irish chemist' to Boyle, 'Croatian mathematician' to Boscovich, and 'Italian physicist' to Volta – which are understandable but would have puzzled these people.[27] Jews formed a diaspora without a state, and debated furiously whether they should aspire to one, or assimilate, but their religious traditions of practice and argument perhaps made them particularly significant in science.[28] Italy was a geographical entity, but after the fall of the Roman Empire had been politically fragmented, with city-states like Genoa, Venice

Figure 21 *Spiral nebula: J.P. Nichol,* The Architecture of the Heavens, *London: Parker, 1850, pl. 12*

and Florence, papal states round Rome, and long-lasting incursions from Byzantium, France, Spain and Austria leading to fluctuating allegiances and boundaries. By Galileo's time, the Tuscan dialect had become the written Italian language, but dialects that could hardly be understood outside their region of origin were the norm. French revolutionary conquests under the young Corsican Bonaparte, later Napoleon, simplified the map for a time, but Nelson's navy propped up the Bourbon Kingdom of the Two Sicilies in the south, and after 1815 the north was largely divided between Savoy and Austria. Later, French troops supported the papal states. As in Germany, nationalism took root, based upon language and the fact that, as a peninsula with the Alps along its north, Italy had reasonably 'natural' frontiers. The support of celebrities like Byron, the fervour of Giuseppe Mazzini (1805–72), the romantic daring of Giuseppe Garibaldi (1807–82) and his redshirts, and the diplomacy of Camillo Cavour (*c.*1810–61) brought about unity under the Savoy monarchy. Science could not be said to have had much to do with it. But Italian unity was exemplified by the career of Stanislao Cannizzaro (1826–1910), who at and after the Karlsruhe Conference of 1860 convinced the world that water was H_2O. Born in Palermo in Sicily, he became a professor in Genoa in the far north, then back in Palermo and finally in Rome, at the centre of things. In 1871 he became a Senator, and devoted himself to questions of public health. Science, prominent in the days of Galileo, was once again becoming something in which Italians could take pride and his Accademia dei Lincei was re-founded.

NATIONAL TRADITIONS

National traditions are a complex aspect of the history of science, involving as they do political, religious, social, economic and linguistic factors.[29] There can be no doubt that as the nineteenth century wore on, national differences in scientific styles became less important. But in our opening years, the contrasts between France and Britain were extreme. On the one hand, there was authoritarian government and centralization in Paris, with an Academy, the École Polytechnique, the museum and its associated garden and zoo, and nationalized industries requiring consultants, all giving something like a career structure. In Britain, on the other hand, there were strong provincial and local loyalties, relatively weak government, voluntary clubs and societies, and a runaway capitalist Industrial Revolution where the man of science had to take his chances and find a niche. As we saw with Davy, in France honour and prizes tended to go for achievement in scientific understanding; in Britain for useful knowledge. British society worked through patronage and impoverished men of science or letters might, on the recommendation of well-placed friends, be given 'pensions' out of government funds as Dalton and Faraday were. In France, Napoleon's sinecures provided for Berthollet and Laplace,[30] but after 1815 governments were less generous and stingy salaries were augmented by accumulating posts, thus restricting opportunities for the young.

As the century wore on, the German model of state support for research universities began to catch on, and travel abroad (very difficult between 1789 and 1815 because of the wars) became steadily easier with steamships and railways. Liebig had gone to Paris to study with Gay-Lussac, and Alexander von Humboldt became his patron, getting him a professorship in Giessen on his return to Germany. There, his research school, training chemists for PhD degrees, attracted students from other parts of Germany – the system, based on the ideas of Alexander's brother Wilhelm, made it easy to migrate between universities. The graduate student, so important in science, had arrived. Soon Liebig, whose writings attracted much attention in Britain and the USA, began to draw students from outside Germany. Previously, prominent men of science like Hermann Boerhaave (1668–1738), Berzelius, Davy, Ampère or Gay-Lussac had welcomed visitors from abroad (especially if they came with a suitable letter of introduction), and sometimes worked with them. But those had been informal arrangements, whereas a qualification in the form of a PhD (in those days always achieved more rapidly than is possible now) and a published paper could be acquired by working in a group alongside Liebig or Bunsen. This degree, 'junior' to doctorates in theology, law and medicine because the philosophy faculty was seen as introductory to these studies, only achieved widespread respectability in Germany about the beginning of our period.[31] In the same way, medical doctorates were awarded for a thesis based upon research (perhaps actually done by the supervisor, as with Linnaeus's pupils), but in theology and law were given for published but unsupervised work. In Britain, the PhD degree was unavailable until after the Great War.

Students from Britain, the USA, Scandinavia and Russia, and even (as with Adolph Wurtz, 1817–84), from France, therefore went to Germany where university life was cheap, the research ethos strong and they were welcomed by powerful professors into a world where science was a collective effort under their firm direction. 'Doctor-fathers' were important figures around whom their students constructed intellectual genealogies. Those arriving in Germany from Britain, such as Frankland and Tyndall, found themselves plunged in at the deep end: they had to pick up the language through total immersion, and were expected to get down to work for long hours and find their feet rapidly. The Russians were in the same position. It was an eye-opening experience. German music, German painting, German philosophy, novels, poetry and drama were all available as well as the chemistry, physiology or physics that were the students' primary concern, and scientists were prominent in taking back to their home countries an admiration for German culture in all its rich manifestations. In different countries (and even different regions or universities), frontiers between sciences were drawn differently, so that chemistry might be an independent science, or seem a province of physics, of medicine or of industrial development. This affected what questions were important, and how they might be answered. Students schooled in a highly empirical Baconian mode might meet bold hypothetico-deductive thinking, or vice versa. By 1860, chemists applying for academic posts in Britain were expected to have spent time in Germany, and to have a testimonial from their doctor-father or some other eminent German.

The traffic was not all one way. Russia had long been a magnet drawing Britons (especially Scots) and Germans to scientific and engineering posts, because there were still too few appropriately educated Russians. There were also many Germans who sought like Prince Albert to make their fortune in England, where industry and academe lacked well-qualified natives. Thus many German chemists came to work in the textile industries, especially on dyes, in the years both before and after the introduction of coal-tar synthetic dyes in the early 1860s.[32] Most of them returned home well-acquainted with the industry by the end of that decade, when German academic research in organic chemistry was supporting the new industry which developed so successfully both there and in Switzerland. Engineers arrived to work in the telegraph and electrical industries and some stayed on in Britain, like Merz and William Siemens (1823–83), whose brother Werner (1816–92) remained in Germany, managing that end of the family business.[33] Inorganic chemicals brought Mond to Britain, where he manufactured soda using the new but tricky Solvay process. He entered into partnership with Brunner, the son of a Swiss pastor who kept a school in Liverpool, and they built up a large and lucrative business. Brunner, elected to parliament as a radical Liberal, was made a baronet, and described as 'the Chemical Croesus'.[34] Their company, in which Mond provided the scientific know-how, was the nucleus of the later giant, Imperial Chemical Industries, rivalling huge German companies like BASF.

Academic chemists also came to Britain: Liebig often, but also Kekulé, who after three years at Giessen and a year in Paris came to London in 1854–5. Working under John Stenhouse (1809–80) at St Bartholomew's Hospital, he met Alexander Williamson (1824–1904), famous for elucidating the mechanism of the reaction in which ether was made from alcohol, and William Odling (1829–1921). He compared their ideas on chemical families, or 'types', with those he had encountered in Germany and France. Supposedly in daydreams gazing into the fire or on the top of a London bus, he imagined atoms forming patterns in space, each kind forming a set number of links to others. Hofmann developed this idea into actual models, and in 1865 Kekulé, dreaming about snakes chasing their tails (an alchemical symbol), came up with the idea that benzene, C_6H_6 – isolated by Faraday much earlier – had a ring structure. These dreams may be authentic, or a picturesque cover for speculative hypothetico-deductive thinking: they led to the prediction (soon confirmed) that there were three possible compounds of the formula $C_6H_4X_2$ (where X can be an atom like Cl or a group like OH or NH_2) called *ortho*, *meta* and *para* as the second X is moved round the ring with the first one kept at the top. These ideas formed the basis for the synthetic work of chemists back home in Germany, where Kekulé after a spell in Ghent became professor at Bonn with its magnificent new laboratory.[35]

Such travels were not unique to chemists. Geologists, like Humboldt and Murchison in Russia, made field trips in conjunction with local *savants*. Earlier, Humboldt, Orfila and others from countries occupied by the French army had gone to Paris, as did the hostile but admiring Davy and Faraday who, following a row about priority (over iodine), went on to the more congenial but less high-powered cities of Italy. Other unhappy and competitive international contacts included the cosmopolitan Tyndall asserting the claims of German scientists, notably Mayer and Clausius, in thermodynamics, when he incurred the wrath of the Scots, particularly Peter Guthrie Tait (1831–1901), who stuck up for Joule and William Thomson.[36] The discovery of the planet Neptune was also contentious. John Couch Adams (1819–92) in Cambridge had predicted its existence from wobbles in the orbit of Uranus, and computed whereabouts it should be, but the observatories at Greenwich and Cambridge were too busy to take serious notice.[37] Soon afterwards, Urbain Le Verrier (1811–77) in Paris independently predicted it and he was successful in finding an observer, J.G. Galle (1812–1910) in Berlin, prepared to drop everything and search for it. There were red faces in Britain, with national pride and hatred of perfidious foreigners being kindled, but fortunately the two protagonists got on well together, honour was satisfied and posts were found for both of them, in Cambridge and at the Paris Observatory.

In medicine, travel was of great value for catching up and seeing things differently, and William Lawrence was only one of the many who profited from a visit to France. The meetings of the German *Naturförscher* provided an important venue where foreign scientists might be met, and so later did meetings of the BAAS, keeping everyone aware of what was going on and diminishing national differences, although language continued to be significant barrier and translation was often patchy. Huxley saw the importance of German embryology and physiology[38] and, like Tyndall, translated papers. After 1860, there was a new development, when that first international scientific congress met at Karlsruhe to put chemistry on a firm footing. As it turned out, the formal sessions were indecisive and tedious, much less important than the opportunities for conversation and the chance afterwards to read Cannizzaro's paper (which he had wisely had printed for circulation) and reflect upon it. Conferences ever since have been the same: the keynote addresses can be great occasions and the papers interesting, but the crucial and most fruitful encounters are generally in the bar or over meals. They have been, and remain, an essential way of making contact and keeping in touch. And they kept alive until 1914, in an era of competition, the belief that science transcended nationalisms.

SCIENCE AND NATIONAL POWER

It was harmless to take pride in the inventions and discoveries of Perkin or Edison, and to celebrate them in international exhibitions. But science and technology were crucial in naked demonstrations of power in warfare. Cook's skills in surveying had been shown on the St Lawrence River, and made possible General Wolfe's capture of Quebec in 1759. In different ways, the scientists and engineers on both sides in 1791–1815 had promoted the war effort. In the Crimean War (1854–6), Cook's successors, trained off South America and Australia, led the Royal Navy into the Baltic, while navvies built a railway to transport supplies to the front in the Crimea itself, and the medical care of soldiers became a major issue. In 1861, in response to reports that the French were building an innovatory battleship, *La Gloire*, the armoured iron ship HMS *Warrior* was launched, powerful enough to take on the world's navies. During the American Civil War (1861–5), and Bismarck's campaigns of unification, troops were transported over great distances by train, the lethal power of modern guns was demonstrated, and armour-plated battleships were seen in action as the lumbering and improvised *Merrimac* and *Monitor* bombarded each other inconclusively. The German empire and the USA both became beacons of modernity, rapidly industrializing. But whereas the Americans had a Wild West to subdue and settle, the Germans lacked an overseas empire comparable to those of Britain or France. Many – notably from 1897 Kaiser Wilhelm II, his new Foreign Secretary Bernhard von Bülow (later Chancellor) and Naval Secretary Alfred Tirpitz – felt that their power and importance were unfairly constrained, and envied Britain and France their 'place in the sun'.[39]

Britain had emerged in 1815 with more warships than the rest of the world put together, and an empire partly acquired in a fit of absence of mind but being steadily augmented. South American states became part of the informal empire, linked to Britain (which had backed their independence from Spain) through trade; the Royal Navy maintained stations to protect trade and expatriates, and British governments recognized that a small trading nation with a seaborne empire required a large navy (but as cheaply as possible).[40] The convention was adopted that the Royal Navy should be kept as powerful as the two next-largest put together, and for a generation it seemed as if the ships that defeated Napoleon's navy could go on and on. By the middle of the century, this was not so. Navies were no longer anything like Nelson's had been. Turner's evocative picture, *The Fighting Temeraire*, showed the enormous wooden ship being towed away in the sunset by a steam tug.[41] The British army had also been run down by the powerful forces of tradition and economy, and by the 1840s Woolwich Arsenal and its workforce were antiquated. John Anderson (1814–86) was appointed in a shake-up, and with enormous energy brought it and the armoury at Enfield up to date, making breech-loading rifled guns from wrought iron (and later steel) in processes invented by Armstrong and his rival Whitworth (whose company Armstrong later took over).[42] Both artillery and small arms were made for the many small wars in which Britain was engaged, while at Harpers Ferry in the USA, and armouries and arsenals elsewhere, continual improvements meant that no nation could neglect spending on armaments.[43]

This aspect was especially acute with warships. They were made of iron, then steel; steam engines, at first auxiliary to sail, propelled them; soon turbines were replacing piston engines, and screw propellers replacing paddle-wheels; breech-loading rifled guns were introduced and armour plate was required to resist their projectiles. Instead of sailing in line, getting in close and bombarding at point-blank range, battleships were designed to engage each other with big guns when almost out of sight. HMS *Warrior* was obsolete within not much more than a decade, and HMS *Dreadnought*, based on the principle of relying upon big guns and equipped with the latest technical innovations, was the world's most formidable ship when launched in 1906, having been built in a year and a day. Existing warships instantly became obsolescent. An arms race began, at huge expense.[44] 'Dreadnought' became a term for ships of her class, and the German Navy League pressed the (supportive) government to build more such leviathans more quickly. The British were determined to maintain a 60 per cent superiority over Germany. The exact point of these swift floating fortresses was unclear, but the existence of the German fleet meant that the Royal Navy could no longer be so widely dispersed – a large home fleet had to be maintained. Hitting a moving ship at long range from another moving ship was very tricky. Scientific and engineering skills were required to make suitable steels, propellants and high explosives, design range-finders and gun-sights and maintain stability, and much training was needed for the crews. Zeppelins and submarines were also developed. The aggressive nationalism that went with this shipbuilding was a factor in the British move to agree an Entente with France, the traditional enemy, in 1904, and the German reaction to it, and in the run-up to war in 1914.

By 1900, Babbage's vision of a world in which the achievements of scientists and engineers were properly appreciated was coming about. The governments of Napoleon I and III, and of the kings who filled the interim between them, ennobled scientists, including Laplace, Berthollet and Cuvier, and the subsequent republican regime honoured them with government posts, decorations and, as with Berthelot, burial in the Pantheon. In Germany, Liebig, Helmholtz, Hofmann, Kekulé and other eminent scientists were ennobled and enjoyed a high status. In Britain, Davy and both Herschels were (like Newton) knighted, but in the second half of the century fuller recognition was given to Lister the surgeon, Armstrong the engineer and William Thomson the physicist (who became Lord Kelvin) when they were raised to the peerage. Faraday refused honours and decorations, but the 'plebeian' Huxley became a Privy Counsellor. Scientists and technologists, including Davy, Watt, Joule, Maxwell and Darwin, were buried in Westminster Abbey, the British pantheon, or commemorated there.[45] Nobody, by the latter part of the century at any rate, could doubt their national importance or think them fuddy-duddies.

IMPACTS OF SCIENCE

Dreadnoughts embodied a lot of science and being bombarded by one could give anyone a jaundiced view of science and technology, but there was nothing new about navies being expressions of national prestige and technical expertise. Those survey voyages round Australia and South America involved a measure of military intelligence, and a determination to keep other European powers out. Cook (provoked by his suspicious reception by the Portuguese authorities) had noted how weak the defences were at Rio; the Portuguese were also anxious to keep Humboldt (exploring under French and therefore hostile auspices) out of Brazil, and gave orders to arrest him should he appear on the Amazon. After 1815, and particularly with Brazilian independence, scientific visitors were more welcome.[46] Darwin landed briefly and marvelled at the richness of tropical vegetation, but the *Beagle*'s duties lay southward. In 1848–50, Bates and Wallace arrived from Britain, collecting specimens commercially.[47] Bates stayed until 1859, ruefully reflecting afterwards that his eleven years had netted him about £800. He moved beyond collecting, and interpreted the patterns on his butterflies in Darwinian terms as 'mimicry' brought about by natural selection. On Bates's return, Darwin persuaded John Murray to publish his travels as a very readable book, and helped him find a post at the Royal Geographical Society.[48]

Wallace was also reflective, but he left Brazil earlier, taking his collections with him – like those of Raffles, they were destroyed by a fire while at sea, so instead of writing them up and selling them, Wallace had to set forth again. In Indonesia this time, he studied the distribution of organisms (and came up in 1858 with the idea of evolution by natural selection). Both Bates and Wallace were free from the racism so common in Victorian Britain, but their researches (and the writings of Humboldt) stimulated economic botany, which could lead to exploitation.[49] Darwin's other ally Joseph Hooker was anxious to promote Kew as an imperial resource, and the Brazilians came to be at the less comfortable, receiving end of science. The Sao Paolo railway, taking coffee to Santos for export, was Britain's most profitable railway enterprise in Latin America.[50] Brazilian logwood had been sent to British mills to be shredded and make a red colour, but this lucrative trade was rendered obsolete by aniline dyes. Rubber, becoming essential for industry, was procured from trees growing wild in Amazonia and collectors were urged to bring seeds back to Kew, although their export was prohibited. James Trail (1851–1919) went out in 1873–5, secretly hunting for them when exploring on behalf of a British shipping company which had acquired large concessions on the Amazon and wanted to know their potential. But he was unsuccessful, and it was Henry Wickham (1846–1928) whose seeds germinated at Kew. The trees were then transported to plantations in Malaya, which made Brazilian rubber uneconomic. Similarly, quinine trees from Peru were grown at Kew, and then planted in India. Trail, under the tuition of Barbosa Rodriguez (1842–1909), became an authority on palm trees, publishing on and naming them when he was appointed professor in Aberdeen on his return. Rodriguez protested, but to no avail: naturalists in places remote from Europe were expected to be deferential.[51] But when, in 1907–9, Carlos Chagas (1879–1934), at Oswaldo Cruz's Institute in Rio, isolated the parasite causing a tropical disease, and identified its vector, nobody could deny that serious science was being done in Brazil.

One reason for the Royal Navy to be active off Brazil was that in 1807 the slave trade had at last been outlawed by the British government, and once the war was over its suppression (along with protection and promotion of honest trade) became a major task for naval ships. Others were stationed off Africa, and land-based expeditions to open up that dark continent were intimately connected with the slave trade and the effort, of government and of pressure groups and missionary societies, to put an end to it. Curiously, explorers such as Mungo Park (1771–1806), Richard Burton (1821–90), David Livingstone (1813–73) and Henry Morton Stanley (1841–1904) had perforce to make use of slavers' assistance and travel with their caravans, in their efforts to map the country, assess its resources, replace the loathsome slaving with respectable trade and bring the Christian Gospel to the people. The maps and information, the supposed riches disclosed, and the resources in ships and guns possessed by Europeans led to the 'scramble for Africa'. Madagascar had been a base for pirates and then evangelized by Protestants from the London Missionary Society,[52] but the suggestion that it might be a British protectorate was, like some other applications, turned down by a government aware of imperial overstretch. The French, who already had an empire in North and West Africa and the Far East, invaded, the Malagasy relying for their defence upon 'General Fever'. He nearly won, but enough French troops survived to bring the huge island with its raw materials into their empire. Medical science was essential in these 'white men's graves'. Cultivating quinine was one thing; the other was the work of Ronald Ross (1857–1932), at last clearly establishing the connection between mosquitoes and malaria in India where he had been born, and where he served in the Indian Medical Service from 1881–99. By the early twentieth century, hospitals for tropical diseases had been set up in European ports including Hamburg, Marseilles, Liverpool and London. European power, suppressing much of the slave trade, brought new possibilities to Africans but made them subject to new masters, primary producers again feeling the impact of science and technology rather than sharing the happier fate of those generating it.

National prestige was involved in exploring and exploiting Africa, taking over vast chunks of it to be appropriately coloured on the map. In this competition the Germans as late starters could sulk over their small share, though they were keen to exploit their colonies, planting sisal in Tanganyika and building a railway. On the other hand, international cooperation continued, notably in the 'magnetic crusade' stimulated by Humboldt in the 1830s. This great geophysical enterprise, arousing particular enthusiasm in Britain where Humboldt was much admired, involved magnetic measurements round the world. Observatories everywhere took part, overland expeditions were sent out, and two specially strengthened ships of the Royal Navy, HMS *Erebus* and *Terror*, under Captain James Clark Ross (1800–62) who had found the North Magnetic Pole, were fitted out as floating observatories. They went to the southern hemisphere in 1839–43, penetrating into the Ross Sea, and visiting Tasmania where John Franklin (1786–1847) was Governor.[53] At the other end of the world, international competition lay behind the more desperate enterprise of seeking the north-west passage round the top of Canada that would bring China and Japan much closer to Western Europe[54] Northern Canada was a barren

Figure 22 *A naval survey voyage – burying a shipmate beneath the Arctic ice, under a portentous 'mock Moon': J.* McClintock, A Narrative of the Discovery of the Fate of Sir John Franklin, *London: John Murray, 1859, p. 74*

region, but one of the spurs to exploration was that the Russians, having got right across Siberia, were already in Alaska and might make further territorial claims to the east or south. Franklin and John Richardson (1787–1865), an Edinburgh-trained doctor, had been the heroic figures in these exploits, navigating and charting the Arctic Ocean in birch-bark canoes, escaping starvation by eating their boots, rescued by kindly Native Americans, and returning for further exploration.[55] Franklin, at the age of fifty-nine, set out again with Ross's ships in 1845, but died in his iced-up vessel in 1847, and his crews, over a hundred men, also died when attempting to escape overland – but they had found the ice-blocked passage. Expeditions sent out to search for them completed the charting of these desolate regions, demonstrating that the passage was impractical for sailing ships. The most successful detective, finding clues and remains, was John Rae (1813–93) of the Hudson's Bay Company, who travelled light and had learned from the Inuit how to survive.[56]

Similarly, Canadians and Americans (the USA had bought Alaska in 1867) explored their own vast territories after the frontier west of the Great Lakes was agreed between Britain and the USA. Nationalism entered into taxonomy when Britons named the sequoia *Wellingtonia* after the Duke, while Americans applied the principle of splitting, sure that American birds and plants must be distinct species from European ones, even if (as with mallards) they looked exactly alike. Railways and telegraphs were essential in nation-building, especially in North America and Russia. Under Perry in 1852–4, the US Navy 'opened up' Japan willy-nilly to the world and the Japanese rapidly adopted Western science and technology, offering foreign teachers limited-term contracts to train Japanese to be their successors. Determined to resist colonial status, they soon built railways and equipped a modern army, as well as visiting Armstrong in Newcastle to order a navy. When war with Russia broke out in 1904, the Russian Baltic fleet was sent round Africa and Malaysia to attack Japan – taking pot-shots at fishermen and others in the English Channel on their way because they took them to be Japanese gun-boats. In the Straits of Tsushima they were annihilated by the Japanese fleet. This defeat of a predominantly European power by an Asian one equipped with the technology of the progressive West successfully demonstrated the place of science in maintaining national pride and identity. Subsequently, an abortive revolution broke out in relatively backward Russia, a sign of things to come. Science had clearly lost its innocence.

Nevertheless, the beginning of the twentieth century was a time of optimism. There is more to national identity and proper patriotism than military power, and more to science than the improvement of weaponry. Alfred Nobel (1833–96), the inventor of dynamite, a relatively safe high-explosive vital for great engineering projects, bequeathed a fortune to fund prizes for scientific achievement, and from 1901 the award has been a matter of national pride, like Olympic Gold Medals, with nations keeping the score in harmless competition. In 1900 the Paris Exposition was an occasion for rejoicing in progress and the availability of electric power. Physicists were discovering all sorts of extraordinary and unexpected phenomena in a new scientific revolution, and chemists producing new substances. In 1901 Marconi sent a radio message across the Atlantic. And in 1903 Wilbur (1867–1912) and Orville Wright (1871–1948) made the first flight in an aeroplane.

CONCLUSION

During the nineteenth century, Europe congealed into major nation-states, with Russia and Austria presiding over empires with mixed populations. Nations were united by folk-memories of their histories, crafted during the Romantic period into powerful myths. Gradually, these were augmented by national pride in the achievements of the scientists and technologists who played a major role in modernizing them and increasing their wealth and power. It was an age of overseas empires, in which colonies were expected to supply raw materials (in the form of data as well as material goods) to be processed in the great metropolitan centres in Europe, and the eastern seaboard of the USA, which had its own backyard to explore. The smaller countries of Europe, where Italy was, like Germany, being united politically and Belgium setting up on its own, looked to their bigger neighbours for models. Travel broadened many minds, as students and mature scientists visited other countries, met new ways of thinking and found things differently arranged – both in academe and in industry. Armies, and particularly navies, upon which national power depended, were transformed with technological changes, while wars became even bloodier (though medical services improved) and an arms race ensued, which was a major factor in precipitating war in 1914. Science had an important impact outside Europe – notably in India, Africa (now brought under European control and subject to new crops and industries) and South America where quinine and rubber were profitable until seeds were smuggled to Kew, and plantations set up in British and other colonies. The Japanese succeeded in avoiding colonial status, adopted Western science and technology, and by 1900 were themselves a formidable power. Amid all this competition and nationalism, cooperative scientific enterprises went on, notably in geophysics. And from the opening years of the twentieth century, Nobel Prizes formed a way of assessing and quantifying national contributions to science. Just what was thought to make a discipline scientific will be the subject of the next chapter.

10

METHOD AND HERESY

It would be grand if good science was the inevitable result of carefully following the right method, and even better if the method could be learned without all the grind associated with actually studying physics, chemistry or biology – like French without tears. Promoters of science have indeed, ever since Descartes' *Discourse of a Method* appeared in English in 1649, emphasized its method, supposed to differentiate it from other and less reliable activities.[1] Some in the nineteenth century even hoped that the method could be applied everywhere, yielding true and tested knowledge, *scientia*, in all fields, in what came to be called positivism or scientism. They saw everything else as the realm of opinion, emotion and subjectivity, hardly worthy of serious consideration except by the antiquarian, psychologist or anthropologist. More hoped that study of scientific method could be part of a liberal education, making those who would never be scientists appreciate science as a cultural force, a route to reliable knowledge about the world and not just an expensive but important aspect of the modern economy. Such thinking goes back a long way: Aristotle had written in his zoological works that the well-educated person ought to be interested in, and able to make sound judgements about, what experts said.[2] It certainly is vital to be able to judge the soundness of inductive and deductive arguments, and not be dazzled by unfamiliar terms or content. It must, however, be easier to be simplistic in thinking about science (for or against) than in doing it – we forget the sheer difficulty and complexity, as well as the excitement, of science.

Hence discussions of method were also very important for insiders who wanted to do science properly or perhaps found accepted canons a strait-jacket. Scientists are cautious conservatives, and one way of dealing with innovators was to brand them as heretics or as speculators, improperly trying to build upon foundations of sand. Thus Cuvier dismissed *Naturphilosophie* and evolution; Henry Brougham (1778–1868), Young's wave theory of light; Huxley, the evolutionary thesis of *Vestiges;* and Wilberforce, Darwin's cumulative, probabilistic arguments. Such polemic is a serious part of science, which is necessarily a critical activity. In writing about method, while practitioners might display their deep convictions about the world, they often paid conventional lip service to accepted norms – which is interesting in a different way to the wary historian. They also sought to understand how great men, notably Newton, had established what seemed to be unshakable truths about nature. He had not been altogether helpful. He was, he said, like a boy on the beach picking up pretty pebbles; he kept problems before him, thinking about them, and little by little the answer came. He left a legacy in his *Principia* of mathematical deductions from what he called 'hypotheses' in the first edition, subsequently dropping this awkward word in favour of 'rules of reasoning' and 'phenomena'. The term 'hypothesis', previously neutral, was coming to mean conjecture, and his were far better grounded than that. In the much more accessible *Opticks* (1704), he followed an inductive, experimental path, cautiously generalizing and using 'crucial experiments' (a Baconian term) to decide between alternatives. In 'queries' at the end, augmented in successive editions, he indicated which side he was disposed to take in open or disputed questions, such as whether light was a corpuscular or wave motion, without committing himself finally. His was the name most respected in the scientific world of 1789, and to follow him was to be on the right lines. But it was not always straightforward.

Partly this is because there are different kinds of scientific minds. Some work by framing hypotheses and testing deductions from them, and by the 1900s this was seen as par excellence the scientific method. Others carefully pile up data and generalize, and in the 1790s this had seemed the safe way to go. Their contemporaries contrasted Davy, keen to discover truth, with Wollaston, anxious to avoid error.[3] The intellectual historian Henry Buckle, lecturing at the Royal Institution in 1858, declared that women possessed a vaulting deductive cast of mind, while men's intellects were plodding and inductive, and that the rapid progress of science would require both.[4] Unfortunately for his argument, his example of female intellect was Newton's. At the end of our period, the French philosopher Pierre Duhem

(1861–1916) contrasted 'ample' (or broad and shallow) English minds with deeply penetrating French ones.[5] This is a rather different distinction, between the synthetic and analytic mind, both again being necessary. Duhem was not a narrow nationalist, for curiously enough Newton's was among his examples of the French mind, and Napoleon's of the English. There must be, and have been, times and places where hypothetico-deductive (often analytic) minds are appropriate, and others where cautious induction and a broad view are appropriate, or indeed requisite.[6]

VARIED TRADITIONS

Thus, in Lavoisier's France, there was a reaction against grand Enlightenment 'systems', broad-brush sketches that could not be rigorously tested and lacked detail. Mathematical and chemical analysis were the order of the day: painstaking experiments, accurate observations and careful measurements were to be expected, in what has been called the 'second scientific revolution' going on in Paris. The philosopher Auguste Comte was a product of the École Polytechnique, and in his 'positivism' he saw knowledge (in individuals and society) progressing through religious and metaphysical stages to a final positive state just being realized in the emancipated and grown-up Europe of the 1820s and 1830s.[7] His six rather indigestible volumes were abridged and translated by the formidably intellectual Harriet Martineau (1802–76), and published by John Chapman. Comte's rigorous philosophy, softened by his 'religion of humanity', appealed in Britain to some of those abandoning orthodox Christianity.[8] British scientists, however, even agnostics, found it constricting. Although one can see Comte's 'three stages' in the structure of Huxley's textbook on the crayfish, for him and for Tyndall to interpret nature, they found it essential to have scope for theorizing and scientific imagination.[9] In France, a Positivistic tone in science was evident to some contemporaries right through the century, markedly in resistance to atomic and evolutionary theory. Though some historians doubt how far this should be taken, philosophical realism was (as Duhem recognized) much stronger in Britain, where physicists like Stokes at Cambridge hoped in Newtonian vein for theories that were true representations of the world, to be believed in rather than just used (and dropped) when convenient.[10]

In Germany and Britain, birthplaces of the Romantic movement, things had developed differently.[11] Despite an admiration for French achievements in science, Germans like Goethe, Humboldt and Oken aimed for a wide world-view, seeking connections in a dynamical world. Britons like Davy thought that way, too, but in reaction to the French Revolution abjured the dangerously radical public science of Priestley, Erasmus Darwin and Beddoes in favour of sound Baconian principles. Thus, they remained in tune with the French, though far behind in the mathematics that was giving a new turn to physics. In 1812, John Herschel had been prominent in that group of young Cambridge mathematicians who appreciated and deplored this gap. They translated a French textbook, and began teaching up-to-date mathematics and testing students on it – bringing them back into the international community.[12] Herschel left Cambridge to work with his father in his observatory, took over on his death in 1822, and in 1829 married Margaret Brodie Stewart. In 1831 he was (like his father) knighted, and after his mother's death in 1832 was financially independent and never needed to specialize or undertake hackwork – the classic 'gentleman of science'. In 1830, as a contribution to a series of popular little books, *Lardner's Cabinet Cyclopedia*, published by Longman, he had written a study of scientific method, *A Preliminary Discourse*.[13] The book became a classic, a template against which to judge claims of scientific status. Especially after his return, as a scientific hero, from a spell in South Africa observing the stars and nebulae of the southern hemisphere, he was regarded as a pundit or sage whose science had brought him wisdom.[14] Though admiring Bacon and emphasizing the role of inductive reasoning, he knew that Newton had gone beyond that, but declared that any hypothesis worth taking seriously as a theory must be a *vera causa*:

> Which we can not only show to exist and to act, but the laws of whose action we can derive independently, by direct induction, from experiments purposely instituted; or at least, make such suppositions respecting them as shall not be contrary to our experience, and which will remain to be verified by the coincidence of the conclusions we shall deduce from them, with facts.

While these criteria would exclude weightless fluids or mysterious effluvia, Herschel believed that by 1830 the wave theory of light had become either: 'An actual statement of what really passes in nature, or that the reality, whatever it be, must run so close a parallel with it, as to admit of some mode of expression common to both, at least as far as the phenomena now known are concerned.' [15] 'Theory' in science meant (and means) something a great deal more solid and testable than it does in common usage or literary and cultural studies: it was an accolade given after a hypothesis had survived critical examination, and had shown its explanatory and preferably predictive power. Herschel's experience in physics and astronomy, and the respect this brought him among men of science, gave his book great authority.

It was picked up by two eminent contemporaries, John Stuart Mill (1806–73) and Whewell. Mill moved in Chapman's circle associated with the *Westminster Review*, and corresponded with Comte. Through the terrifying educational programme devised for him by his father James, he had acquired a utilitarian and inductivist philosophical outlook, modified by his reading of Coleridge, but he was never a practitioner in science. In his *Logic* he emphasized inductive reasoning, playing down the hypotheses and deductions that Herschel had allowed (or even encouraged). He ran up against the problem that induction had to be judged inductively: it was reliable because it had worked in the past. This could not satisfy those who accepted David Hume's argument that we can never be sure that the past will be a guide to the future. While scientists have not on the whole been very interested in that kind of certainty, it has been a source of much philosophical discussion. Nevertheless, for English readers, Mill thus made scientific reasoning (or one aspect of it, the commonsensical empiricist tradition) central and relevant – along with the moral and political thought for which he was famous as writer and Member of Parliament.

Whewell was involved with his friend Herschel in the 'Analytic Society' at Cambridge, and made his career there, becoming a very powerful figure in its affairs as it slowly modernized. He was a polymath, prominent in scientific societies, working on mineralogy and the tides as well as what we would call philosophy, and in Cambridge was a great upholder of the value of applied mathematics as better training for the mind than the pure branch, which might all too easily lead into deductive habits remote from the real world (and its God).[16] Despite that, what he admired in Herschel was the emphasis upon the mental leap going beyond generalization. For him, getting the right end of the stick, the fundamental idea, in any science was crucial in transforming a mass of data into a science. Experimental test for him, as for Herschel, was vital. Nevertheless, not only experiment, but also observation, or the collection and ordering of materials, was only sensible and scientific when there was an end in view, something to test. In his multi-volume *History of the Inductive Sciences*, so-called because they had an empirical component, he showed how fundamental ideas had emerged and been developed.[17] The book was dedicated to Herschel, while the companion volumes on philosophy of science were dedicated to Sedgwick, his geologist colleague at Trinity College, Cambridge.

There were strong connections in physics and mathematics between Cambridge and Scotland, and Britons everywhere thus got ideas about method stemming from Herschel, but going in rather different directions. In Oxford, the mathematician Baden Powell (1796–1860, father of the founder of the Boy Scouts) wrote about inductive philosophy, but made little impact on the university during the great religious upheaval there in the 1830s and 1840s. With other liberal theologians, he was a contributor to *Essays and Reviews*, but died in 1860 in the midst of the rows associated with it.[18] By then, Mill was becoming widely admired in Oxford as it became increasingly secular – to be displaced in turn, in the latter years of the century, by Kantian Idealists.

German universities taught philosophy more formally than was customary in England, with Germans at the centre of exciting debates in philosophy, and German scientists better informed and more self-conscious about the status of their disciplines than those in Britain. In Germany, debates about teleology, atomism, space and time, and the nature of knowledge were prominent, and Goethe's polemical work on colours (where he challenged Newton's conclusions, and indeed truthfulness) was a source of national pride or embarrassment.[19] Looking through prisms rather than at spectra formed at a distance from them, Goethe believed colours were generated by light and shade. He was interested in visual illusions, and in artists' use of colour to suggest depth and mood. Forty years on, in 1845, his book (toned down) was translated into English by Charles Eastlake (1793–1865, Director of London's National Gallery), and studied by Turner, who made use of some ideas from it. Not long afterwards, in 1853, Helmholtz devoted one of his great public lectures to Goethe's science, respectful of the great man's literary reputation and observational skill, but well aware also of the mistakes a cobbler can make when deserting his last. For Helmholtz, another polymath, Goethe's system of optics was really a part of psychology rather than physics, and he could therefore be sympathetic to the enterprise rather than simply condemn it.[20] In the USA, Johann Bernhard Stallo (1823–1900), who had emigrated from Germany and became a judge, gave up *Naturphilosophie* and wrote a powerfully positivistic book, *The Concepts and Theories of Modern Physics*, but this does not seem to have created much stir.[21] Generally, Germans were explicit about their position, whether idealistic, positivistic, vitalistic, materialistic or whatever, compared to their English-speaking contemporaries.

Others, too, came to see *Naturphilosophie* as excessively a priori and based upon weak analogies – though we can see that through Goethe and Oken it pointed towards evolution, and through Ørsted towards conservation of energy.[22] Liebig had been highly critical, and in the latter years of the century, Ernst Mach (1838–1916) strenuously opposed 'metaphysics' in science, which for him should be concerned only with relationships between observables. Writing in 1883 about the science of mechanics, he was thus critical of Newtonian axioms about space and time, which he believed (like Leibniz earlier) could not be 'absolute'. In the twentieth century, his writings were to influence Einstein, and then the Vienna Circle of 'logical positivists', and he had as an ally Wilhelm Ostwald (1853–1932), the eminent physical chemist who at the end of the Great War published Wittgenstein's *Tractatus*. Heinrich Hertz (1857–94) had earlier declared similarly that, in field theory, Maxwell's equations were what mattered; and that any hypotheses or models that he had used were no more than scaffolding. But many would have sided with Arthur Schuster (1851–1934), who was brought up in Germany and retained links there, but made a career in academic science in Manchester. In his fascinating account of the transformation in physics in his lifetime, he attacked positivism as cowardly:[23]

> I have . . . contrasted . . . the former tendency to base our theoretical explanations of natural phenomena on definite models which we can visualize and even construct, with the modern spirit which is satisfied with a mathematical formula . . . I believe [this] to be fatal to a healthy development of science. . . We all prefer being right to being wrong, but it is better to be wrong than neither to be right nor wrong.

When truth ceases to be the object of the search, science may seem rather like an intellectual game.

Most philosophy of science has concentrated upon physics, but Swainson wrote a companion volume to Herschel's, a *Preliminary Discourse* on natural history.[24] This is a curious volume, because Swainson, a well-known and talented illustrator, set out a system of classification based upon triads of circles, perhaps connected to his High Church Trinitarian principles. Each third circle was replaced by three small ones, making five (two large and three small) on each taxonomic level: it was thus called the quinary system. Fitting, and often forcing, animals and birds into this scheme, Swainson went on to follow his *Preliminary Discourse* with a series of books in the *Cabinet Cyclopedia* on different aspects of zoology. The idea of an intelligible pattern seduced a number of biologists, among them Darwin, Huxley and

Wallace, but they all gave it up, and returned to the messy 'natural' system of family grouping based on multiple criteria, which in the hands of the Jussieu dynasty at the Parisian Jardin des Plantes had from 1789 gradually displaced the Linnean classification.[25] Descriptive science was a serious business, and classification an important part of scientific method, so Hugh Strickland's demolition of the quinarians at the BAAS, and drafting of the code of naming that became the basis of international agreement, was crucial.

Evolutionary theory did promise to bring some structure and reason into biological classification, though millions of years of natural selection yields an untidy pattern of survivors. But in the 1860s taxonomic science scored a notable victory, with the coming of tables into which all the known chemical elements could be placed, thus bringing order into chemistry. The most successful one was Mendeleev's Periodic Table, from 1869, because he not only grouped the elements into families with similar properties, but also left gaps for undiscovered elements whose nature he could predict.[26] Swainson had hoped to do that with his circles, but in chemistry it worked. Gallium, scandium and germanium duly turned up, with properties astonishingly like those forecast. Even when a whole new family of elements – the inert, rare or 'noble' gases – were isolated at the end of the century, a place was duly found for them.[27] Different authors tried, and still try, different ways of setting out the table, sometimes in three dimensions. They also tried to account for it. It led to a revival of William Prout's and Thomas Thomson's idea that the elements were all polymers of hydrogen, or maybe helium – and with the success of Darwinism it was inevitable that inorganic evolution would be suggested. Crookes's speculations, especially about the group of extremely similar 'rare earth' metals (our lanthanides) tantalized contemporaries, and seemed to fit with spectroscopic study of the Sun and nearby stars that indicated they had histories of development in which elements might be generated, and would decline.[28] But down to 1914 no fully satisfactory explanation of the way the elements were grouped could be found. It was a taxonomic scheme that made learning the facts of inorganic chemistry much easier, and worked brilliantly. When an explanation came, in terms of quantum mechanics, it meant that chemistry might be seen as 'reduced' to physics, but in the meantime here was a highly developed science of huge economic importance which was still in the 'natural history stage' as far as devotees of method were concerned. Clearly, there was no single method appropriate to all times, places and sciences.

GETTING IT WRONG

If even sophisticated discussions of method failed to produce agreement, how, then, were outsiders or tyros to be enrolled, and how were lines to be drawn between what was real and what pseudo science? Schuster's contemporaries had found that intuition about *verae causae* no longer worked in a world where atoms could split, where the atoms of an element were of different weights, where in radioactivity one element could change into another, where invisible radiation abounded, and where the crucial experiments that had demonstrated light was a wave motion were contradicted by others proving it to be particulate. All these paradoxes led to established Nobel Prize-winning science: rule-breaking and intellectual opportunism paid off. Already in other spheres of science, as with the coming of statistics and with Darwin's theory, great scientists had stretched the boundaries of the acceptable, rather as great poets, novelists, painters or musicians had defied and transcended the rules current in their day. It was not easy to tell what was going to turn out well.

Great rows and rivalries were a feature of nineteenth-century science: Cuvier and Lamarck, Sedgwick and Murchison, Owen and Huxley, and at the end of our period, the chemists James Dewar (1842–1923) and William Ramsay were not on speaking terms. Partly this was a feature of the state of science. Recognized qualifications like the PhD were slow in coming and so it was not clear when scientific apprenticeship was over. This led to 'father and son' rows like Davy's with Faraday, and to Huxley's anger when Owen's patronage became patronizing. Then, because the rules were unclear, publication might be slow, and international communications were uncertain, priority disputes could be sharp and wounding, and intrusion on intellectual territory bitterly resented: Faraday had some troubles on both these scores. The introduction in the early 1800s of 'offprints', copies of papers supplied to authors to send to friends and rivals, helped in establishing priority. Gentlemen of independent means did not need to worry so much about being anticipated, though for Darwin it was a horrid shock, even though his livelihood did not depend upon success in science like Wallace's (who was in the event magnanimous). But gentlemen of science, for example in the Geological Society, were competitive individualists and might easily fall out,[29] for it was a feature of the time that leading men of science, striking out on their own, easily became prima donnas.

There could also be issues of scientific method that were deeply and genuinely divisive. Thus Norman Lockyer (1835–1920) and Richard Proctor

(1837–88) fell out over the scope of astronomy and how science should be communicated.[30] Chemists agonized over whether to build up their science inductively, or more boldly to imagine molecular structures and test them. Hermann Kolbe (1818–84) thus famously denounced van't Hoff as a charlatan over his three-dimensional atomic models. Somewhat similarly, George Carey Foster (1835–1919), who had studied with Kekulé, Wurtz and Bunsen, teased John Newlands (1837–98) about his arrangement of elements in a table of 'octaves', asking whether he had tried arranging them in alphabetical order instead.[31] Where Chambers and Darwin saw the value of a theory of the origin of species, to critics like John Phillips at Oxford it seemed like turning away from careful study into hypothesizing.[32] Darwin and Phillips were irenic men, but for their quarrelsome allies Huxley and Owen this provided something else to fight about. Huxley at the famous BAAS meeting at Oxford in 1860 accused Owen of incompetently anatomizing the brain of a gorilla, because he did not want to see how similar it was to a human's.[33]

BAD SCIENCE

Sometimes such rows might lead to accusations of scientific misdemeanour. We saw Thomson and Berzelius fall out over atomic weights, with Thomson seeing exact ratios masked by experimental uncertainty, while Berzelius stuck defiantly to the figures he painstakingly obtained, accusing Thomson of tidying up or making up his results.[34] Thomson was doing what Dalton had done in reaching his Laws of Chemical Combination, and what we do at school when verifying Boyle's Law: looking for the message against a background of noise. Misinterpretation, inadvertence, contamination of samples and deficiencies of apparatus are always around, but deliberate falsifications did and do happen also. In 1830 Babbage, poking a finger at various contemporaries, described different modes of scientific fraud: trimming, cooking, forging and hoaxing, pointing out portentously and chauvinistically, echoing Julius Caesar's comment on his wife, that the scientist must be above suspicion.[35]

Trimming was a matter of cleaning up observations and omitting those that seemed to diverge from the proper result. Everybody knows that we get poor readings sometimes, and until Gauss's work on the error (or bell) curve became well known through Quetelet and others, it was very tricky to decide which readings from a large number were best.[36] In analysing the data from the many observations of the Transit of Venus across the

Sun in 1769 (some from Tahiti on Cook's first voyage), different astronomers could not agree which were best, and might well decide just to adopt
those made by their fellow-countrymen.[37] The observations turned out to
be trickier than expected, so uncertainty about the Earth's distance from
the Sun, which could be worked out from this data by number-crunching
astronomers, therefore remained. It can turn out, as Babbage pointed out,
that observations cast aside were actually the best. Trimming, then and
since, is bad science indeed, but easy and tempting to do. Indeed, scrutiny of
Mendel's classic paper on peas and the factors governing their inheritance
indicates that his results are too good to be true, and that some trimming or
sampling must have gone on.[38]

Cooking is much more serious. Some results are dreamed up, perhaps as
in school where an experiment has gone wrong, calculated back from the
desired result. As Babbage noted, a reputation for high accuracy may be
attained by the cook: he hoped that it would be temporary, that science is a
self-correcting enterprise in which claims are independently tested. But the
cook can also slide into the forger, making the whole thing up, so Babbage
saw a slippery slope from trimming through cookery to downright dis-
honesty. We have in our own day seen forgery, where stakes are high, and
ethics committees have to investigate possible cases, distinguishing honest
mistakes, negligence or inadvertence from fraud, where their verdict must
lead to utter disgrace. It happened in the nineteenth century, too (though
in science formal ethics committees came much later than in older professions, including medicine). Suggestions of forgery were made when the
missing-link toothed bird fossil *Archaeopteryx* turned up so conveniently for
Huxley in a slate quarry – but it was genuine. Davy kept laboratory notes
on odd pieces of paper, while Faraday was much neater, but it became very
important indeed to keep exact records in order to claim priority or avoid
charges of cookery.

Plagiarism, the other major scientific misdemeanour, was also present
in the nineteenth century. After the distinguished anatomist Sir Everard
Home (1756–1832) died, it was found that he had worked up material left
unpublished by John Hunter and published it as his own work in papers in
the Royal Society's *Philosophical Transactions*. It seems he then destroyed the
original notes, which had been lovingly preserved by Hunter's assistant,
William Clift (1775–1849), who had fortunately made and kept copies of
some of them.[39] Simultaneous discovery, that fascinating topic that seems
to make science so different from the creative arts, was also often associated with accusations of plagiarism. For example, when Stephenson and

Davy clashed over the invention of the miners' safety lamp, and Urbain Leverrier's (1811–77) supporters showed John Couch Adams (1819–92) surreptitiously copying their hero's calculations through a telescope.[40]

Hoaxes are, as Babbage also noted, more problematic. In a curious episode of 1803, Wollaston put out an anonymous advertisement that palladium, a previously unknown metal, could be bought at a certain shop. Richard Chenevix (1774–1830), a distinguished Irish chemist, suspected something, bought it, and declared at the Royal Society that while it was an alloy so stable that he could not analyse it, he had synthesized some from platinum and mercury.[41] This would upset understanding of elements and compounds. The Royal Society awarded Chenevix the Copley Medal, but attempts to repeat the synthesis were unsuccessful. Wollaston informed Banks confidentially that he was the discoverer of palladium, and that Chenevix was in error, but refused until 1805 to come into the open. Meanwhile, Chenevix had gone on working, but admitted that the synthesis had worked only four times in about a thousand experiments. Curiously, the two men seem to have remained friendly, though we might expect that Chenevix would have been livid. It is hard to work out why Wollaston behaved as he did: he was perfecting his process for making platinum in malleable metallic form, and may therefore have felt the need to be secretive about discovering another very similar metal – but normally scientists delight in discovery, and bask in the glory of it.

In 1835 there was a more straightforward kind of hoaxing, when the *New York Sun* published a scoop – John Herschel in South Africa with his magnificent telescope had revealed the inhabitants of the Moon, including creatures rather like people with wings.[42] This created a huge sensation, and no doubt sold lots of copies of the paper. Its interest for us is that it reveals the general expectation that there must be life elsewhere in the universe. This was contrary to the general run of Church teaching, notably the creation story in Genesis, but for many it had become inconceivable that in an enormous if not infinite world, our little planet could be the only seat of life or intelligence. This idea was prominent in *Vestiges*, and was later acrimoniously debated by Brewster and Whewell. Whewell perceived that 'plurality of worlds' went with speculative evolutionary and materialist views, and argued (on astronomical grounds) for our uniqueness, while to Brewster (ever the scourge of Cambridge men) this seemed obtuse and absurd.[43] That argument rumbles on.

Then at the very end of our period, in 1912, the fossil 'Piltdown man' was discovered in Sussex: a very early human skull with an ape-like jaw, duly named as *Eoanthropus*. It went to the Natural History Museum in London, and became one of its treasures. At the Festival of Britain in 1951 a recon-struction of this earliest Briton was on view, and he glared memorably at young visitors like me. Two years later, the skull was shown to be an arte-fact, a fake – a human cranium, painstakingly broken and stained, with the jaw of a modern orang-utan. The scientific community should have been wary. After all, Lyell had remarked in 1863 that if we have a common ances-tor with the apes, then hominid fossils might be expected in Africa or East Asia where they live.[44] Lyell had taken a paternal interest in the exploration, by the extremely scrupulous Joseph Prestwich (1812–96), of caves in Britain where there were signs of humans coeval with extinct creatures like mam-moths. Such momentous science had to be above board. By 1912 things must have become laxer, the diggers were distinguished, their conclusions were welcome, and although there were doubters few people seem to have scrutinized the original material, being satisfied with reconstructions. What is not clear is what the point of it was: whether a joke got out of hand when it was taken so seriously, or whether there was a successful conspiracy to deceive. The Piltdown man who never was now has websites, and there is as long a list of suspects as in any mystery story.

Discussing the evidence for strange manifestations in seances, the celebrated conjurer and entertainer John Nevil Maskelyne (1839–1917) remarked somewhere that because science depends on honest and open communication, scientists are easy to deceive compared to streetwise mem-bers of the general public.[45] They trust each other. Philosophers of science have urged critical vigilance, a less-agreeable climate of suspicion and attempted falsification, and since the very beginnings of modern science in the 1660s scientific journals have (at least in principle) adopted policies of peer review, where submitted papers are checked for accuracy and experi-ments repeated. The scientific enterprise ought to be self-correcting if its methods are correctly followed; nevertheless, given the uncertainties about just what the method is (or methods are) it should not surprise us that there were false starts – and on the whole, as we see in health scares, the credu-lity of the scientific community is and was considerably less than that of the general public. But trust and truth-telling are important and attractive aspects of science, and excessive suspicion sad.

FAILED SCIENCES

Not only do individuals make mistakes (or play jokes), but would-be sci-ences are launched amid fanfares, only to founder after showing initial promise. Thus phrenology seemed to have a great deal in its favour. Launched by an eminent Viennese doctor, based upon empirical findings in prisons and hospitals, it fitted well with a psychology in which human 'faculties' were distinct, with everyone having a different mixture (rather than a compound). Phrenology seemed also to be making real testable science out of intuitions that everyone felt (and willy-nilly we still feel) about physiognomy and character. Spurzheim's lectures and his big book, heavily revised in a second edition in 1815, brought the science to public attention in Britain, particularly in Edinburgh with its prominent medical school. Students were attracted to lectures there and a phrenological society was formed, and published a journal. George Combe, the educationalist, was enthusiastic about phrenology in his writings, which had enormous circulation, and it was taken up in schools, in Mechanics' Institutes, and by artists anxious to get their historical scenes right by giving the brave, the saintly, the clever and the villainous the right shape of head.[46] There were some poor fits, where villains had prominent 'bumps of benevolence', but that happens in all sciences (and especially with prognoses in medicine) and was taken to mean that further refinement (perhaps modifying the idea that faculties were so separate) was needed. As earlier with astrology, if the predictions turn out to be wrong, one could change one's phrenologist (while he could report the case in the journal, and suggest it proved the need for further research and investment). Falsifying 'pseudo-sciences' is more complex than some suppose.

In the first half of the nineteenth century, as we can see in the career of the combative Hewett Cottrell Watson (1804–81), pioneer of the study of plant distribution, phrenology was respectable, while evolution was not.[47] Watson, and Chambers in *Vestiges*, happily combined them. But after 1850 or so, it was becoming clear that phrenology was not advancing. This had happened with alchemy in the eighteenth century. Sciences do not usually go out with a bang, as it were, when disproved in some great public test: instead, they become irrelevant as practitioners turn to more promising topics. Rather than clash about method and heresy, scientists often become adept at knowing when to move out of a 'degenerating research programme' and take up topics closer to what others are doing more fruitfully.[48] Phrenology had never quite managed to become established, and there was never a section devoted to it, for example, at the BAAS. By the 1850s, it became discreditable, out of date and mildly ridiculous. In contrast, after 1859 evolution opened up new avenues for research, among fossils, animal and plant populations, and classification, and went with a zeitgeist in which progress was discerned everywhere. The criticisms directed against Darwin and Huxley by Wilberforce, for whom evolution was neither good induction nor rigorous deduction, and Whewell, for whom it was not appropriate as a fundamental idea, failed to squash it. Darwin indeed thought that he had been following the methodology advocated by Whewell. Had he not seemed to be challenging Christian faith, Whewell might have been proud of him.

There are aspects of fashion about the choice of both topics and method. Thus Mendel was advised to stop working on peas and do something more directly concerned with evolution, and there were allegedly some in the late nineteenth century who advised the young that physics was almost complete and they would do better in other sciences. The rise of museums had to some extent devalued field work, because the serious business of taxonomy was done indoors. Then taxonomy, with its careful literature searches, categorizing and naming, began to look old-fashioned with the rise of evolutionary studies, and the newer sciences of physiology and ecology that were based in the laboratory and the field. Naturalists began to carry binoculars rather than guns in the later years of the nineteenth century, and by the early twentieth century were looking more carefully at the behaviour of the creatures they were studying, such as mating rituals.[49] They noticed the importance of territory in birds' lives (curiously, a new idea), and the way that colours can conceal or warn – studies useful for camouflage (including dazzle painting, to break up shapes) in the Great War.[50]

BACKLASH

Vaccination had been one of the great medical successes of the nineteenth century, an example of (highly empirical) science saving lives through the conquest of smallpox.[51] It was duly made compulsory in Britain and other countries. The older process of inoculation with matter from the pustules of a patient had provoked unease about inflicting a disease (in a mild form, if all went well) upon a healthy person on the hypothesis that they might otherwise catch it in a severe, maybe fatal, version. The infection might also spread from the inoculated to the unprotected, starting an epidemic. Given that smallpox was endemic in the eighteenth century, the first objection had little force and the second was an argument for more general inoculation. With vaccination, the 'cowpox' that developed on the site was generally a source of discomfort rather than actual illness, but as smallpox receded, the fact that some (albeit very few) of those vaccinated suffered distressing symptoms became more significant. In a generation unfamiliar with the disease, civil libertarians became concerned about compulsion, and opposition to vaccination became a rallying point for those uneasy about the arrogance of scientists, their materialism and machismo.

Diseases do wax and wane in virulence, and some argued that vaccination was not the cause of the disappearance of smallpox. Wallace, ever the maverick, joined in the campaign against it, publishing in 1898 a pamphlet that he considered irrefutable, *Vaccination a Delusion*.[52] He was sure 'That the time is not far distant when this will be held to be one of the most important and most truly scientific of my works.' The Antivaccination League included among its members many of the people who were also resolutely opposed to vivisection. Living dogs had been used in experiments in the seventeenth and eighteenth centuries, but during the nineteenth century increasing numbers of people in Britain, notably from the expanding middle classes, became outraged by cruelty to animals, and the (later Royal) Society for the Prevention of Cruelty to Animals was founded in 1824. Bell had done his work on the nervous system without vivisection; but after 1815 medical visitors to France found that vivisection was a customary procedure there, taken for granted in physiology where rapid advances were being made. Britain would fall behind in this important science unless it adopted it, too. Magendie's pupil Claude Bernard, who published his *Experimental Medicine* in 1865, was the great name in mid-century French physiology.[53] Suspicious of statistical evidence, a firm determinist and critic of 'vitalisms', he emphasized the 'internal milieu' which living organisms maintain constant as far as possible. The laboratory, rather than the clinic, was for him where discoveries and new treatments would be found. And in his, animals were in constant use. He noted that whereas rabbits' urine (like that of other herbivores) was cloudy and alkaline after feeding, when they had fasted for twenty-four or thirty-six hours, it became clear and acid, like that of carnivores. The fasting, hungry rabbits were in effect eating rabbit. Such observations led to experimental interventions of a kind to make the modern reader queasy. Modern medical science, in the hands of Bernard and Pasteur in France, and Müller's disciples, Brücke, Du Bois-Reymond, Helmholtz and Ludwig in Germany, required physiology as its basis; and that meant vivisection, though to a lesser extent among the Germans (who tended to use parts from already dissected animals) than the French. Medical research everywhere came to mean laboratory work, and frequently with living animals rather than just test-tubes.[54]

Huxley, who hated doing vivisection, perceived this, and trained Michael Foster, who at Cambridge from 1870 built up a laboratory able to compete with work on the continent. A powerful and noisy anti-vivisection movement developed, horrified at the vision of the brutal vivisector relishing or indifferent to animal agony while allegedly advancing science and medicine, which could be done better by other means. They demanded a ban on animal experiments. Frances Power Cobbe (1822–1904) was a moving spirit in this and other causes.[55] She came from a landowning family in Ireland, and aroused her oppressive father's ire by becoming an agnostic, though subsequently a deist, moving after his death in 1857 to England and publishing an edition of the writings of Theodore Parker, who had found Boston Unitarianism too constricting and glum.[56] Nevertheless, in London she attended the Unitarian chapel of the eminent James Martineau (1805–1900), Harriet's brother. Though a staunch Conservative, she was an energetic and prominent feminist, becoming a journalist, supporting the admission of women to university, and campaigning for married women to retain control of their property, and for domestic violence to be a ground for judicial separation. She saw men everywhere exerting power over the defenceless, and her feminism readily went with her intransigent opposition to vivisection, against which she began to campaign in 1863. In the face of what became a powerful movement, bringing bad publicity for science, Huxley and Darwin rather unhappily stressed the need for vivisection in medical research. Eventually, after parliamentary debate, it was permitted but only under licence from the Home Office.

In the light of her campaign, Frances Power Cobbe came to see scientists and medical men as arrogant, brutal and atheistic. Given Bacon's aphorism that knowledge is power, this need not surprise us. The rhetoric of science was and remains macho: there are images of warfare against ignorance, superstition and disease; and of conquest and interrogation of nature, involving perhaps torturing her in the laboratory until she gives the required answer. There were women who played important roles in sciences, not only as supportive wives, daughters and assistants, but also as writers.[57] Davy, who had urged the importance of women's education, nevertheless went in for sexy macho rhetoric:[58]

Not content with what is found upon the surface of the earth, [the chemist] has penetrated into her bosom, and has even searched the bottom of the ocean for the purpose of allaying the restlessness of his desires, or of extending and increasing his power.

Elsewhere, he wrote in more 'female' vein of the scientist being 'visited' by discovery, and as well as the urge to penetrate there are stories of lying back and dreaming. But in general the widespread exclusion of women from science was no accident: it was man's work. And that could provide a good reason for being against it.

Love of nature often went with the desire to control her, and dominion did not necessarily entail oppression. But technology and pollution, with dark satanic mills, could also seem a consequence of the brutal, arrogant amorality of science. Novels by Dickens, Disraeli and Gaskell in Britain, and Zola in France, revealed more tellingly than statistical surveys the miseries of industrial city life. Nostalgia for unspoilt nature grew. Frances Power Cobbe refused to wear feathers in her hats, in a widely copied gesture that led to the founding of the Royal Society for the Protection of Birds (in defence originally of great crested grebes) in 1860; and the later years of the century also saw the establishment of National Parks in the USA, with Yellowstone in 1872, and the founding of the National Trust for the protection of the countryside and historic buildings in Britain in 1895. These were practical steps to conserve what was valuable. But those to whom science seemed a ruthless and sinister power, opposed to all that was spiritual, could also embrace the occult and the irrational, and the late nineteenth century saw huge interest in spirits, spooks and magic which, curiously enough, was one route into 'modernism'.[59]

RESPONSES

None of this necessarily excluded science. Love of nature, appreciation of the fine arts, and sensitivity to social injustice and to the spiritual were all fully compatible with science. Field clubs were out regularly looking at the animal, vegetable and mineral realms, and doing valuable surveys on their local distribution as pioneered by Watson. Ornithologists were studying birds and their behaviour with the assistance of binoculars. There was wide interest in Reichenbach's auras surrounding sensitive souls,[60] as well as in spooks. And the Society for Psychical Research, founded by Henry and Nora Sidgwick (another pioneer feminist), set out to investigate the phenomena reported by spiritualists,[61] using the methods of experimental science. Crookes's attempt to get the Royal Society to publish his account of a seance in which there were spectacular manifestations of 'psychic force' before reliable witnesses was blocked.[62] Nevertheless, he, J.J. Thomson, William James (1842–1910), Pierre Janet (1859–1947), Lord Rayleigh,

Balfour Stewart (1828–87), John Couch Adams (1819–92), Oliver Lodge (1851–1940) and other prominent scientists continued to be interested and were prominent in the SPR. This was another case where the barrier between proper and improper science was not obvious, but in the event the phenomena were hard to reproduce, control or explain (even in a world of X-rays and other invisible radiations) and further investigation gradually seemed not to be worthwhile. Although parapsychology is still with us, the high hopes that psychical research would turn into a real science were not fulfilled, as hardly any of the pioneering investigators found evidence of spirits that was above suspicion.

Darwin's cousin, Francis Galton (1822–1911), went to South Africa and in 1855 wrote a book for travellers filled with advice based upon his own improvisations and derring-do.[63] Shakespeare's Prospero had meditated upon Caliban:[64]

A devil, a born devil, on whose nature
Nurture can never stick; on whom my pains,
Humanely taken, are all lost, quite lost;
And as with age his body uglier grows,
So his mind cankers.

On his return, Galton became fascinated with this balance between nature and nurture in human development, and studied men of science, and then other professions. He found that scientists were on the whole a vigorous and manly lot, with good heredity, and came to the conclusion in his *Hereditary Genius* (1869) that success ran in families. Using a Gaussian curve for the first time to indicate mental ability, he argued that, statistically speaking, inheritance was the most important factor in success.[65] He saw different races having different intellectual capacities, with Europeans at the top because they were furthest on in the Darwinian progress of civilization.[66] He became very interested in quantifying people's measurements as well as their attributes, notably fingerprints, and was increasingly concerned with the poor physique of working-class city dwellers. This perception was confirmed by recruitment for the British army for the Boer War in 1899, when many volunteers were found to be unfit for service. For Galton, the medieval period had been so barbaric and lasted so long because the most intelligent and sympathetic in each generation had gone into monasteries or nunneries and left no descendants: society had selected against the highest qualities, and must not be allowed to do so again.

Galton studied photographs of criminals, but concluded (unlike his Italian contemporary Cesare Lombroso (1836–1909), or phrenologists) that the criminal mind was not visible in the face.[67] But he noted with alarm that although the practice was regarded as immoral, professional couples were planning their families, and having less children. Among slum dwellers, large families remained the rule, and he foresaw the superior offspring of the middle classes being swamped by the prolific underclass. Late Victorian society was therefore also selecting against excellence – and decline would be inevitable. He thus promoted eugenics, another science that has not worn well. But it flourished in Edwardian Britain, where Galton was knighted in 1909, and beyond – notably in the USA and Sweden – before being enthusiastically taken up in Nazi Germany. As a result, we associate it with sterilization and extermination, but positive eugenics went with child allowances for professionals – teachers and university lecturers, for example (I was myself a beneficiary, in the pay-scale for academics in the 1960s). Despite the ill repute of eugenics, those who see science as a social construction point out that the austere statistics with which students have to wrestle were considerably advanced by Galton in the context of this dicey science.[68]

CONCLUSION

Since Descartes, method has seemed the key to good science, and many in the nineteenth century hoped that it might be separable from the actual practice of science, and applicable in other intellectual realms. A close look, however, indicated that there were different kinds of scientific minds, and a range of methods applicable to different fields of study, times and places. Throughout the century, the French tended towards the positivism formalized by Comte, seeking laws and treating theories as tools rather than true or false statements about the world. In Britain, where John Herschel's writing was especially influential, scientists sought to interpret the world using Newtonian *verae causae*, entities and mechanisms that really represented nature. German scientists, better trained in philosophy, were more conscious of what they were doing when adopting or opposing teleological, atomistic or dynamical views. Goethe's radical critique of Newtonian physics, Schelling's and Oken's wide-ranging *Naturphilosophie*, Liebig's down-to-earth chemistry, and then Mach's positivism indicate a wide range of possibilities. As well as explanation, taxonomy was a major concern, as not only plants and animals but also chemical elements were tabulated and ordered, following rules of method painstakingly agreed internationally.

Some science was bad because it broke the rules of method, trimming or cooking results, and there were also the occasional hoaxes and frauds. Some promising sciences failed to grow, notably phrenology in the first half of the century and psychical research in the second. Like alchemy and astrology before them, they were not straightforwardly falsified, but as experiments proved inconclusive and explanations ad hoc or vague, scientists moved on to something more profitable and law-abiding. Even good science, done in full accordance with rules of method, was not without its critics. It struck some outsiders as bad because it was amoral: compulsory vaccination seemed a gross violation of personal liberty, and vivisection of animals mere indulgence in cruelty. Science seemed to such critics to go with machismo, its practitioners arrogant, materialistic and narrow-minded. At the turn of the century, eugenics looked like a form of class warfare.

Nevertheless, the success of Darwinian theory in promoting evolutionary explanations, the power of science-based industry to change society and the place of science in philosophy all meant that scientists were becoming regarded as sages, widely looked to for wisdom. Science had become a great cultural force, an important aspect of the triumph of the West. What had been a hobby in 1789 was in 1914 a profession, or a number of professions. Not everyone was by 1900 exposed to science as part of their education (it was frequently unavailable to girls), but the option of a scientific education was there.[69] It is to scientists' aspirations and claims to cultural leadership that we shall next turn.

11

CULTURAL LEADERSHIP

Science in 1900 was a crucial part of 'Western' culture, perceived as the key to its power, wealth and dominance. Scientists were an important group. How had this come about? In most places, science in 1789 had been a hobby or avocation for gentlemen, or craft skills employed in a profession or trade. It was not in itself a profession, with criteria of admission and ethical standards, like the Church, the law or medicine, and there was no scientific career structure, except in medicine where scientific learning separated physicians from surgeons or apothecaries. Britain had its Royal Society as a focus, where interested gentlemen (some of them practitioners) met during the winter season in London; other countries had academies on the French model, departments of state having advisory roles and laying down what was, and was not, real science. Banks, Lavoisier, Humboldt and Franklin were important in the culture of their countries, and their science gave them authority and prestige. But even the years of 'Enlightenment', ending in revolutions, made reason rather than empirical science the key to knowledge. The gentleman, and sometimes the lady, might – as observer rather than participant – appreciate science as a component of culture, as they would music, painting or drama, and some might think also of applying it as useful knowledge to the improvement of their estates. Churches had always been responsible for almost all education, and clergy retained a leading role, notably in universities that remained conservative institutions where tradition and scholarship were valued. Elsewhere, connoisseurship, discrimination and wisdom were more important culturally than scientific knowledge.

REVOLUTIONARY AND ROMANTIC REACTIONS

The French Revolution undermined the Church, and with it the theology of nature in which a sustaining Creator maintained the world in order. In his published lectures in Germany in 1799, Friedrich Schleiermacher (1768–1834) made religion a matter of feeling rather than reason.[1] Meanwhile, romantic thinkers valued the poet as the seer, putting creative, imaginative reason above mere understanding, and looking back to heroic medieval or pre-Christian times. Painters and musicians joined writers, and in Germany philosophers, as cultural arbiters as the old order crumbled. And when the University of Berlin was founded on Wilhelm von Humboldt's plan, it was in the spirit of neo-humanism with philology, the study of language, at its heart.[2] What we call science, *Naturwissenschaft*, was only a part of what they saw as science, *Wissenschaft*, which included disciplines we would describe as arts, humanities and social sciences; neither in Germany nor in Britain were natural sciences privileged in those revolutionary years. In Britain indeed they were seen, notably in Erasmus Darwin and Priestley, and in Beddoes's circle in Bristol, as possibly subversive.

That was because the French Revolution, as it gathered momentum and horrified British spectators, was perceived as the forbidden fruit of the tree of knowledge, undermining hard-earned experience and tradition and replacing social order with anarchic violence and terror. In France 'two cultures', scientific and humanistic and with their different Academies, had begun to emerge by 1789, and a number of young men, fired with ambition for election to the Academy of Sciences and a life of research, had specialized and networked for posts there.[3] Most men of science, including Lavoisier, were indeed supporters of the Revolution, especially in its earlier phases, and as revolutionary France was beset by enemies, they patriotically devised ways of melting church bells into cannons and otherwise advancing the war. The engineer Lazare Carnot was the survivor of the triumvirate (with Robespierre and Saint-Just) in the Committee of Public Safety that oversaw the Terror, and was the patron of the young Napoleon Bonaparte. As First Consul and Emperor, Napoleon was gratified by election to the Academy, and reciprocated by choosing scientists for both sinecures (as Senators in his tame legislature) and active administrative posts. The tradition of prominent scientists going into politics remained strong in France right through the nineteenth century, examples being the applied mathematicians Fourier and Arago, and the chemists Jean Baptiste Dumas

(1800–1884) and Berthelot, all of whom held high office. Science was usually but not exclusively identified with the secular, modernizing tradition,
though for Ampère his Catholicism was central.[4] Comte saw science as the
acme of knowledge, and the scientist as having outgrown and put away the
phoney consolations of religion and the empty wordiness of metaphysics.[5]
By the early nineteenth century, scientists in France could truly be said to
aspire to cultural leadership.

In Britain, Davy, who unlike Banks owed his status entirely to his scientific talent, sought both in his lectures at the Royal Institution as a young
man, and as President of the Royal Society in the 1820s, to occupy the
position of a sage or cultural icon along with the poets who had been his
friends from the 1790s.[6] For him, science could be a route to wisdom. In
his poetry, and then at the end of his life in his little books *Salmonia* (1828;
2nd expanded edition, 1829) and the posthumously published *Consolations in
Travel* (1830), he sought in series of dialogues to expound a scientific world-
view with room for poetry, enjoyment of scenery, and deistic or pantheistic
religion.[7] He did not survive to become a sage, as John Herschel and Huxley
in different ways were to do, and his death (in 1829, at fifty) was an occasion
for deploring the decline of science in England. Davy and Wollaston, both
eminent for chemical researches that had also (in the miner's lamp, and in
platinum apparatus) proved very useful industrially, died within a year of
one another. Young, the polymath remembered for his wave theory of light,
and work on Egyptian hieroglyphs, died shortly after Davy. Babbage took
the occasion to bemoan the decline he perceived, and was joined by others
who compared England unfavourably with Scotland as well as with France
and Germany.[8]

RAISING THE PROFILE OF SCIENCE

Babbage's proposals for more honours and government money for men of
science fell on stony ground, but invocation of the example of Germany
proved very fruitful. French science was concentrated in Paris, the capital,
but fragmented Germany, a nation but not a state, had no capital. Berlin,
Munich, Dresden, Weimar, Hamburg and many other cities were capitals
of states large and small, many of them with universities (being modernized on the Berlin plan) and academies. Oken's plan of calling together the
Naturforscher from all over Germany to annual meetings, to be held in a
different state each year, was as we saw a great success. After initial misgiv-
ings, the states welcomed the men of science of the Gesellschaft Deutscher

Figure 23 *Lecture theatre of the London Institution, 1828: C.F. Partington,* A Manual of Natural and Experimental Philosophy, *London: Taylor, 1828, frontispiece*

Naturforscher und Ärzte (GDNA), and laid on a good show for them. British visitors commentated favourably upon these open meetings, enormously stimulating to science in the city where they were held, and overcoming the loneliness of men of science isolated in small places. Babbage attended the Berlin meeting in 1828, J.F.W. Johnston (1796–1855) that in Hamburg in 1829, and their accounts of these occasions enthused Brewster, who became a great advocate for a similar society in Britain. Scotland had strong scientific and medical institutions, but it was small. England had strong provincial traditions, local pride being connected with booming trade and industry (and often dissenting religion) in cities like Bristol, Birmingham, Manchester, Liverpool, Leeds, Derby and Newcastle, and with cathedral-city culture in Exeter, Norwich, Lichfield, York and Durham. By 1830, institutions like Literary and Philosophical Societies or Athenaeums, subscription libraries and Mechanics' Institutes promoted science as an important aspect of culture – and, moreover, more accessible to the *nouveau riche* manufacturer than gentlemanly knowledge of fine art or literature.

The German model seemed to provide a way to overcome the supposed decline, perceived as connected to the metropolitan and genteel character of the Royal Society. In 1831 the well-connected William Venables Vernon Harcourt (1798–1871), son of the Archbishop of York and himself a clergyman, convened a meeting of all those interested in science, in the summer in York – a city both respectable and provincial, and the place where rich collections of fossils from the moors and coast around Whitby were held and displayed. He had attended Buckland's lectures in Oxford, and had afterwards worked with both Davy and Wollaston, but he was a gifted administrator rather than researcher and that was just what was required to put science on the map.[9] At that first meeting of this British Association for the Advancement of Science, the pattern (common to conferences ever since) of having some plenary sessions, and some time in specialized sections, was established.[10] Each year the Association would meet in a different place, bringing to a succession of cities its vision of sweet reasonableness freed from dogma and prejudice through empirical study of nature. It was not only applied science that was in focus. Subsequent meetings in Oxford and Cambridge, Edinburgh and Dublin, were followed by very successful visits to commercial and industrial cities.[11] The annual itinerary was planned well in advance, the President being appointed the year before and supported by an administrative team: Harcourt was General Secretary for the first five years, and President in Birmingham in 1839. Vice-Presidents would be chosen from eminent men of the neighbourhood who were interested in science and technology, and local organizers would plan excursions and events to make the occasion enjoyable and memorable. Cities wanting to host a meeting began to prepare bids, undertaking to found a museum, a library or a college in order to tempt the Association to come: it brought business in the slack summer season, and also ministered to civic pride in its promotion of intellectual culture. Reporters from national newspapers and journals would be there, and the proceedings were well-publicized, taking place in the 'silly season' when parliament and the law courts were not sitting and there was not much other news.

Although the running of the Association soon, and necessarily, fell into the hands of elite 'Gentlemen of Science' (usually Oxford or Cambridge graduates based there or in London), anyone could offer a paper to one of the sections, and the organizers tried to attract particularly local men of science as participants and congregation. They would have the chance to see and hear, maybe even speak to, men like Faraday, John Herschel and

Huxley. Visitors from overseas were there as they had been in Germany, and distinguished foreigners might be specially invited. Women were welcomed, though (as in other respectable mixed environments) it was for many years very rare for them to be speakers. Indeed, the stuffy Royal Society continued until 1945 to reject women as Fellows, though wives, daughters and guests might be brought to the occasional intellectual parties there. Whewell's word 'scientist' was coined at the BAAS's Cambridge meeting. Though it was long before the term caught on, the BAAS succeeded in creating a self-conscious group concerned with science in our restricted sense. They were choosy about what to include, and their list came to define science as against other intellectual activities. Huxley wrote of the 'Church scientific', and the BAAS came to stand for it, its meetings functioning rather like the annual Assemblies of the Church of Scotland.

The BAAS might legislate like a synod or parliament: at its Dublin meeting Berzelius's notation for chemical elements (C for carbon, O for oxygen) was adopted in place of Dalton's hieroglyphics; subsequently, Strickland's code for naming in taxonomy was adopted and promoted internationally (if gradually). The BAAS produced reports on the state of science, and set up committees to report at later meetings on matters of controversy; eventually, it recommended the use of metres, grams and litres in British science. Meetings almost always made a profit, and this was used in making grants to individuals whose research was deemed worthy of support. Presidential addresses were carefully prepared and widely noted. They called attention to the achievements of British scientists and engineers, and the shortcomings of government in support of them. Often the President would be part of a delegation mandated to call upon government ministers to promote a particular scientific activity or institution. For his year in office, culminating in the annual meeting, the President enjoyed an official position as spokesman for British science, alongside the President of the Royal Society, and in their own fields the Astronomer Royal and the Presidents of the Royal Colleges of Physicians and Surgeons. And because of the open nature and public profile of the BAAS, he could represent a much wider cross-section of the scientific and technological community. Huxley had jokingly coined the phrase the 'Church scientific' to stand for organized science: but the best jokes are serious, the metaphor is illuminating, and the meetings of the GDNA, the BAAS, the American Association for the Advancement of Science and such bodies might be taken (with a pinch of salt) not only as festivals, but also as synods debating doctrine.[12]

THE CHURCH SCIENTIFIC

This 'Church scientific' might usurp or replace the cultural authority of the clergy (a long-established profession) and their stranglehold on education, with a secular clerisy (to use Coleridge's word).[13] This was part of the project of Chapman and his associates on the *Westminster Review*, and then of the X-club.[14] Anti-clerical feeling was not confined to Britain: Samuel Butler (1835–1902), the writer and would-be man of science, reported in 1881 that in newly united Italy people said to him, 'For the future, let us have professors and men of science instead of priests.'[15] Many in the BAAS would have been happy with a more moderate version of this programme, giving them authority alongside the clergy, but such aspirations drove scientists in their search for status and their frustration with the cultural dominance of classics and theology. Outside Roman-Catholic Italy, and especially in the English-speaking world, nineteenth-century churches were intensely competitive. Though (or because) there was religious toleration, different denominations were at each other's throats, sure that only they knew the way to heaven. There was warfare for souls, and Christians hated each other. In a world of such skirmishes, it was easy to see organized science, long presented as waging war against ignorance and superstition, as in conflict with organized religion. But whereas churches had creeds and ordained and licensed ministers, there was no clear credo or qualification to distinguish and authorize the real scientist – that was Butler's problem, ostracized as he was by the scientific establishment. The Royal Society steadily became in effect an academy (though with a subscription rather than a salary), but outside that august and metropolitan body the boundary of organized science was not clear.

By the middle of the century, the specialized societies that had begun with the natural historians and the geologists had increased and multiplied: astronomers, geographers, zoologists, chemists, physicists, microscopists and others had their own group, usually publishing a journal. Aspiring scientists like young Huxley faced problems with paying the fees required to take one's place in the scientific community, with admission fees to societies commonly between six or ten pounds (or guineas) and annual subscriptions not much less.[16] If the eminent figures who ornamented the platform at BAAS meetings might be the scientific equivalent of the bench of bishops, then those with a published contribution to science or membership of a prestigious society might constitute its clergy, or clerisy. But clergy are formally ordained or commissioned: what distinguished the real scientist? Some of them were involved in teaching science. But others worked in industry, or were gentleman amateurs, and most science teachers had never published a scientific paper. Gradually, formal education in science provided a criterion, but right through this period there was no clear demarcation available in Britain, like the German PhD, that indicated scientific training and qualification. That would have eased Faraday's emergence from Davy's shadow, for example, as an approved scientist rather than an assistant.

Churches deplore those who rock the boat, but rows and schisms are a feature of Church life: thus in 1843 Thomas Chalmers (1780–1847), famous as preacher and science lecturer, led 470 ministers out of the Assembly of the Church of Scotland in the Great Disruption leading to the formation of the Free Church. The split, over patronage, lasted more than half a century, and even then was not fully healed. Similarly, rows (less drastic) happened in science, sometimes in public. Sharp scientific debate was attractive to outsiders, but the leading figures in the BAAS were always determined that their meetings, dedicated to empirical science, must not be an arena for conflicts that might split the body and weaken the authority of science. The electrifying moment in Oxford in 1860 when Huxley accused Owen of lying about his dissection of a gorilla's brain, and the debate which followed that meeting when Huxley clashed with Wilberforce, were deplored and played down. Tyndall's presidential address at Belfast in 1874, when he brilliantly advocated scientific naturalism, was widely perceived as an unfortunate departure, in poor taste, into a speculative realm in which quarrels were inevitable.[17] The prima donna side of eminent scientists was curbed as far as possible in public.

PROFESSION?

We think of scientists as professional, but in the nineteenth century it was more complicated. Science was set against practice, and learned societies dedicated to science set against professional ones, which were in effect more like guilds or white-collar trade unions. The Colleges of Physicians and Surgeons had always been like that, and in engineering various institutions were formed whose most important function was to set standards, license qualified practitioners and ensure that interlopers did not take jobs appropriate to their members. Depending on one's viewpoint, these bodies and later ones like the (Royal) Institute of Chemistry were gallantly protecting everyone against quacks and incompetents, or were a conspiracy against the public, maintaining high fees. Within the sciences, these professional

bodies made clear boundaries, and still do. The qualified doctor, the chartered chemist, architect or engineer, can perform services that nobody else
is allowed to do. These might then count as clerisy, and indeed sometimes
do, but when we think of great scientists, such professionals are not usually
the people we have in mind – and our ancestors thought the same way. The
professional, a doctor or lawyer perhaps, was not a tradesman. His fees were
in guineas, he might perform services gratis for the poor, and in the social
scale he ranked above the manufacturer but below the gentleman (who
might be much poorer than either).[18] Natural philosophy, applied curiosity, had since the seventeenth century been appropriate to the gentry, and
though many like Davy and Faraday were of humble origins, they became
nature's gentlemen.

George Johnston (1797–1855) was an Edinburgh-trained doctor who
practised in Berwick, and was careful to distinguish his professional from his
scientific activities. He was a little jealous of his landed friends whose time
was their own, grumbling that:[19]

> I am compelled sometimes to a seeming neglect of my friends by profes-
> sional, social and other engagements which none but *gentlemen* can well
> disregard. You know it does not become a doctor, who is, or should be,
> a staid animal, to move about like a *Pecten* or a gentleman. . . I am lost
> in despair at the riches that surround me. There lie volumes to be read,
> letters to be answered, specimens to be named and described, bottles full
> of funny little fellows sporting away and too seldom looked at. Had I the
> leisure of a *gentleman*, it would not do more than enable me to put all to
> rights.

But in his intervals of leisure, he got more done in science than most of
them. He wrote about local natural history, and had the idea of the field
club, making excursions to look for animal, vegetable and mineral curiosities rather than just hearing lectures and looking at specimens in a museum,
and naturalists proved keen to form communities and networks.[20] In 1831
Johnston founded the Berwickshire Naturalists' Club, which also published
a journal – and which continues. As a way of spreading enthusiasm and
knowledge, these were very effective and soon sprang up all over Britain
and beyond, sometimes giving themselves fancy names like the Innerleithen
Alpine Club in Scotland, which made local excursions and published a
report illustrated with photographs taken by members.[21]

Johnston also had the idea of a publishing club for natural history, where

Figure 24 *A field naturalist at work, pond-dipping: E. Newman,* A Familiar Introduction to the History of Insects, *London: Van Voorst, 1841, p. 96*

illustrated books were necessary but expensive. If a definite market were guaranteed, then they could be much cheaper. Publishers had appreciated this with subscription publishing, where advance orders made printing economic, and had also published sumptuous natural histories in parts, again normally by subscription. Johnston's idea led to the Ray Society being formed in 1844 to publish works, mostly but not exclusively of British natural history: among many other works, the society published Darwin's exhaustive study of the living barnacles. His volumes on extinct barnacles were published by a sister society, the Palaeontographical, founded in 1847 to publish books on fossils. Meanwhile, in 1846, the Hakluyt Society had been founded, to publish on similar principles scholarly editions of voyages and travels for subscribers with geographical interests. The Ray Society's work in propagating natural history augmented that of commercial publishers, notably John van Voorst (1804–98). Usually the society published (and still does) what became standard descriptive and taxonomic works (including translations), but exceptionally and controversially it published a translation of Oken's speculative work, *Elements of Physiophilosophy*, in 1847.[22] Associated with all this were the crazes for ferns to put in pots in conservatories (or perhaps sealed in glass 'Wardian cases', as used for transporting plants), and for making up aquaria as part of the newly popular seaside holiday – as promoted by Charles Kingsley (1819–75), and by Philip Gosse (1810–88) who organized educational holidays by the sea to learn natural history.[23] Such dissemination of science was very important in raising its cultural profile, though it blurred the distinction between the 'professional' practitioner and the interested amateur. Local natural history was to become very important in ecology, but in the nineteenth century it brought science into the jolly activity of getting fresh air and exercise on holiday, and into the romantic nostalgia for a pre-industrial world.

ORGANIZERS OF VICTORY

Lazare Carnot was named the 'organizer of victory' for his crucial role in saving the French Revolution from its enemies, but his name is less familiar than those of the generals he promoted and supported. So it is in science: the change in scale and pace of science that went with the second scientific revolution depended upon the organizers and administrators, the grey eminences, whom we tend to forget in our admiration for the star performers, the original thinkers and discoverers whose careers they made possible. They might be executives, administrative secretaries, like Walter White at the Royal Society whose genially sceptical journal casts wonderful light upon its great men,[24] or might be themselves scientists, like Arthur Aiken (1773–1854) at the Society of Arts or Bates at the Royal Geographical Society. The biologist William Benjamin Carpenter (1813–85) became in 1856 Registrar of the University of London, as well as being prominent in the BAAS and the Royal Society. Less formally perhaps, in their various ways, Johnston, Banks, Harcourt and Oken exemplify the importance of administrators, but there were many others. Lord Northampton (Spencer Compton, 1790–1851) was President of the Geological Society and of the BAAS, and from 1838 to 1848 of the Royal Society, which was effectively modernized during his regime. In the USA, Joseph Henry directed the Smithsonian Institution from its inception in 1846, and made it a powerhouse of American science and technology. His ally Alexander Dallas Bache (1806–67) superintended the US Coast Survey, mapping the entire coastline and promoting surveys of the West.[25] From Germany, Humboldt promoted international geophysical measurement, notably the 'magnetic crusade' involving measurements around the world. Through the BAAS, Edward Sabine (1788–1883) promoted this to the government, which duly funded the strengthening and fitting out of two ships, *Erebus* and *Terror*, as magnetic observatories for Ross's Antarctic expedition (1839–43).[26] Under John Barrow (1764–1848), the British Admiralty had undertaken numerous such expeditions, such as Henry Foster's,[27] and the tradition continued, culminating in the oceanographic voyage of HMS *Challenger* (1872–6).[28] Meanwhile, at home, Henry de la Beche (1796–1855) inaugurated the Geological Survey in 1832.[29] Big, expensive and sometimes dangerous, science that requires teams rather than solitary geniuses was not an invention of the twentieth century. Sending ships and crews on voyages lasting years, or overland parties carefully surveying and mapping unknown territory, was a very costly business, necessarily involving governments. Such promotion of science brought them credit, with the promise of power and prosperity, and the feting of the heroes on their return was good publicity for science.

Among the various forms that scientific organizations took, the most common were open societies where all who were interested were welcome – usually subject to the agreement of the membership or of a committee. Would-be members were proposed and seconded, and (as at the Royal Society) others might add their names in support before the matter came to the vote, when the unsuitable would be 'blackballed' and rejected. The *Naturforscher*, the BAAS and the societies modelled upon it in France, the USA and Australasia were all more accessible than this, being open at a price to all (suitable) comers; after all, their aim was to disseminate science. But 'working men' were not expected to join, and got their lectures separately from the main sessions. The Royal Institution, the Literary and Philosophical Societies in Manchester and Newcastle, and the Subscription Libraries, Athenaeums and Lyceums elsewhere in Britain and North America, all playing an important part in the cultural life of their cities, were similarly open to all interested and suitable persons and increasingly to both men and women. Their subscriptions paid for a library, and perhaps also a laboratory with a resident researcher, and a journal. Because of its fashionable London location, its exalted membership and the exciting discoveries made there, the Royal Institution was of national importance and the lectures given in its theatre disseminated science among powerful people, and particularly their children.[30] Science was not only improving, but also entertaining.

While these societies did not confine their lectures to science in the

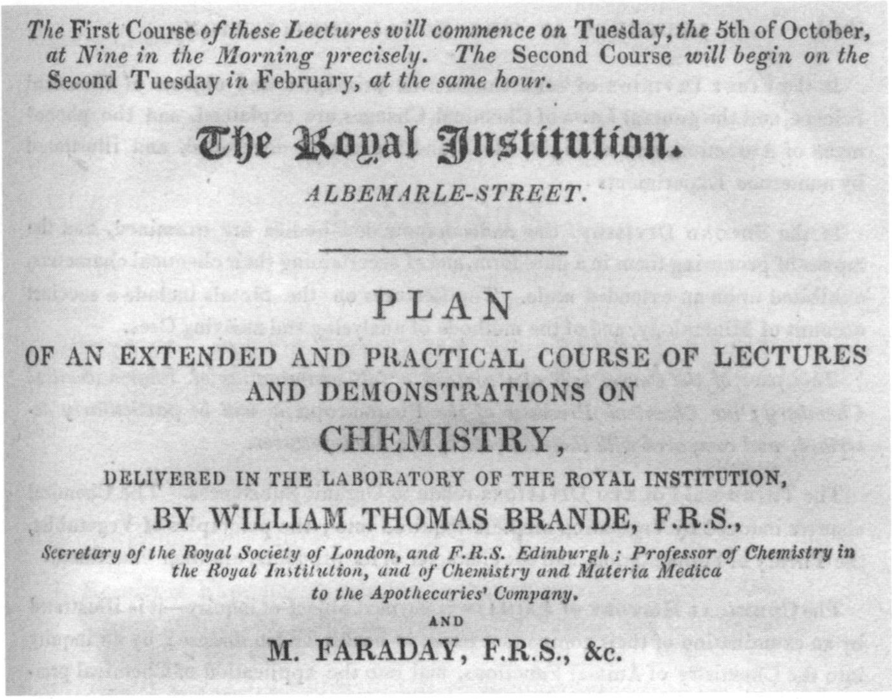

The First Course *of these Lectures will commence on* Tuesday, *the* 5th of October, *at Nine in the Morning precisely. The* Second Course *will begin on the* Second Tuesday *in* February, *at the same hour.*

The Royal Institution.

ALBEMARLE-STREET.

PLAN

OF AN EXTENDED AND PRACTICAL COURSE OF LECTURES AND DEMONSTRATIONS ON

CHEMISTRY,

DELIVERED IN THE LABORATORY OF THE ROYAL INSTITUTION,

BY WILLIAM THOMAS BRANDE, F.R.S.,

Secretary of the Royal Society of London, and F.R.S. Edinburgh ; Professor of Chemistry in the Royal Institution, and of Chemistry and Materia Medica to the Apothecaries' Company.

AND

M. FARADAY, F.R.S., &c.

Figure 25 Lecture prospectus, Royal Institution, 1825, Quarterly Journal of Science, Literature and the Arts 18 (1825): 199

narrow sense to which the BAAS had confined it, they also reckoned that all the sciences were their province. Others were narrower. In France, the informal Societé d'Arcueil (with membership by invitation) had been concerned with the sciences of its patrons, the chemist Berthollet and physicist Laplace.[31] The BAAS with its sections provided forums where chemists, physicists or mathematicians might get together in their 'section' and discuss abstruse and technical questions, and by 1831, when it was founded, metropolitan societies specializing in particular sciences were beginning to flourish. These were generally open, because after all there were few possible formal qualifications that distinguished the expert, and few committed practitioners – who needed all the support they could get. Gradually, societies began requiring some evidence of specific knowledge rather than general interest in their science, especially as the prestige of science increased and people began to use membership of a society as a sign of expertise when applying for jobs or offering services. Thus the Chemical Society of London, founded in 1841, which published a journal full of formulae forbidding to any outsider, broke into two societies in 1871, with the dissidents forming the Institute of Chemistry to act as a professional body.[32] It sought like the Royal Colleges in medicine (but with only moderate success) to control qualifications, fees and fields where only qualified chemists could practise. Meanwhile, with the expansion of higher education, the rump remaining in the Chemical Society closed ranks and became a learned society, publishing and promoting research and teaching, open only to graduates in chemistry.

Already by the mid-century, the Royal Society was becoming closed: under Northampton and his successors, the fellowship was limited, and each year nominations were scrutinized by a committee who selected the set number of those to be admitted. This process brought the kind of authority that academies had enjoyed in France, Prussia, Russia and other countries. But it also brought problems as a generation of scientists grew up less familiar with the culture of liberal education, and as scientific journals became increasingly impenetrable to outsiders. Moreover, we saw that in the French Revolution the Academy of Sciences had been criticized and briefly suppressed as elitist, and the later nineteenth century was a time when, in the fine arts, academies were disdained by the avant-garde as centres of dreary conservatism and conformity where originality was stifled. 'Academic' became a dirty word, and as part of the 'establishment' scientists were likely to be distrusted, no longer as Faustian necromancers but as feathering their own nests at the expense of the public, foisting sclerotic official views, vaccination and other aspects of big government upon everyone. Nevertheless, such negative perceptions were themselves a sign of the recognized importance of science. Thus Hobhouse greeted the twentieth century with his remarks about the efficacy and problems of specialization. It brought rapid advance, but also made science duller, less imaginative and less valuable in general education.[33] As science became more specialized and technical, requiring great numbers of textbooks, opportunities arose for publishers: they might go in for science publishing, working with one or more societies, and perhaps publishing an independent journal. Taylor and Francis in Britain did both these things.[34] Macmillan was induced by Lockyer to publish *Nature*, which for many years lost money but brought prestige, respectability and sales to the textbook series the firm also produced. There were also opportunities for writers, journalists and publishers for popularizing science at all levels, from children's books to those like Mary Somerville's, written for experts in one branch of science wanting to understand other branches in the field.[35] Crookes published both the *Quarterly Journal of*

To be Published in October, 1869, *in* 8vo, 24 pp. *Weekly, No.* 1 *of*

NATURE:

AN ILLUSTRATED JOURNAL OF SCIENCE.

The object which it is proposed to attain by this periodical may be broadly stated as follows. It is intended—

FIRST, to place before the general public the grand results of Scientific Work and Scientific Discovery; and to urge the claims of Science to a more general recognition in Education and in Daily Life;

And, SECONDLY, to aid Scientific men themselves, by giving early information of all advances made in any branch of Natural knowledge throughout the world, and by affording them an opportunity of discussing the various Scientific questions which arise from time to time.

To accomplish this twofold object, the following plan will be followed as closely as possible:—

Those portions of the Paper more especially devoted to the discussion of matters interesting to the public at large will contain:

I. Articles written by men eminent in Science on subjects connected with the various points of contact of Natural knowledge with practical affairs, the public health, and material progress; and on the advancement of Science, and its educational and civilizing functions.

II. Full accounts, illustrated when necessary, of Scientific Discoveries of general interest.

III. Records of all efforts made for the encouragement of Natural knowledge in our Colleges and Schools, and notices of aids to Science-teaching.

IV. Full Reviews of Scientific Works, especially directed to the exact Scientific ground gone over, and the contributions to knowledge, whether in the shape of new facts, maps, illustrations, tables, and the like, which they may contain.

In those portions of "NATURE" more especially interesting to Scientific men will be given:

V. Abstracts of important Papers communicated to British, American, and Continental Scientific societies and periodicals.

VI. Reports of the Meetings of Scientific bodies at home and abroad.

In addition to the above, there will be columns devoted to Correspondence.

The following eminent Scientific men have already promised to contribute Articles, or to otherwise aid in a work which it is believed may, if rightly conducted, materially assist the development of Scientific thought and work in this country:—

ABEL, F. A., F.R.S.	MASKELYNE, N. S., F.G.S., *British Museum.*
BASTIAN, PROF. C., F.R.S., *University College.*	MIVART, ST. GEORGE, F.R.S.
BRODIE, PROF. SIR BENJAMIN, F.R.S., *Oxford.*	MURCHISON, SIR RODERICK I., BART., F.R.S., *President Geo-*
CLIFTON, PROF., F.R.S., *Oxford.*	*graphical Society; Director Geological Survey.*
COOKE, J. P., Jun., *Cambridge, U.S.A.*	ODLING, PROF., F.R.S., *Royal Institution.*
DARWIN, C., F.R.S.	OLIVER, PROF. D., F.R.S., *Kew Gardens.*
DAWKINS, PROF. W. BOYD, F.R.S., *Owens College, Manchester.*	PENGELLY, WILLIAM, F.R.S.
DAY, G. E., M.D., F.R.S.	PHILLIPS, PROF. J., F.R.S., *Oxford.*
ETHERIDGE, R., F.R.S.E., *Geological Survey.*	PRESTWICH, J., F.R.S.
EVANS, J., F.R.S., *Secretary Geological Society.*	PRITCHARD, REV. C., F.R.S.
FARRAR, REV. F. W., F.R.S., *Harrow School*	QUETELET, M., *Sécrétaire perpetuel de l'Academie des Sciences*
FLOWER, PROF. W. H., *Royal College of Surgeons.*	*de Belgique.*
FORBES, DAVID, F.R.S.	RAMSAY, PROF. A., F.R.S., *Geological Survey of England.*
FOSTER, PROF. MICHAEL, *Royal Institution.*	ROLLESTON, PROF., F.R.S., *Oxford.*
FOSTER, PROF. CAREY, F.R.S., *University College.*	ROSCOE, PROF., F.R.S., *Owens College, Manchester.*
FRANKLAND, PROF. E., F.R.S., *Royal College of Chemistry.*	SCLATER, P. L., F.R.S., *Secretary Zoological Society.*
GALTON, DOUGLAS, F.R.S.	SMITH, ARCHIBALD, F.R.S.
GEIKIE, A., F.R.S., *Geological Survey of Scotland.*	SORBY, H. C., F.R.S.
GLADSTONE, DR. J. H., F.R.S.	SPOTTISWOODE, W., V.P.R.S.
GRAHAM, THOS., F.R.S., *Master of the Mint.*	STAINTON, H. T., F.R.S.
GRANT, PROF. R., F.R.S., *Director Glasgow Observatory.*	STEWART, BALFOUR, F.R.S., *Director Kew Observatory.*
GROVE, W. R., F.R.S., Q.C.	STONE, E. J., F.R.S., *Royal Observatory, Greenwich; Secre-*
HOOKER, DR. W., F.R.S., *Kew Gardens.*	*tary Royal Astronomical Society.*
HUXLEY, PROF. T., F.R.S., *President Geological Society.*	STUART, JAMES, *Trinity College, Cambridge.*
JEFFREYS, J. GWYN, F.R.S.	TAIT, PROF. P. G., *Edinburgh University.*
JONES, DR. BENCE, F.R.S., *Secretary Royal Institution.*	THOMSON, PROF. SIR W. F.R.S., *Glasgow University.*
KINGSLEY, REV. PROF. C.	TYNDALL, PROF. J., F.R.S., *Royal Institution.*
LEWES, G. H.	WALLACE, A. R., F.L.S.
LUBBOCK, SIR JOHN, BART., F.R.S.	WILLIAMSON, PROF. A., F.R.S., *President Chemical Society.*
MACMILLAN, REV. H.	WILSON, T. M., *Rugby School.*
MARKHAM, CLEMENTS, *Secretary R.G.S.*	WOODWARD, H., F.G.S., *British Museum.*

The Secretaries of Scientific Societies and Authors of Memoirs and Books on Scientific subjects will confer a great favour by transmitting their publications for notice as soon after publication as possible, to the Editor, care of the Publishers.

MACMILLAN & CO., LONDON.

Figure 26 Prospectus for the new weekly journal, Nature, *1869*

Science, with broad-ranging review articles, and *Chemical News*, a weekly directed at chemists both in academe and in industry.[36]

SCIENCE AND SPIRITUALITY

Science proved to be genuinely useful knowledge, bringing prosperity, but for some such as the surgeon and poet Keats, it had disenchanted the world, turning the rainbow into an example of refraction and reflection of different frequencies. Scientists might be perceived as technicians, manipulating the world but indifferent to beauty and truth, or as materialistic and reductive philistines, anxious to squash finer feelings in the name of realism. Science might be dreaded as a conspiracy to turn the world grey in the name of a crusade against ignorance and superstition, as well as a ruthless exploitation of nature (including humans). Priestley and Paley in different ways had used natural theology in response to such feelings: the book of nature carefully studied aroused awe and wonder and turned the mind towards God the creator. Theology of nature, in which God's existence and benevolence is taken for granted rather than argued for, was part of the framework not only explicitly in works like the *Bridgewater Treatises* or the series the Religious Tract Society published, but also in lectures and popular writings generally in the English-speaking world.[37] Elsewhere, this tradition had been greatly weakened, but everywhere it was necessary to establish that science was not indifferent or opposed to the higher things in life.

Priestley's vision of a dynamical science bringing together chemistry, electricity and optics was realized in the early nineteenth century in electro-chemistry and electromagnetism. Powers rather than brute matter seemed the key to the world, and electricity akin to spirit. Davy wrote pantheistic verses, and Ørsted, the scientific sage of Denmark, published at the end of his life a series of poems, lectures and essays, translated into English in 1852 with the title *The Soul in Nature*.[38] Aware like Keats how our bodies continually replace themselves, he believed that matter cannot constitute persons: we endure as waterfalls do, through a flux of material components. In Britain, the book was one of the few flops that the energetic Bohn published, and privately described by Darwin as 'dreadful', but as an attempt to get a dynamical science across, in which apparent rest was equilibrium of forces, and as an exercise in the scientific sublime, it was important in Denmark and in Germany. In his version of *Naturphilosophie*, Ørsted promoted science as a spiritual activity.

The deistic *Vestiges* of 1844 had vividly presented another vision, which has been called the evolutionary epic.[39] In France with Lamarck, and in English verse with Erasmus Darwin, this was a grand vision of solar systems coagulating out of undifferentiated and chaotic matter, the Earth cohering and cooling, an electric spark kindling life in the cosmic soup, and progressive evolution leading when the times were ripe to the appearance of human beings – able ultimately to interpret it (as in John Herschel's poem). For many, the same unchanging laws would certainly have led to rational beings elsewhere in this awe-inspiringly vast but uniform world – we need not be terrified of these vast spaces, and we were probably not alone. As archaeologists worked out a framework of stone, bronze and iron ages, artists began to portray our remote ancestors no longer as Adam and Eve, perfect specimens at ease in paradise from which they had subsequently fallen, but as hairy savages, leading lives that were nasty, brutish and short.[40] Mankind's history was not a decline from a golden age, but a story of progress. This epic might be taken as a realization of God's plan through the vast ages of unrecorded time, but it also allowed for God to be left out of the story. Either way, it provided a magnificent story vindicating the imaginative power of science,[41] and one that provoked a new genre, the novel of religious doubt, such as Mary Ward's best-selling *Robert Elsmere* (1888), which was astonishingly popular in the middle years of the century.[42]

Of those who left God out in recounting this epic story, one of the most eminent was Haeckel at Jena, who provoked much controversy in stuffy Wilhelmine Germany with his embryological studies, his famous drawing of the tree of life and his portrait of our ape-men ancestors on the wall of his study, now preserved as a museum.[43] In Britain, John Tyndall notoriously left God out, but his triumphalist Belfast address of 1874 was prepared amid the Alps, and he was moved to pantheistic reveries among the summits. In 1870 at Liverpool he had addressed the BAAS on the use of the imagination in science.[44] Controlled by experiment and observation, he perceived its role as essential: the favoured Baconian interpretation of science, even his friend Huxley's view of it as trained and organized common sense, was not enough. Collecting facts and generalizing without some imaginative leap would be impossible. Well read and eloquent, he quoted Emerson, Kant and Goethe, demonstrating that for him at least literary and scientific cultures were fully compatible. So for Ørsted science was spiritual, revealing the soul in nature, while for Chambers it was epic, and for Tyndall imaginative. There was yet another humanistic epithet that could perhaps be applied: mystical. Thus Charles Piazzi Smyth (1819–1900), the eccentric Astronomer Royal for Scotland, made careful measurements of the pyramids of Egypt, seeing arcane knowledge (and British units of length) embodied in them.[5]

SCIENCE SPREADS OUT

How science was received is as striking in its way as how it was achieved. The signs that science was assuming a leading position in culture became steadily more numerous as the century wore on. Whether disenchanting or inspiring, science found its way into fine art, literature and even music. John Constable was indebted in his cloud studies, so important in his big landscapes, to Luke Howard's meteorology and his distinguishing cumulus, nimbus and so on. Goethe's work on colours was a stimulus to artists, including Turner, who like modern photographers could use Goethe's perception that depth can be achieved using red to suggest nearness and blue distance, as well as his more sophisticated ideas about shadows.[46] Photography was an offshoot of chemistry and optics, and developing plates brought chemical reagents and processes into improvised darkroom laboratories in homes and gardens of enthusiasts, launching Crookes's career, among others.[47] The war in the Crimea (1854–6) had its photographer, Roger Fenton, an unsuccessful landscape painter who had turned to the new medium, and who had taken his van equipped as a darkroom for developing his glass plates: exposures were long and pictures had to be posed, but wounds, squalor and immediacy were evident, and the days of painting glorious victories and battlefields were over.[48] People began to look at paintings differently. A prosaic German eye-doctor, lecturing at the Royal Institution, raised the question whether distortion in paintings was pathological. He suggested that artists (like El Greco) who painted elongated people suffered from astigmatism and saw them that way, and that Turner's later razzle-dazzle works, seeking to capture light in paint, were the result of failing vision.[49] Artists saw things otherwise. The greater accuracy of photography did not as predicted speedily kill realistic academic painting, but it did make artists and critics think about visual representation, which need not after all depend upon the kind of perspective that went with geometry and the camera obscura. Appreciation of older painting unconcerned with the rules of perspective, such as Greek and Russian icons and medieval miniatures, became easier. And no longer having to do what the photographer can do better, painters (notably in France) could experiment with impressionism and pointillisme, picking up ideas from Helmholtz and others on optics and the physiology and psychology of vision.[50]

In literature, the story is fuller. To appreciate Goethe's novel *Elective Affinities* without some knowledge of chemistry would have been difficult, and the same applies to the poetry of Percy Shelley, a devotee of science.[51] His wife Mary picked up enough science from him and from Davy to write *Frankenstein*, now perceived as pioneering science fiction.[52] It was resonant with symbolism: God had placed Adam in a garden, and made Eve to keep him company and be his wife, bearing children the proper way. Excluding women, playing God, Frankenstein in contrast fled from the monster

A LITTLE CHRISTMAS DREAM.

Figure 27 Popular science and its nightmarish possibilities, Punch, *Christmas 1868*

made in his image, left him to fend for himself among glaciers and denied him a mate. Cutting himself off from his friends, with no moral compass, Frankenstein had worked alone among sickening body-parts from the char-nelhouse and the abattoir. No wonder it turned out badly. And the book has come to stand for a cautionary tale about overweening macho science, recklessly pursuing its amoral agenda. The story seems almost plausible: we have come some way from the stories of the medieval rabbi making a golem, or Faust a homunculus.[53] Erasmus Darwin's evolutionary ideas, the electrical experiments of Galvani and Volta, the horrid twitchings of the corpse when terminals were applied in Andrew Ure's (1778–1857) ghoulish public dissection of a hanged criminal in Glasgow, and the nasty milieu of grave-robbing in Edinburgh, all made Frankenstein's experiment conceivable.[54] Less suspension of disbelief was required than for the earlier stories involving magic or alchemy. Real science was now becoming powerful and dangerous.

By the optimistic years of the mid-century, most people saw science in a more positive light, as reflected in the novels of Jules Verne (1828–1905). While *Round the World in Eighty Days* (1873) celebrated the steamships and railways which (as well as the resourcefulness of the heroes) made the feat possible, his other books might be seen as a reaction to the way that North America, Siberia, Africa and Australia had been opened up by explorers. Even though there were still large blank spaces on maps, there was less scope for thrilling voyages and travels into the unknown on the surface of the globe. So Verne's heroes made a *Journey to the Centre of the Earth*, or *Twenty Thousand Leagues under the Sea*. Like Frankenstein, but more prudently, they made use in each case of extensions of available science. Their adventures made good stories and must have recruited young readers for science in ways that textbooks could never do: the future looked bright, especially for young inventors and scientists. By the end of the century, for jaded observers all too aware of degeneration rather than improvement, things looked bleaker. Wells's *Time Machine* (1895) is no longer optimistic about the future of the human race, which is portrayed evolving into two species (one preying upon the other), and ultimately becoming extinct.

Science plays an important part in Tennyson's great poem *In Memoriam* (1850), written about the death of his great friend from Cambridge, Arthur Hallam (1811–33). One of the most able undergraduates of his day, and engaged to Tennyson's sister, Hallam had after graduation gone for an extended holiday on the continent, and died in Vienna. The poem, in 131 sequences of four-line stanzas, traces Tennyson's gradual coming to terms with this blow.[55] In it, the abyss of geological time and the evolutionary science of *Vestiges* raise frightening questions about Nature, 'red in tooth and claw', and her indifference to the death of individuals: 'So careful of the type she seems,/so careless of the single life'. The poet tells us he had, like St Paul with beasts, fought with Death, and refused to believe that his friend and we were merely material, and would utterly perish:

Not only cunning casts in clay:
     Let Science prove we are, and then
     What matters Science unto men,
At least to me? I would not stay.

For Tennyson, reasoning like Paley's from 'eagle's wing, or insect's eye' did not lead to faith, but out of the darkness of doubt and sorrow came hands, like a father's comforting a crying child, 'that reach thro' nature, moulding men'.

Huxley was (like his wife) a great admirer of Tennyson, quoting his poems in his public lectures and making sure that the Royal Society was officially represented at his funeral (by four of its Officers as well as nine other Fellows) because he was:[56] 'the only poet since the time of Lucretius who has taken the trouble to understand the work and tendency of the men of science'. He believed that *In Memoriam* showed an insight into scientific method 'equal to that of the greatest experts'. The haunting cadences of the prophet of doubt resonated in the founder of agnosticism, who himself struggled to make sense of the death of his eldest son, Noel, and suffered periodic assaults of the Blue Devils of depression. Just as Tennyson rejected the materialism of *Vestige's* evolutionary epic, so Huxley (to the horror of some of his associates, notably Spencer) came to reject an ethic based upon evolution. He came to believe that the survival of the fittest in the struggle for existence, leading to Social Darwinism, could not be the guide to civilized life. Whatever is, is not necessarily right, and in his Romanes Lectures, *Evolution and Ethics* (1894), delivered in Oxford at the end of his life, Huxley argued that morality represents a defiance of nature.[57] Instead of jostling our way into the lifeboat, we should let women and children get in first. Science was not for him the complete guide to life that others believed. But his lecture was a sign that science could no longer be left out of moral discussions, even if like Huxley one felt that it must be transcended. And by then, science, with eugenics, armaments, medicine, technology, and the 'acclimatization' of plants like rubber and quinine, and of animals including rabbits, grey squirrels, cattle and goats, raised a host of ethical questions.

Francis Hueffer (1845–89), music critic and champion of Wagner, remarked in 1887 about the new enthusiasm for music that was spreading from Germany to Britain:[58] 'With the exception of natural science . . . there is no branch of human knowledge, or of human art, in which the change that the half-century of the Queen's reign has wrought, is so marked as it is in love of music.' Like Davy, Tennyson was unmusical, to the surprise of Huxley – who noted, however, that men of science seemed more often than poets or men of letters to appreciate music. They brought some new understanding into it: Helmholtz had written on music as well as colours – *Sensations of Tone* (1863; 4th edition, 1877) – a great book seeking its physiological basis, more sensitively than Richard Liebreich (1830–1917) had for painting in his Royal Institution lecture.[59] The text includes both music and equations: Helmholtz used the adaptable Fourier series here, as his friend Lord Kelvin had used them in energy calculations. A tour de force, it brought out the mathematical and physical basis of music (something Pythagoras had begun), and included work on the anatomy and physiology of the ear, and a discussion of aesthetics. He described his process of discovery:

> I was like a mountaineer who, not knowing his path, must climb slowly and laboriously, is forced to turn back frequently because his way is blocked but discovers, sometimes by deliberation but often by accident, new passages which lead him onwards for a distance. Finally, when he reaches his goal, he finds to his embarrassment a royal road which would have permitted him easy access by vehicle if he had been clever enough to find the proper start. In my publications, of course, I did not tell the reader of my erratic course but described for him only the wagon road by which he may now reach the summit without labor.

Just as his work on optics and the eye was important for innovatory artists at the end of the century, so his musical studies played a part in the new music.

NEW DEVELOPMENTS

Helmholtz wrote of his reader as 'he', and the rhetoric of science as we have encountered it was macho, but in his lifetime women were moving into science at last, no longer as helpmates but in their own right. Any claim for cultural ascendancy could hardly exclude them. In astronomy, especially in the USA, women proved themselves to be excellent observers, and they had long played a big part in natural history and its illustration. In J.J. Thomson's Cavendish Laboratory at Cambridge there were women researchers, doing fundamental physics: notably Rose Paget (1860–1951), daughter of a medical professor at Cambridge, whom Thomson married in 1890. But in science as in other activities, women were expected to give up their careers on marriage, and she duly did so. Then they would revert to the traditional maternal role. A scientific career for a married woman was difficult: Marie Curie (1867–1934) was the great exception, both scientist and mother. Women who did not marry, or perhaps whose marriage was childless or who were widowed like Nora Sidgwick, could aspire to

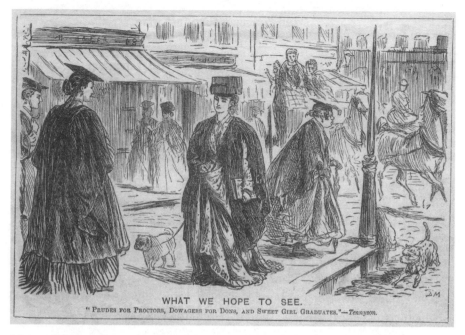

WHAT WE HOPE TO SEE.
" Prudes for Proctors, Dowagers for Dons, and Sweet Girl Graduates."—*Tennyson*.

Figure 28 *Facing up to the astonishing prospect of women graduates*, Punch Almanac, 1866

an academic career in science, especially in women's colleges, and might teach science in schools, although science was less readily available to girls than to boys.[60] In the 'maternal' tradition, a number of women had written scientific books, and several scientists had in fact picked up science at their mother's knee at the outset of their careers.[61]

Married women were expected to be consumers, and nineteenth-century advertisements are a rich and entertaining source for social history. They also reveal the increasing prominence of science.[62] Trade cards, illustrating processes and products, go back to the eighteenth century. Nineteenth-century books often advertise the publisher's list at the back, but those aimed at emigrants, to the goldfields or diamond mines or to less dramatic life in the colonies, advertise supposed necessities for the voyage and for life afterwards. Scientific journals from the beginning of the century also sometimes carried articles that were in effect puffs for apparatus or other inventions. Magazines and newspapers published increasingly illustrated advertisements, notably for patent medicines and for devices in which electricity allegedly played a part in restoring or maintaining health: pseudo-scientific claims, making use of learned language, are a very old feature of the trade. Informal scientific publications, like the delightfully titled

CADBURY'S COCOA, being genuine, does not thicken in the cup; no chemicals are used, as in many of the so-called "pure" foreign Cocoas. The *Analyst* says: "It is the typical Cocoa of English manufacture—absolutely pure."

"CADBURY'S COCOA has in a remarkable degree those natural elements of sustenance which give the system endurance and hardihood, building up muscle and bodily vigour, with a steady action that renders it a most acceptable and reliable beverage."—*Health*.

CADBURY'S COCOA is entirely free from objectionable alkalies, giving a false appearance of strength by producing a dark colour. The Editor of *Braithwaite's Retrospect of Medicine* writes: "Of absolute purity and freedom from alkali; may be prescribed without hesitation, with the certainty of obtaining uniform and gratifying results."

CADBURY'S COCOA is absolutely pure, therefore best. The *Medical Annual* says: "The name 'Cadbury' on any packet of Cocoa is a guarantee of purity."

NO CHEMICALS USED

(AS IN MANY OF THE SO-CALLED PURE FOREIGN COCOAS.)

Figure 29 The prestige of science equivocally indicated in an advertisement for cocoa,
Science Gossip, *December 1895*

Science Gossip (1865 onwards) depended upon advertising revenue:[63] books, lectures, cocoa, cabinets for curios, mounted insects, birds' eggs, shells, microscopes, soap, indigestion pills, steel pens and spectacles all featured. In advertisements for soap, cocoa and standard remedies, the emphasis was upon purity, and sometimes the message might be reinforced by an image of the doctor or chemical analyst.

CONCLUSION

In 1789, wisdom and connoisseurship were generally more valued than scientific knowledge. Science was often perceived as a somewhat comical pastime, and culturally the clergy were dominant. That changed in continental Europe with the French Revolution, when science assumed a prominent place in the new order there. But in Germany philosophers and philologists, and in Britain poets, claimed the cultural high ground, to the chagrin of scientific men with their bold vision of nature interpreted and of technical progress. In the 1820s, the GDNA in Germany, and peripatetic meetings of *Naturforscher*, provided the example for the BAAS, and these open and well-reported gatherings turned men of science into a self-conscious group. Like a church, they had their rows and schisms, but gradually scientists began to turn themselves into a kind of profession. They formed increasingly specialized societies, some learned and some professional, including publishing clubs and field clubs. With booming industry and more specialized education, scientists became a kind of professional group, able to live by science. The process required administrators, important figures less remembered than the stars of science whose careers they made possible. By the end of the century, scientists were much respected, but like other academicians might be suspected of stuffy conservatism. Science had important implications for spirituality, its methodological doubt being infectious in agnosticism, and its epics weakening the need for a First Cause. Fine art was revolutionized by the invention of photography from 1839; in literature, science fiction and novels of religious doubt were prominent; science entered musical theory; and ethical questions raised by science, practically in technology and theoretically in evolution, were pressing. And at last, women began to be able to play a visible part in science. They were still restricted by limited educational opportunities, conventions about women not speaking in public when men were present, and the expectation that on marriage a woman would devote herself to the family, but they became steadily more important.

By 1900, then, in the West, electric light, telephones and telegraphs, steamships, railways and the first motor cars, with a consumer society buying things made in factories, were all features of middle-class life. The message of Davy, Huxley and other apostles of applied science could no longer be doubted, and applied science was believed to depend upon the pure kind, the interpretation of nature. In the late medieval Church, a butcher's son, Thomas Wolsey (c.1475–1530), could rise to be a Cardinal and one of the most powerful men in the land. The Church had been a meritocracy, admittedly imperfect. In the nineteenth century science everywhere functioned in much the same way. To think of it as a kind of church, displacing the Christian churches from their cultural eminence, can be useful. And as the Church tried to be international (even while ministers blessed the artillery on their nation's side), so within the sciences there was a belief that they were above the squabbles of governments. We shall end this saga with a look at the opening years of the twentieth century, a time that can seem like a lull before the storm, a peaceful afternoon, but which was beset with turmoil, intellectual and social, in which science played its major part.

12

INTO THE NEW CENTURY

The twentieth century was to be an epoch of big science, of the 'military industrial complex', and of research carried out by teams of researchers with extremely expensive apparatus in huge laboratories funded by governments and by big companies.[1] The decades either side of 1900 were still a time of individual heroic discoveries in France, Britain and Germany that dramatically changed science. But the twentieth century, especially after 1933 when Hitler came to power in Germany, was to be America's. Just as poorly educated nineteenth-century England had depended for much of her science and technology upon well-qualified immigrants from Scotland and Germany, so the USA benefited hugely from immigrants, many of them political refugees as Priestley had been in 1794. Coming in waves, after the failure of the 1848 revolutions, after pogroms and revolutions in Russia, and then from the Nazi regime, they found there a familiar milieu, a tradition of precision measurement and instrumentation, and powerful innovative industries. American universities had by the late nineteenth century enthusiastically adopted the research ethos from Germany, and their graduate schools became incubators for scientists, native-born, refugee, or coming down the 'brain drain' from poorer or more hierarchical countries and often tempted to stay by the equipment and the environment.

In 1900, German was the language of chemistry, and undergraduate students in Britain and the USA were usually made to learn it: a practice that survived into the 1950s, although after 1945 English had become the international language for science and much else, as Latin had been in the seventeenth century. Problems of translation were no longer as acute as they had been in the nineteenth century, when scientists speaking one language could be unaware of research published in another. Germans in the nineteenth century had been better served by translations in their journals than the French or the British, but ignorance played a part in nasty rows about priority, with nationalism raising its ugly head, as happened with Archibald Scott Couper (1831–92) and Kekulé over structural chemistry, and when Adams and Leverrier predicted where the new planet (Neptune) would be found. Since simultaneous discovery, and the importance of being first and sadness of being second, are features of science, such unedifying quarrels inevitably continue. By the end of the nineteenth century, rapid publication in weeklies like *Nature*, and the 'abstracts' published by scientific societies, made it easier both to claim priority and to keep abreast. But the flood of publications made it impossible to read everything relevant, and the practice of sending offprints and proofs (preprints) meant that it could be hard for outsiders to penetrate 'invisible colleges' where data was exchanged between acquaintances (and rivals). As well as publishing abstracts, societies (and commercial publishers) brought out reviewing journals, which instead of original research carried weighty surveys of recent work in a particular field – writing such papers was an important and worthwhile task for active scientists, making them and their readers think about where they were going.

INTERNATIONALISM

Despite or because of these problems connected with the explosive growth of science, the nineteenth and early twentieth centuries were a time of international cooperation in science. The nation-states which we regard as the normal or proper way to organize ourselves politically were invented and consolidated in Europe during the long nineteenth century, though outside Europe those same states built up multi-ethnic empires wherever they could.[2] Rivalry (boosted by technology) was a constant feature of their relations, but might be managed through diplomacy and channelled into cultural competition, for example, in science. Because no state could be self-sufficient in science, there was free trade subject to the exigencies of language and access. The 'tariff' barriers of military and governmental secrecy that beset science in the twentieth century were less of a problem in the long nineteenth, though becoming apparent in, for example, the naval arms race of the 1900s, but the commercial barriers were always there, and industrial espionage was an important feature of the international spread of science right through our period. International agreements on patents and copyrights had come about only in the second half of the century – both of these presented obstacles to the free spread of knowledge, balanced in different degrees by the benefits to inventers and authors.

There was a long history of international cooperation. The Swedish botanist Daniel Solander (1736–88) had accompanied Banks on Cook's first voyage, and the Germans Johann (1727–98) and George Forster (1754–94) had gone as naturalists on the second voyage.[3] Well-prepared foreigners were not excluded from such government-funded voyages of discovery, and imperial science in India, Africa, the Far East and Australia depended heavily on them (as well as on locals, often poorly rewarded and recorded).[4] Alexander von Humboldt was fired with enthusiasm by Forster, and his own expedition to Spanish America took place, and was written up, under French auspices. In turn, he became a great enthusiast for European, and indeed worldwide, cooperation in geosciences, promoting the 'magnetic crusade' in which measurements of terrestrial magnetism were made simultaneously over a period by observatories round the world. His British admirers, notably Herschel and Edward Sabine (1788–1883) – whose wife Elizabeth (1807–79) translated Humboldt's books – pressed his proposals upon the British government. The crusade involved 'big science' bringing in men of science and governments from all over Europe and North America, then setting up properly equipped observatories around the world, organizing voyages like that of Ross to Tasmania and the Antarctic, and overland expeditions to suitable sites around the world, generating great quantities of data.[5]

Humboldt himself made a second expedition, this time into central Asia. He was interested in everything: the animals and plants of a region, and their distribution, the minerals and their uses, the climate, the people and their culture, as well as the latitude and longitude, and the magnetic variation. Going beyond description into generalization, he was also sensitive to the beauties of forests, mountains and the night sky.[6] Such wide-ranging global science, summed up in his volumes written in reflective old age, *Cosmos* (1845–62), required international collaboration and has been called Humboldtian by historians.[7]

Survey voyages with astronomers and naturalists aboard, like Nicholas Baudin's and Matthew Flinders's to Australia, HMS *Beagle's* to South America and subsequently to Australia, and of the American Charles Wilkes in the Antarctic and the Pacific, were national enterprises that aroused great international interest,[8] but the oceanographic voyage of HMS *Challenger* (1872–6) was explicitly international in its scientific aspects. Charles Wyville Thomson (1830–82) led the team of six civilian scientists on board, which included the German Rudolph von Willemoes Suhm (1847–75) who unfortunately died of erysipelas during the voyage.[9] Photographers were also funded, and a darkroom provided on board, so a splendid record of the voyage survives. When the ship returned, Thomson persuaded the government to continue funding for an office to coordinate and publish the results, and involved experts abroad in the project, to the indignation of some of the Britons that he deemed less qualified. His vigorous assistant John Murray (1841–1914) took over on Thomson's death, and the *Report* came out in fifty volumes between 1880 and 1895, a splendid example of international collaboration in science.[10]

Other publication projects had international aspects. Many scientific journals carried reports of meetings abroad, and some translations of papers, but in a new departure, the publisher Richard Taylor (1781–1858), with the support of the BAAS, had launched *Scientific Memoirs* in 1837. This was to consist entirely of translated papers; it did not sell as well as he had hoped, though containing much important science, but after five volumes had come out, the young and fiery Huxley and Tyndall, well aware of the importance especially of German science, each helped edit a volume, on natural history and natural philosophy, in 1852–3.[11] After that, the series expired and insularity reigned once more. But the Ray Society published Louis Agassiz's great bibliographical guide to papers on geology and zoology in 1848–54, edited and translated by Strickland,[12] and the Royal Society in 1867 inaugurated a magnificent project, the international *Catalogue of Scientific Papers, 1800–1900*, which appeared under its aegis in nineteen volumes ending in 1925.[13] By then German scientists were still not fully rehabilitated and readmitted to the international scientific community, internationalism was at a low ebb and the project was discontinued. The *Catalogue* is not only a very valuable guide, but also interesting because it indicates the journals regarded as respectable by the compilers. Papers that the historian might think important, and that might have been widely read because in a 'popular' periodical, may well not be listed. Clearly, in 1914, science was an international enterprise, though firmly based in (almost confined to) Europe and North America. And there was a scientific revolution in progress, embracing scientists from many countries.

ANOTHER SCIENTIFIC REVOLUTION?

It is plausible to say that there was a scientific revolution centred in France at the end of the eighteenth century, when science emerged as a career, and new and tighter criteria for being scientific were agreed. A century later, there was another, this time with centres in France, Britain, Germany and the USA, when long-accepted tenets of scientific method, and axioms about the world, had to be revised in the light of new evidence. It involved major revisions in physics and chemistry, with knock-on effects on other sciences (notably geology). Lord Kelvin, marking the turn of the century with a lecture at the Royal Institution, spoke of clouds over the confident classical physics of the late nineteenth century, and was proved right.[14] The relationship of matter and energy, the nature of atoms and elements, conceptions of time and space, and the idea that crucial experiments can settle theoretical problems, were all called in question.[15]

In 1896 Antoine Henri Becquerel, son and grandson of eminent scientists, found that, although packed in light-proof wrapping, photographic plates (by then ubiquitous) in his laboratory were blackened by nearby samples of uranium salts. This was the phenomenon of radioactivity, and his work was followed up in the doctoral dissertation of Marie Curie, from Warsaw but studying in Paris, who found that thorium was also active. She went on with her husband Pierre (1859–1906) to isolate in 1898 the much more active polonium and radium through long series of careful recrystallizations from the mineral called pitchblende. For this, they and Becquerel shared a Nobel Prize in 1903, and after Pierre's death in 1906 she was elected to his chair at the Sorbonne – and awarded a second Nobel Prize in 1911 for her further work on radium and its properties that was eventually to cause the leukaemia that killed her. She had demonstrated, to a generation used to scientific arguments to the effect that women were intellectually inferior,[16] that women could be scientific geniuses.

Her interests were chemical: a particular and measurable intensity of radioactivity was a property of some elements, just like electrical conductivity or the formation of an insoluble chloride. The New Zealander Ernest Rutherford, coming 'home' to study in a Britain he had never seen, on a scholarship funded by the profits of the Great Exhibition, and then appointed to a post in Montreal, was a physicist not satisfied with such categorization, and wanted to know what was going on in the process. He distinguished the alpha and beta rays that made up the radiation. In 1903, working with the chemist Frederick Soddy (1877–1956), he came up with the idea, and the experiments to support it, that radioactivity was 'sub-atomic chemical change'.[17] His researches earned him a Nobel Prize in 1908, curiously enough for chemistry. Uranium was decaying into lead at a steady rate, and other radioactive elements were undergoing similar changes – but in a curious way that seemed random although completely regular, and undermining simple understanding of the axiom that every event must have a cause. Richards, the eminent analytical chemist at Harvard, confirmed that lead found in the neighbourhood of uranium had a slightly different atomic weight from lead found elsewhere, as Rutherford's calculations predicted. Soddy called atoms of the same element that differed in weight 'isotopes', and suggested that they were common. Dalton's axiom had been that all atoms of an element weigh the same.[18]

Although Dalton sometimes used the word 'atom' for carbon dioxide and other compounds, he did not believe that atoms of the elements could be split; and chemists, after atomic theory was generally accepted in the 1860s, mostly agreed. That after all is what the word means. But the first subatomic particle, the electron, was identified in 1897 at the Cavendish Laboratory in Cambridge, to which in 1895 the young Rutherford had gone to work with J.J. Thomson. Helmholtz, giving a lecture in London in memory of Faraday in 1881, had suggested that since definite quantities of elements are deposited by definite quantities of electricity, if matter is atomic, so must electricity be,[19] and earlier George Johnstone Stoney (1826–1911) had had the same idea. This hypothetical atom of electricity was called an 'electron'. But in the light of Maxwell's electrodynamics, field theory and the investigation of electromagnetic waves preoccupied physicists.

J.J. Thomson distinguished physicists, deductive thinkers who like travellers had a destination in mind, from chemists who were like explorers, inductive and interested in everything they found.[20] It was Crookes the chemist who took up Faraday's researches on the discharge of electricity through gases at low pressures, which with improved pumps could now be taken further. He discovered and investigated the cathode rays, projected in straight lines from the negative terminal, casting shadows and turning little propellers, and performed wonderful demonstration lectures illustrating these extraordinary effects.[21] At the end of 1895, William Konrad Röntgen (1845–1923) investigated the fogging of his photographic plates near a cathode-ray tube, named the radiation responsible X-rays, and took the famous photograph of his wife's hand. These amazing X-rays seemed to be electromagnetic waves, and most contemporary Germans believed that cathode rays must be similar.

Thomson, though rather ham-fisted, had at his disposal a very well-equipped laboratory in Cambridge – inherited from his predecessor Lord Rayleigh – with pumps more efficient than those elsewhere, along with graduate students and expert assistants.[22] Like Crookes, he believed that the

CHEMICAL NEWS,
Feb. 20, 1891. *Electricity in Transitu: from Plenum to Vacuum.*

Radiant Matter and " Radiant Electrode Matter."

In recording my investigations on the subject of radiant matter and the state of gaseous residues in high vacua under electrical strain, I must refer to certain attacks on the views I have propounded. The most important of

by Puluj on " Radiant Electrode Matter and the So-called Fourth State." Dr. Puluj's paper concerns me most, as the author has set himself vigorously to the task of

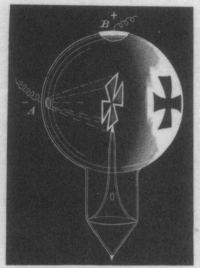

FIG. 20.—P. = 0·00068 m.m., or 0·9 M.

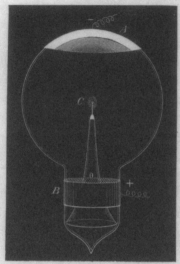

FIG. 22. P. = 0·000076 m.m., or 0·1 M.

FIG. 21.—P. = 0·001 m.m., or 1·3 M.

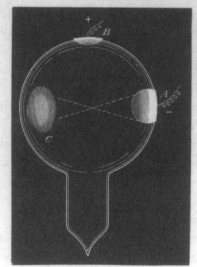

FIG. 23.—P. = 0·00068 m.m., or 0·9 M.

these questionings are contained in a volume of " Physical Memoirs," selected and translated from foreign sources under the direction of the Physical Society (vol. i., Part 2). This volume contains two memoirs, one by Hittorff on the " Conduction of Electricity in Gases," and the other

opposing my conclusions. Apart from my desire to keep controversial matter out of an address of this sort, time would not permit me to discuss the points raised by my

Figure 30 Cathode rays: W. Crookes, 'Electricity in Transitu', Chemical News, 29 February 1891: 91

cathode rays were a stream of particles with a negative charge, but in that
case it was odd that they could not be deflected by an electric field. This
ceased to be a puzzle when he found that when the pressure is low enough,
they are deflected, and can be deflected back again by magnets. Elegantly
balancing measured electric and magnetic forces in this way, he could calculate the ratio of mass to charge of his particles, or 'corpuscles' as he called
them, echoing the term used by Boyle and Newton for fundamental particles. The ratio was about 1/1800 that of a hydrogen ion, H+, so they were
tiny. He envisaged the atom as a 'plum pudding' of positive matter with
electrons embedded in it like currants, and he realized that the hypothesis of
Prout and others, that the elements were polymers of hydrogen, was looking plausible again.[23] Between 1909 and 1912, Robert Millikan (1868–1953)
in the USA directly estimated the charge on the electron by following the
movements of single charged drops of water and then oil in an electric field.
Meanwhile, Rutherford came back to Britain, to Manchester, in 1907 and
with his group there found that when alpha rays were directed on to thin
films of metal, most went straight through but a few suffered large deflec-
tions.[24] This was quite impossible if they were ploughing through positive
dough and hitting a few electrons. He inferred that the atom was mostly
empty space, a tiny, massive, positively charged nucleus with electrons cir-
cling it in orbits.[25] That was fine and dandy, except that it was inherently
unstable and the electrons would speedily spiral into the nucleus.

The model (a term that was coming into use in analysing theories) was
saved when Niels Bohr (1885–1962) left Denmark for Britain, initially to
work with Thomson but then in 1912 joining Rutherford. He was well aware
of recent German physics. Max Planck (1858–1947), wrestling with one of
Kelvin's 'clouds' – the nature of radiation from 'black bodies' – suggested
that theory and experiment could be brought into line if radiation was not
continuous, but came in packets that he called 'quanta'. In the Zurich patent
office, Albert Einstein (1879–1955) took this model seriously and in 1905
published a paper explaining why red light, however strong, did not affect
photographic plates or eject electrons from potassium, whereas blue light,
however weak, does both these things. If light were wave motions, that
would not happen. But if Newton had been right about particles of light,
and radiation comes in the form of 'photons', the red ones carry less energy
than the blue – insufficient to bring about these effects. Shining red light on
potassium would be like shooting at a rhinoceros with small shot, whereas
blue would be like rifle-bullets. In the hundred years since Thomas Young
(1773–1829) did his experiment with a double slit, and Augustin Jean Fresnel

(1788–1827) had subsequently worked out the physics of transverse waves, every crucial experiment had confirmed that light was a wave motion. By 1900, it was taken as an established truth. Einstein's Nobel Prize was given him for this upsetting piece of work. Light was not just a wave motion, but nor was it just particulate – life and the universe were no longer as simple as they had seemed. Bohr came up with the idea of 'complementarity' to cover incompatible models that between them accounted for the phenomena.

Bohr also applied quantum numbers to the hypothetical orbits of the electrons in Rutherford's atom, which could thus move only in certain orbits and nowhere in between. Energy given to an atom would be absorbed in quanta that would make electrons jump between orbits; jumps back again would emit energy in the form of radiation of definite frequencies. This explained one of the mysteries of spectroscopy. The Sun's spectrum was continuous from red to violet, as we see in rainbows and Newton saw with his prism. The black lines that Wollaston and others noticed in it were in 1821–2 shown by Josef Fraunhofer (1787–1826) to be really there at definite places, and not generated by defects in prisms. Chemists were familiar with 'flame tests' for elements such as sodium that colours flames yellow, but these were at best a guide rather than a real test. When in 1860 Bunsen and Kirchhoff heated substances intensely and looked at their spectra, they found them discontinuous, consisting of bright lines that turned out to be characteristic of elements (and puzzlingly coincided with Fraunhofer's dark lines). Chemists, at first sceptical, were content to use this convenient method of analysis. But physicists asked what the lines might mean, launching solar physics and raising question about inorganic evolution. For Bohr, the lines represented the jumps electrons made between the orbits that quantum theory would allow them to occupy. By 1914, quantum mechanics, with its unsettling uncertainties about the nature of light and matter, and about causality, was an inevitable part of physics.

In that amazing year 1905, Einstein also laid sceptical doubts about atomism at rest with his analysis of the motion of minute suspended particles that Brown had noticed decades earlier: they were due to bombardment by invisible molecules. Matter really was atomic. Chemists had been right to use models like those of Hofmann and van't Hoff rather than try to explain their science in terms of thermodynamics, as Ostwald (another Nobel Prize winner) would have had them do. Einstein also proposed special relativity, later generalized, in which long-accepted ideas of time, space and the geometry of the universe (the axioms of Newtonian mechanics) were called into question. Nothing seemed safe. Chemists, who in Priestley's time had thought that theirs was the fundamental science, found their science being 'reduced' to physics with the rise of thermodynamics and then the new theories of the atom.[26] They duly resisted, noting the difference between explanation in principle and in real detail; boundaries between sciences are to do with convenience rather than with the way the world is, and chemistry has in the event retained its autonomy. With spectroscopy and then X-ray crystallography, 'physical methods' even began to take over from the long-established craft techniques of analysis used, with gradual improvements, by Lavoisier, Berzelius and Marie Curie.[27] When dirty ores are reduced, the result is a shining ingot of metal, but when a science is reduced, the effect is to make it into a service discipline, less attractive to really innovating minds. In fact, synthetic chemists have flourished mightily in the twentieth century, giving us all sorts of things our ancestors never dreamed about. They have been no more restrained by physics than architects are: both are constructing within a framework of natural laws.

WHAT THE WORLD SAID

In the 1870s, it seemed plausible to Tyndall and his associates to contrast science (the realm of reason) with faith (irrationally believing six impossible things before breakfast). Positive knowledge was, it seemed to them, coming to replace the religious and metaphysical stages through which humans, collectively and individually, were growing – and the important thing was to hurry the process along, through education in its various forms. Scepticism was fiercely applied to religion, through the logic of Hume and Mill, the textual analyses of biblical scholars, and the moral criticisms of utilitarians like Sidgwick whose *Methods of Ethics* (1874) was published in the same year as the BAAS meeting in Belfast where Tyndall delivered his famous address. One of Sidgwick's students in the new 'moral sciences' degree at Cambridge was Balfour. His uncle was the prominent Conservative politician Lord Salisbury (1830–1903), a rather gloomy intellectual who became Chancellor of Oxford University in 1869, was President of the BAAS in 1894 (where he welcomed Darwinism in a teleological form), and who became three times Prime Minister, leading Britain into the twentieth century.[28] After graduation, Balfour was elected to parliament, and in 1878 accompanied his uncle, then Foreign Secretary, to the Congress of Berlin where the European powers sorted out the Balkans – temporarily, as it turned out. The following year he published his first book, *A Defence of Philosophic Doubt*. He had intended to use the word 'scepticism', but this was a term generally applied to religion, and his target was science. The book was accessible and witty, and its thesis was that science (like all knowledge) depends upon belief: its axioms are not self-evident. Like his uncle, he admired science, knew scientists and recommended deferring to their expert opinions, but we were all willy-nilly believers, and he was not, he wrote,[29] 'acquainted with any kind of defect to which systems of belief are liable, under which the scientific system of belief may not properly be said to suffer'. Its metaphysical foundations need careful and sceptical examination. That was something that Mach and then Ostwald had also urged, but their extreme, root-and-branch scepticism took them towards a new and more austere version of Comte's philosophy, which became the logical positivism of the Vienna Circle, so influential in the mid-twentieth century.

Balfour's was a genial scepticism; he was not gazing into an existential abyss. He allowed for inconsistencies, for science, religion, morality and things of beauty. But against the prevailing tide, and from a man perceived as unserious or lightweight, the book had little impact. By 1894, when he published his second book, Balfour was famous. His party was a staunch opponent of Gladstone's policy on 'home rule' for Ireland, and the motion on which Gladstone was defeated was drafted in Balfour's home. In his uncle's subsequent government, he had been put in charge of Irish affairs, and had done very well in this awkward portfolio. He was being recognized as a future leader whose opinions mattered. In *Foundations of Belief* (1894), he essentially repeated the arguments in the earlier book but included a famous denunciation of scientism or 'naturalism' as empty and incapable of accounting for the important things in life:[30]

> Man will go down to the pit, and all his thoughts will perish. The uneasy consciousness, which in this obscure corner has for a brief space broken the contented silence of the universe, will be at rest. Matter will know itself no longer. 'Imperishable monuments' and 'immortal deeds', death itself, and love stronger than death, will be as though they had never been. Nor will anything that *is* be better or worse for all that the labour, genius, devotion, and suffering of man have striven through countless generations to effect.

This was indeed the world of H.G. Wells's *Time Machine*, and Huxley, laid low with influenza on what turned out to be his deathbed, struggled to answer Balfour but died unsatisfied with what he had tried to write.[31] Balfour's idea that science rested upon faith was beginning to look much more plausible in the light of current developments. In the 1860s, the Metaphysical Society had in London brought together intellectuals seeking common ground between their various beliefs, and publishing in James Knowles's (1831–1908) review, *The Nineteenth Century*.[32] But the members (including Huxley, Tyndall, Tennyson and Gladstone) did not find very much to agree about, and as they got older the society closed down. Following publication of Balfour's book, the Synthetic Society was founded with the same hope, exploring his idea that knowledge rested on belief, and it published for the members a good record of their deliberations.[33]

In 1904, Balfour, who had succeeded his uncle as Prime Minister, was invited to preside also over the BAAS, that year meeting in Cambridge. Always eloquent, he spoke of how much of the science he had picked up there about 1870 had been falsified, and especially about how electricity, a matter of parlour tricks a hundred years before, was now perceived as a crucial feature of matter. Well-briefed, he used examples from the latest science to vindicate his view that it rested upon beliefs. Had he or his audience known what Einstein was up to, they would have had further reason to agree. In 1906 Balfour lost the general election in a landslide. During the Great War, he was back in the Cabinet, and proclaimed support for Jewish settlement in Palestine. And as the war ended, the Council of the Royal Society hoped that, by now an elder statesman, he might agree to be President in the new and uncertain world. He declined on finding that it was more than a figurehead position, but during the 1920s he was in effect the first Minister for Science in a British government.[34] Meanwhile, he helped promote philosophy and liberal theology, and his view of knowledge was important both for scientists questioning the status of theories and models, and for outsiders comprehending science as a cultural activity.

Balfour belonged to the Society for Psychical Research that Sidgwick and his wife Eleanor (Nora) founded in 1882, as did J.J. Thomson and other scientists from Britain, the USA and continental Europe. While the society's investigations into ghostly manifestations, and even their experiments on telepathy, proved disappointingly uncertain, Myers, one of its prominent researchers, became very interested in the threshold of consciousness and what lay below it, the 'subliminal', perhaps to be revealed in dreams, under hypnosis or in a seance, or in cases of multiple personality.[35] These interests led him to Freud whose clinical studies in Vienna and then in Paris (where he went in desperate search for a fruitful specialism) with Charcot took him into psychoanalysis. Myers then introduced his work to the English-speaking public. In the years after 1914, Freudian theory was to be of huge importance both in general culture and in clinical psychoanalytic practice, comparable to phrenology a hundred years before,[36] but in psychology, the experimental approach of Wilhelm Wundt was far more influential. Especially in the USA, psychologists tried to follow the methods of experimental inquiry and statistical inference that would make the discipline 'scientific', with links to evolutionary biology and physiology. The prestige of science was at its height.

UNCERTAINTY AND OPTIMISM

It was not only in the physical sciences that apparent certainties were shaken. In Britain, the USA and Germany, evolution had long before 1900 been accepted by professionals in the Earth and life sciences, and indeed by lay people like Lord Salisbury and Archbishop Temple. Even in France, where the clashes of Lamarck, Cuvier and Saint-Hilaire were recalled and exactitude rather than probability was demanded, initial rejection there had been overcome by 1900.[37] Henri Bergson (1859–1941) very influentially embraced an evolutionary world-view in his *Creative Evolution* in 1907. But the number of people anywhere who, like August Weismann (1834–1914), a professor in Freiburg, were strict Darwinians, attributing most or all evolution to natural selection and the struggle for existence, was small. When the fiftieth anniversary of the publication of *Origin of Species* came round in 1909, the elderly Wallace and Hooker were guests of honour at the celebrations, but they were almost the only true believers. Darwinism, strictly speaking, was in eclipse.[38]

Even in Darwin's circle, Huxley had never been happy with natural selection, favouring bigger and more sudden changes, while Asa Gray promoted a theistic Darwinism, in which evolution was the means whereby God created the animal and vegetable kingdoms.[39] Gray was followed by Salisbury and many others, whose vision of the world was more benign and optimistic than Darwin's. Darwin had been unable to explain inheritance, though it played a crucial part in his theory. He thought it might be the result of minute gemmules from all over the body contributing to the reproductive material in each parent, which is blended in their offspring. He could not answer the objection of the engineer and friend of Robert Louis Stevenson, Fleeming Jenkin (1833–85), that because an animal with an unusual and pronounced characteristic would usually mate with one without it, their progeny would have it in less marked degree, and in two or three generations it would effectively be swamped.[40] Weismann had specialized in microscopic study of invertebrates until his eyesight began to fail. Subsequently, he had to become more reflective and he began to investigate inheritance. His observation that germ cells could be distinguished from body cells very early in development made him question Darwin's gemmules, and instead he postulated the 'continuity of the germplasm', a line of descent through a line of germ cells unaffected by what happens to the body. He utterly rejected the widely held 'Lamarckian' idea that characteristics acquired in life could be passed on, and he held that the chromosomes (visible through the microscope) contained the hereditary material. He saw that in conception half the chromosomes from each parent are involved, so that their permutations will give rise to variation.[41] Weismann's works were translated into English by two eminent scientists, *Studies in the Theory of Descent* (1882) by the dye chemist and friend of Darwin, Raphael Meldola (1849–1915), and *Essays upon Heredity and Kindred Biological Problems* (1889–92) by the expert on animal coloration, Edward Poulton (1856–1943).

Chromosomes, unlike gemmules, were visible entities, and now there was a definite theory of inheritance to test. The stage was set for the recovery in 1900 of the work of Mendel, a monk in then Austrian Brünn, now Brno in the Czech Republic. Trained in mathematics and physics, pursuing ideas going back to Linnaeus that hybridization might explain how the multitude of species had arisen from fewer originally created plants, and with the improvement of agriculture in mind, he experimented in the monastery garden on pure-breeding varieties of pea. Isolating particular characteristics, like being smooth or wrinkled, green or yellow, he found these were not blended when varieties were crossed and re-crossed. Some characteristics, like yellowness, were dominant, so that in the next generation all had them: but breeding from this generation gave a 3:1 ratio in favour of the dominant kind in the second generation. He proposed that there were factors coming from each parent, say Y and G for colour. The thoroughbred parents would have had YY and GG, so the first generation will all have YG, and look yellow, but in the second there are four possibilities: YY, YG, GY, and GG. Three of these will look yellow, and one green. He found these ratios with remarkable exactitude.[42] Indeed, there seems no doubt that he worked like a physicist, testing a hypothesis rather than generalizing from gardening experience. His paper was read to the local natural history society in 1865, and published in its journal. Copies found their way into libraries of learned societies, but nobody took any notice. To publish fundamental work and have it ignored is a nightmare that haunts scientists, but Mendel was elected Abbot, had no more time for research, and died unknown to the wider scientific community.

In 1900 three botanists, Hugo de Vries (1848–1935) in the Netherlands, Carl Correns (1864–1933) in Germany and Erich von Tschermak (1871–1962) in Austria, independently came across Mendel's paper, in an intellectual context now very different from his. They called attention to it, and as evolutionists creatively misunderstood it because it seemed to conflict with the strict Darwinism they disliked. If variation was a matter of the shuffling of factors, then the sort of slow incremental change envisaged by Darwin seemed to be impossible, and a jerky evolutionary mechanism based on 'mutations' was proposed instead. The work of Mendel and his rediscoverers was enthusiastically taken up by William Bateson (1861–1926) at Cambridge, who coined the term 'genetics', and wrote *Mendel's Principles of Heredity: A Defence* in 1902. Darwin had relied on information from stock-breeders and pigeon-fanciers, but horses, cattle and birds breed slowly and have relatively few progeny. That was why peas were such a good choice (they also happened to give much better results than most garden plants); and in 1908 Thomas Hunt Morgan (1866–1945) in New York began rigorous experimental study of fruit flies and their mutations. His team demonstrated that the factors, or genes, resided in the chromosomes, and in due course their study of minute mutations paved the way for a reconciliation of Darwinian and Mendelian theory. Galton and others had done much statistical work on inheritance, but the recovery of Mendel set genetics in a new direction and meant that the twentieth century was to be, among other things, the age of the gene.

THE NEW WORLD

When I heard Harold Macmillan reminisce at the Oxford Union about the decade before 1914, it sounded like a long and glorious summer afternoon before the storm of war broke. Hindsight made it seem like that. It was full of snobbery and anti-Semitism, social and industrial unrest, disputes over welfare legislation, crises and diplomatic incidents, with nationalism cracking the Austro-Hungarian empire, revanchist anti-German sentiment in France, and an Anglo-German naval arms race in full swing.[43] Nevertheless, there was much to be optimistic about, and most of that was connected with science and technology. The campaign to endow research was having some success, notably with the foundation in Germany of the Kaiser Wilhelm Institutes (subsequently renamed after Max Planck) in

1911.[44] Such institutions, privately funded and separate from universities and industries though linked to them, were set up in other countries, too, sometimes, as in Russia, associated with Academies of Sciences. They played an important part in twentieth-century science, pure and applied. The Irish intellectual historian William Lecky (1838–1903) had written in his *History of Rationalism* (1865) that:[45] 'It is impossible to lay down a railway without creating an intellectual influence. It is probable that Watt and Stephenson will eventually modify the opinions of mankind almost as profoundly as Luther or Voltaire.' And in another work, on *European Morals* (1869), he devoted a long final chapter to the (unfair) position of women, where any supposed consensus on separate spheres was clearly breaking down by 1900.[46] The International Exposition in Marie Curie's Paris in 1900, with its emphasis upon electricity, reinforced a scientific and technical optimism. It was and remains impossible to divorce science from technology and medicine: whether we like it or not, knowledge is indeed power. Understanding has increasingly been aligned to usefulness, as applied science, a visionary prospect in 1800, became a reality by 1900.

This book has been concerned with the rise of the scientist as the interpreter of nature, and we get some impression of that from the census returns for England and Wales. The first census was taken in 1801 and everyone was asked about their occupations, but only agricultural, trade and manufacturing, and 'others' were recorded. By 1831, there were many more specific trades included, and 'professional men' were in a category of those who do not labour with their hands – but although there must have been many that we would call technicians and some scientists among these categories, they are not specific enough to tell us anything. What we can be quite sure about is that the number living by science at these dates was very small. By 1911, the distinctions between the various crafts and trades are elaborate and fascinating, with the survival of old ones and the appearance of new: wheelwrights were in decline, the motor and bicycle industries expanding. The introduction to the data on occupations and industries reviewed the methods and distinctions used in past and present censuses, and described how a system of punched cards had been used to sort the returns. By 1911, there were 6,246 males 'engaged in scientific pursuits', and 145 females of whom 20 were married. This was about double what it had been in 1901.[47] There were also 7,398 male scientific instrument makers, and 788 females, 49 married. Engineers and surveyors, and physicians and surgeons, will have included technologists and medical scientists, but it is not clear how many in the chemical, gas or

electrical industries, or how many writers, teachers and lecturers, might
be counted as members of the scientific community. What is apparent
is that a substantial number of people thought of themselves as engaged
full-time in science. By that time there really was a clerisy in Britain, and
in some other countries it will have been proportionally larger, though by
then polytechnics in Britain were closing the gap with Germany.[48] Science
had come of age, and history could not properly be written or understood
without it.

Our story began with a bloody revolution, and the execution of Lavoisier
in the Reign of Terror. It ends with a terrible and bloody war.[49] The revolutionary and Napoleonic Wars that dragged on for so many years had little
connection with science, and natural philosophers were awarded prizes by
academies on the other side, and in Davy's case granted permission to come
to Paris to collect it. Although Louis XVI had been guillotined, this was a
war of kings rather than of people, and hatred of the enemy was limited.
There was distrust of things French among British conservatives, political
and medical, but also liberal enthusiasm: Napoleon had his British admirers, and there were French Anglophiles. In succeeding wars, science played
a greater role. Steam ships, rifled guns, photography, medical treatment
and a railway were important in the Crimea, and the course of the Indian
Uprising, or Mutiny, was affected by the telegraph because vital requests
for reinforcements got through rapidly to the Punjab. In the USA, the
bloody battles of the Civil War involved new technology, including the
first engagement between two armoured steam-powered warships. That
war, and the rapid mobilization of the Prussian army in 1870, showed the
strategic importance of railway networks. But the Great War turned out
to be different, with the wanton destruction in Belgium followed by the
terrible stalemate bogged down in trenches. Accurate rifles, machine-guns
and artillery, now engaging targets at long range whereas before they had
normally been in sight, were a horrific combination; and they were joined
by minefields, Zeppelin bombers, aeroplanes and tanks. At sea, ships were
torpedoed in the unrestricted submarine warfare that brought the USA into
the war. Huxley had said that science (unlike religion) never did anyone any
harm, and Clifford became lyrical:[50]

It is idle to set bounds to the purifying and organizing work of science.
Without mercy and without resentment she ploughs up weed and briar;
from her footsteps behind her grow up corn and healing flowers; and no
corner is far enough to escape her furrow.

But in 1914–18 science lost that innocence. What had begun in wondering and delighted interpretation of nature had resulted in Dreadnoughts, heavy artillery and world war.

The conflict was described as 'the chemists' war'. The synthetic dye industry was by then in effect confined to Germany and Switzerland, and there were no dyes for British army uniforms until a crash programme to develop the chemical industry was put in hand. British chemists were recruited for the military, but their status was at first low and they were given the rank of sergeant. Meanwhile, Fritz Haber (1868–1934) had in 1908 invented the process whereby nitrogen and hydrogen were combined into ammonia, using a catalyst in a vessel under enormous pressure, and by 1913 the chemical engineer Carl Bosch (1874–1940) at BASF had successfully scaled-up the pilot venture into a works making 36,000 tons of ammonium sulphate per year. The impetus for this had been the need for fertilizers, but in the event, until 1919 all the output went into armaments. Haber was also responsible for the attempt to break the deadlock on the Western Front by using poison gas, Davy's chlorine, in clouds and then in artillery shells. He seems to have seen this as relatively humane, promising a quick end to the slaughter, but it filled contemporaries with horror. As things worked out, it was not even very effective, and the allies soon retaliated with greater quantities and more lethal substances.[51] Those who had been gassed, or witnessed gassing, acquired a jaundiced view of science – or at least realized that it was not simply a force for good, or a harmless form of curiosity. Prominent German scientists publicly endorsed the behaviour of their army in Belgium, and nationalistic fervour gripped scientists on both sides as they put their knowledge at the disposal of their governments in devising weapons and means of detecting enemy submarines and artillery-positions. Germans were expelled from societies and academies, and only slowly re-admitted to the international scientific community after 1918. An era had ended.

It would be wrong to end our story on that melancholy note. We have after all been following a success story, the story of the 'Age of Science', in which an ugly duckling – a hobby chiefly for inquisitive if dotty professors, gentlemen of leisure and country parsons – turned into a beautiful and powerful swan, the cynosure of all eyes.[52] This process had its intellectual, practical and social aspects. Science is about discovery, interpreting nature, learning more about the world and how it works, through observation, experiment and reasoning to the best available explanation. In the long nineteenth century, an enormous amount of knowledge was gained. Logical positivists would have us pipe down if we have questions science cannot answer, but for most of us there is more to life than that, as there was for Huxley and Balfour. Science is not just about insatiable curiosity, childishly asking questions when grown-ups have shifted their attention to important matters like sex, status and money. It is important because it is practical, its usefulness something that became manifest at last in the nineteenth century, when applied science became a reality. Earlier, devices such as cannons, waterwheels and steam engines had promoted science, as Galileo, John Smeaton (1724–94), Lazare and Sadi Carnot, and Watt[53] pondered about how they worked. But in the nineteenth century, especially in the chemical and electrical industries, science preceded technology. Society was duly transformed: the long nineteenth century began with turnpikes, horses, picks and shovels, low levels of literacy, pikes and muskets, and heroic medicine and surgery; and ended with railways, motor cars, steam shovels, cheap postage, photography, telegraphs, telephones and wireless, universal schooling, modern weapons and hospitals.

The able and socially mobile people who were attracted into science would no doubt in earlier times have become clergy or judges, but science is not just a business of geniuses making discoveries and engineers applying them. Chemistry, a science not always given its due, exemplifies this particularly collective, intellectual and utilitarian enterprise. Science is a social activity, carried on by a group, sometimes indeed a chorus supporting a soloist or two, but by the nineteenth century it had developed into something much bigger than that. Its growth required people to convene meetings, exchange information, hear, criticize and publish papers and books; to raise money and consciousnes; network, debate, arrange conferences and exhibitions; administer museums and laboratories; and spend time in the field and the laboratory, sometimes in danger of their lives. The growth of a 'Church scientific', a major cultural force, was one of the major features of the long nineteenth century, and should be the prime focus of any study of its history. It is a complex story, and different people will seek different threads to guide them through the labyrinth. Some may think they have found a royal road that takes them straight through without diversions and wrong turnings. But we should recall the wise advice: 'when things begin to look clear historically, odds are you are missing something'.[54]

TIMELINE

1789 Lavoisier, *Elements*; A. de Jussieu, *Genera plantarum*; E. Darwin, *Loves of the Plants*; Bastille falls in French Revolution; George Washington inaugurated.

1790 First steam-rolling mill.

1791 Priestley's house sacked; his ally Kirwan converted to oxygen theory.

1792 Priestley defends phlogiston at Hackney Academy; Arkwright dies.

1793 Dalton, *Meteorological Observations*; France and Britain at war; Board of Agriculture established; Parisian medical schools reorganized; Pinel at the Bicêtre Hospital; Paris, Museum of Natural History.

1794 École Polytechnique; execution of Lavoisier; Priestley flees to USA.

1795 Bramah's hydraulic press; London Missionary Society set up.

1796 York Retreat.

1797 Bichat's lectures, and Proust's analyses, begun.

1798 *Philosophical Magazine*; Malthus, *Essay*; Coleridge and Wordsworth, *Lyrical Ballads*; Jenner begins vaccinating; Rumford, on heat.

1799 Schleiermacher, *On Religion*; metric system; laughing gas; Royal Institution; Volta, electric battery.

1800 College of Surgeons; Baudin to Australia, W. Herschel, infra-red.

1801 Britain and Ireland united; Flinders and Brown to Australia; Jacquard loom; Ritter, ultra-violet; Young, wave theory of light.

1802 Paley, *Natural Theology*; Peace of Amiens; Louisiana purchase; gas lighting; Davy's inaugural lecture.

1803 E. Darwin, *Temple of Nature*; Berthollet, *Chemical Statics*; Napoleon, emperor; war resumes.

1804 Gay-Lussac, balloon ascent to 23,00 feet; British and Foreign Bible Society.

1805 Percival, code of medical ethics; Battle of Trafalgar.

1806 Marcet, *Conversations on Chemistry*; continental blockade; Britain seizes Cape colony.

1807 Gmelin, *Handbook of Chemistry*; Davy, potassium; Geological Society of London; slave trade abolished.

1808 Dalton, *New system, I*; Jenner, *Memoir*; Gay-Lussac, gas laws.

1809 Lamarck, evolutionary theory; Charles Darwin born.

1810 Davy, 'chlorine'; Hahnemann, homeopathy; University of Berlin.

1811 Avogadro's hypothesis; Cuvier on fossils of Paris; Luddite riots; Regency.

1812 Laplace, probabilities; US enters war; Napoleon retreats from Moscow.

1813 *Annals of Philosophy*; Davy and Faraday to France; iodine.

1814 Wollaston, chemical equivalents.

1815 Spurzheim, phrenology (2nd edn); Prout's hypothesis; Raffles founds Singapore; safety lamp for coal miners; Apothecaries Act; Battle of Waterloo.

1816 Post-war depression: Spa Fields riots.

1817 Parkinson, 'shaking palsy'.

1818 Silliman's *American Journal of Science*; Institute of Civil Engineers; Mary Shelley, *Frankenstein*.

1819 Berzelius on chemical proportions; Lawrence's lectures; Peterloo massacre.

1820 Accum, *Coal gas*; Ørsted, electromagnetism; death of Joseph Banks.

1821 Bell on nerves; Fraunhofer lines.

1822 Fourier, equations for heat flow; Magendie on nerves; Mitscherlich on isomorphism; Congress of Vienna.

1823 Mechanics' Institutes, Glasgow and London.

1824 (Royal) Society for Prevention of Cruelty to Animals; death of Byron.

1825 Liebig to Giessen; Faraday, benzene; Stockton and Darlington railway.

1826 Ohm's law of electricity; London Zoo; Society for Diffusion of Useful Knowledge.

1827 Audubon's *Birds of America* (to 1838); Walker's friction matches.

1828 Wohler, urea synthesis; Thomas Thomson and Berzelius quarrel; University College London, founded.

1829 Burke and Hare murder in order to dissect.

1830 Herschel's *Preliminary Discourse*; Lyell's *Principles*; Babbage, *Decline of Science*; Liverpool and Manchester railway; revolution in France; Comte, positivism.

1831 British Association for the Advancement of Science; HMS *Beagle* sails; cholera reaches Britain.

1832 Lear on parrots; Geological Survey; Liebig's *Annalen*; Durham University; Reform Bill.

1833 *Bridgewater Treatises* and *Tracts for the Times* begin.

1834 Somerville, *Connexion of the Physical Sciences*; Babbage's difference engine; slavery abolished in British empire; New Poor Law (workhouses).

1835 Académie des Sciences, *Comptes rendus*; Struve to new Pulkova observatory; Quetelet on statistics.

1836 HMS *Beagle* returns.

1837 Whewell, *History of the Inductive Sciences*; Victoria becomes Queen; registration of births, deaths and marriages in UK.

1838 *Annals of Natural History*; Brunel's *Great Western* steamship crosses Atlantic.

1839 Murchison, *Silurian System*; Schwann, cell theory; Ross's magnetic voyage; Daguerre and Fox Talbot, photography.

1840 Kane, *Elements of Chemistry*; Liebig, agricultural chemistry; Hess, law of heat change in chemical reactions; Victoria and Albert married; penny post.

1842 Chadwick, *Report*; Mayer on heat and work; Nasmyth's steam-hammer.

1843 Doppler on sound; Lawes, superphosphate; Scottish Church disruption.

1844 Chambers, *Vestiges*; Morse's telegraph line from Baltimore to Washington.

1845 Humboldt begins *Cosmos*; Adams and Leverrier on Uranus; *Scientific American*.

1846 Famine in Ireland; corn laws repealed; railway mania; ether anaesthesia; chemistry lab at UCL; Smithsonian inaugurated under Joseph Henry.

1847 Lectures by Joule and Helmholtz on energy; Semmelweiss on puerperal fever; chloroform anaesthesia; railway time; ten-hour factory act; Armstrong, Elswick works.

1848 Smithsonian, *Contributions to Knowledge*; Owen, on archetype and homologies; revolutions all over Europe; American Association for the Advancement of Science.

1849 Herschel, ed., *Admiralty Manual of Scientific Enquiry*; Elizabeth Blackwell qualifies as a doctor in USA; Fizeau, velocity of light;

1850 Williamson on alcohols and ethers; Tennyson, *In Memoriam*.

1851 Great Exhibition, Crystal Palace; Helmholtz's ophthalmoscope; Owen's College, Manchester.

1852 Napoleon III emperor; Perry's squadron opens up Japan.

1853 New York international exhibition; Whewell and Brewster on plurality of worlds; Snow gives Queen Victoria chloroform.

1854 Laurent, *Chemical Method*; Spencer, *Principles of Biology*; Snow closes Broad Street pump (cholera); Crimean War – first war correspondent, and photographer, Florence Nightingale, Mary Seacole.

1855 *Pacific Railroad Reports*; Berthollet on total synthesis; Faraday magnetizes light; Wilson Professor of Technology; London's Great Stink.

1856 Bessemer steel.

1857 Perkin's mauve; Indian Uprising (Mutiny).

1858 Gray, *Anatomy*; Virchow, *Cellular Pathology*; Kekulé on structural chemistry; Societé chimique de France; Atlantic cable.

1859 Darwin, *Origin of Species*; Whitworth, steel artillery.

1860 Karlsruhe conference; Bunsen and Kirchhoff on spectra; *Chemical News*; *Essays and Reviews*.

1861 MIT founded; US Civil War; Italy united; HMS *Warrior*; death of Prince Albert.

1862 Nightingale Nursing School; Thames Embankment begun; Gatling gun; Krupp adopts Bessemer process at Essen.

1863 Huxley, *Man's Place in Nature*; Lyell, *Antiquity of Man*; Helmholtz, *Sensations of Tone*; Alkali Act; Solvay soda works.

1864 *Quarterly Journal of Science*; London, underground Metropolitan railway; Guldberg and Waage on mass action.

1865 Bernard, *Experimental Medicine*; Mendel on peas; Hooker director of Kew; Hofmann's molecular models; Lister, antisepsis; Sprengel, vacuum pump; Lincoln assassinated.

1866 Boltzmann on second law of thermodynamics; frozen mammoth found in Siberia.

1867 Second Reform Bill; USA buys Alaska from Russians.

1868 US National Academy of Sciences.

1869 Galton, *Hereditary Genius*; alizarin synthesis; Suez Canal; *Nature*; Mendeleev's Periodic Table.

1870 Franco–Prussian War, and siege of Paris; Devonshire Commission.

1871 Darwin, *Descent of Man*; Chemical Society splits; Stanley finds Livingstone.

1872 Darwin, *Expression of Emotions*; HMS *Challenger*; *Popular Science Monthly*.

1873 Maxwell, *Treatise on Electricity and Magnetism*; Wundt, *Outlines of Psychology*; Carnegie begins steel-making.

1874 Tyndall, *Belfast Address*; Lockyer, *Solar Physics*; van't Hoff and le Bel on molecular structure; Johns Hopkins University; first Impressionist exhibition.

1875 Draper, *Conflict between Religion and Science*; London Medical School for Women; London main drains completed, population self-sustaining.

1876 Wallace, *Geographical Distribution of Animals*; Bell, telephone.

1877 Edison, phonograph.

1878 Gilchrist-Thomas steel process.

1879 Tay Bridge disaster; London telephone exchange; Edison and Swan, light bulbs; full public admission to British Museum.

1880 Marsh on extinct toothed birds.

1881 Helmholtz, Faraday memorial lecture; Natural History Museum, London.

1882 Rowland, diffraction gratings; Society for Psychical Research; death of Darwin.

1883 Mach, *Science of Mechanics*; W. Thomson on sizes of atoms; Arrhenius on electrolytes.

1884 BAAS in Montreal, and Johns Hopkins seminars; Emil Fischer on sugars; van't Hoff on chemical dynamics; Balmer on spectra; Maxim gun.

1885 Pasteur, vaccine for rabies.

1886 Stevenson, *Jekyll and Hyde*; Anglo-German agreement on East Africa.

1887   Michelson and Morley's experiment; Hertz detects radio waves.

1888   Paris, Pasteur Institute; Dunlop invents pneumatic tyre for bicycles.

1889   Paris Exhibition, Eiffel Tower; Benz exhibits motor car there.

1890   James, *Principles of Psychology*; Rydberg on spectra.

1892   Pearson, *Grammar of Science*.

1893   Armstrong and Whitworth amalgamation.

1894   Huxley, *Evolution and Ethics*; Balfour, *Foundations of Belief.*

1895   Wells, *Time Machine*; argon discovered; Röntgen, X-rays.

1896   Becquerel, radioactivity; Sunday opening of British national museums; Marconi, wireless telegraphy.

1897   J.J. Thomson, electrons; Ross on malaria; Diesel engine.

1898   Curies isolate radium and polonium; Liverpool School of Tropical Medicine.

1899   London School of Hygiene and Tropical Medicine (Hamburg, 1900; Marseilles, 1907).

1900   Paris Exposition; Planck, quantum theory; Mendel's work 'rediscovered'.

1901   De Vries, mutation theory; death of Queen Victoria.

1902   Bateson, 'genetics'; Pavlov, conditioned reflexes.

1903   Rutherford, 'new alchemy; Wright brothers' flight.

1904   Merz, *European Thought in the Nineteenth Century*; motor taxi-cabs; Russo–Japanese War.

1905   Einstein, quantum theory, atoms (Brownian motion) and special relativity.

1906   HMS *Dreadnought.*

1907   Imperial College, London.

1908   Morgan, genetics of fruit flies.

1909   Millikan, electrons; Perrin, Brownian motion; Blériot flies the Channel; Ford Model-T; Bakelite.

1910   Ehrlich, salvarsan for syphilis.

1911   Rutherford, nuclear atom; Marie Curie, second Nobel Prize; Kaiser Wilhelm Institutes; Amundsen, South Pole.

1912   Braggs on X-ray crystallography; Piltdown Man; *Titanic* sinks; Bohr joins Rutherford in Manchester.

1913   Moseley on X-ray spectra; Soddy, 'isotopes'; Haber-Bosch ammonia process.

1914   Huxley, on courtship of grebes; BAAS in Melbourne; world war.

NOTES AND REFERENCES

Preface: The Age of Science

1 J. Issitt, *Jeremiah Joyce: Radical, Dissenter and Writer*, Aldershot: Ashgate, 2006.

2 W.H. Brock, *Justus von Liebig: The Chemical Gatekeeper*, Cambridge: Cambridge University Press, 1997.

3 W. Clark, *Academic Charisma and the Origins of the Research University*, Chicago, IL: Chicago University Press, 2006; D.M. Knight and H. Kragh, eds, *The Making of the Chemist*, Cambridge: Cambridge University Press, 1998.

4 K.T. Hoppen, *The Mid-Victorian Generation, 1846–1886*, Oxford: Oxford University Press, 1998, p. 598.

5 Ibn Khaldûn, *The Muqaddimah: An Introduction to History*, tr. F. Rosenthal, London: Routledge, 1958, vol. 1, pp. 59–60 and *passim*.

6 J. Molony, *The Native-Born: The First White Australians*, Melbourne: Melbourne University Press, 2000.

7 T.S. Kuhn, 'The Function of Dogma in Scientific Research', in A.C. Crombie, ed., *Scientific Change*, London: Heinemann, 1963, pp. 347–69.

8 J.H. Brooke and G. Cantor, *Reconstructing Nature: The Engagement Between Science and Religion*, Edinburgh: T. & T.Clark, 1998; D. Lindberg and R. Numbers, eds, *Where Science and Christianity Meet*, Chicago, IL: Chicago University Press, 2003; D.M. Knight, *Science and Spirituality: The Volatile Connection*, London: Routledge, 2004; D.M.

Knight and M.D. Eddy, eds, *Science and Beliefs: From Natural Philosophy to Natural Science*, Aldershot: Ashgate, 2005.

9   M. Shelley, *Frankenstein* [1818], ed. D.L. Macdonald and K. Scherf, 2nd edn, Peterborough, Ontario: Broadview, 1999.

10 B. Bensaude-Vincent and C. Blondel, eds, *Science and Spectacle in the European Enlightenment*, Aldershot: Ashgate, 2008, p. 2; D.M. Knight, *Public Understanding of Science; A History of Communicating Scientific Ideas*, London: Routledge, 2006.

11 M. Fichman, *An Elusive Victorian: The Evolution of Alfred Russel Wallace*, Chicago, IL: Chicago University Press, 2004; A. Owen, *The Place of Enchantment: British Occultism and the Cult of the Modern*, Chicago, IL: Chicago University Press, 2004.

12 D. Pick, *Faces of Degeneration: A European Disorder, c.1848–c.1918*, Cambridge: Cambridge University Press, 1989.

13 A. Tennyson, *In Memoriam*, ed. S. Shatto and M. Shaw, Oxford: Oxford University Press, 1982, p. 115 (section 97).

14 J.F.W. Herschel, *Essays from the Edinburgh and Quarterly Reviews, with Addresses and Other Pieces*, London: Longman, 1857, p. 737.

Introduction: Approaching the Past

1   B.T. Moran, *Distilling Knowledge: Alchemy, Chemistry and the Scientific Revolution*, Cambridge, MA: Harvard University Press, 2005, p. 65.

2   A. Lundgren and B. Bensaude-Vincent, eds, *Communicating Chemistry: Textbooks and Their Audiences*, Canton, MA: Science History Publications, 2000.

3   I. Hargitai, *Candid Science: Conversations with Famous Chemists*, London: Imperial College, 2000.

4   J. Priestley, *The History and Present State of Electricity* [1775], intr. R.E. Schofield, New York: Johnson, 1966; *A Scientific Autobiography*, ed. R.E. Schofield, Cambridge, MA: MIT Press, 1966; M.E. Bowden and L. Rosner, eds, *Joseph Priestley: Radical Thinker*, Philadelphia, PA: Chemical Heritage, 2005.

5   W. Paley, *Natural Theology* [1802], ed. M.D. Eddy and D.M. Knight, Oxford: Oxford University Press, 2006.

6   C.C. Gillispie, *The Edge of Objectivity: An Essay in the History of Scientific Ideas*, Princeton, NJ: Princeton University Press, 1960.

7   H. Davy, *Collected Works*, London: Smith, Elder, 1839–40, vol. 7, p. 15.

8   S.A. Smith and A. Knight, eds, *The Religion of Fools? Superstition Past and*

Present, Oxford: Oxford University Press, *Past and Present* Supplement 3, 2008.

9 C.J. Fox, *A History of the Early Part of the Reign of James the Second*, London: Miller, 1808.

10 H. Butterfield, *The Whig Interpretation of History*, London: Bell, 1931.

11 See the essays on historiography in the *TLS*, 13 October 2006; H. Kragh, *An Introduction to the Historiography of Science*, Cambridge: Cambridge University Press, 1987.

12 H. Butterfield, *The Origins of Modern Science, 1300–1800*, London: Bell, 1949.

13 A.R. Hall, *The Scientific Revolution, 1500–1800: The Formation of the Modern Scientific Attitude*, London: Longman, 1954; revised as *The Revolution in Science, 155–1750*, 1983; D.C. Lindberg and R.S. Westman, *Reappraisals of the Scientific Revolution*, Cambridge: Cambridge University Press, 1990.

14 C. Chimisso, *Writing the History of the Mind: Philosophy and Science in France, 1900–1960s*, Aldershot: Ashgate, 2008.

15 A. Koyré, *From the Closed World to the Infinite Universe*, Baltimore: Johns Hopkins University Press, 1957.

16 W. Whewell, *History of the Inductive Sciences*, 3rd edn, London: Parker, 1857.

17 A.G. Debus, *Chemistry and Medical Debate: Van Helmont to Boerhaave*, Nantucket, MA: 2001; B.T. Moran, *Distilling Knowledge*, Cambridge, MA: Harvard University Press, 2005, pp. 157–81.

18 K. Thomas, *Religion and the Decline of Magic*, London: Weidenfeld and Nicolson, 1971; O. Mayr, *Authority, Liberty & Automatic Machinery in Early Modern Europe*, Baltimore, MD: Johns Hopkins University Press, 1986; D. Freedberg, *The Eye of the Lynx: Galileo, His Friends, and the Beginning of Modern Natural History*, Chicago, IL: Chicago University Press, 2002; A.G. Debus, ed., *Alchemy and Early Modern Chemistry*, London: Society for the History of Alchemy and Chemistry, 2004; O. Hannaway, *The Chemists and the Word*, Baltimore, MD: Johns Hopkins University Press, 1975; W.R. Newman, *Promethean Ambitions: Alchemy and the Quest to Perfect Nature*, Chicago, IL: Chicago University Press, 2004; L.M. Principe and A. Grafton, eds, *Transmutations: Alchemy in Art*, Philadelphia, PA: Chemical Heritage, 2002; L. Abraham, *A Dictionary of Alchemical Imagery*, Cambridge: Cambridge University Press, 1998.

19 T.S. Kuhn, *The Structure of Scientific Revolutions*, 2nd edn, Chicago,

IL: Chicago University Press, 1970; I.B. Cohen, *Revolutions in Science*, Cambridge, MA: Harvard University Press, 1989.

20 A. Hallam, *A Revolution in the Earth Sciences: From Continental Drift to Plate Tectonics*, Oxford: Oxford University Press, 1973; W. Shea, ed., *Revolutions in Science: Their Meaning and Relevance*, Canton, MA: Science History, 1988.

21 J. Morrell and A. Thackray, *Gentlemen of Science: Early Years of the British Association for the Advancement of Science*, Oxford: Oxford University Press, 1981; M.P. Crosland, *Science under Control: The French Academy of Sciences, 1795–1914*, Cambridge: Cambridge University Press, 1992.

22 K. Popper, *Conjectures and Refutations: The Growth of Scientific Knowledge*, London: Routledge, 1963; *The Logic of Scientific Discovery*, London: Hutchinson, 1972.

23 J.C. Thackray, *To See the Fellows Fight: Eye Witness Accounts of Meetings of the Geological Society of London and its Club, 1822–1868*, BSHS Monograph 12, 1999; 24 I. Lakatos, *The Methodology of Scientific Research Programmes*, ed. J. Worrall and G. Currie, Cambridge: Cambridge University Press, 1978, pp. 102–38.

25 T.R. Wright, *The Religion of Humanity: The Impact of Comtean Positivism on Victorian Britain*, Cambridge: Cambridge University Press, 1986; T. Dixon, 'The Invention of Altruism: Auguste Comte's *Positive Polity* and Respectable Unbelief in Victorian Britain', in D.M. Knight and M.D. Eddy, *Science and Beliefs*, Aldershot: Ashgate, 2005, pp. 195–211.

26 The British Society for the History of Science organized a conference to commemorate this on the seventy-fifth anniversary, in September 2006.

27 M. Berman, *Social Change and Scientific Organisation: The Royal Institution, 1799–1844*, London: Heinemann, 1978; contrast F.A.J.L. James, ed., *The Common Purposes of Life: Science and Society at the Royal Institution*, Aldershot: Ashgate, 2002.

28 D.M. Knight, *Sources for the History of Science, 660–1914*, London: Sources of History (later Cambridge, Cambridge University Press), 1975, p. 26; Japanese translation by Hazime Kasiwagi, Tokyo: Uchida Rokakuhu, 1984.

29 M. Shortland and R. Yeo, eds, *Telling Lives in Science; Essays on Scientific Biography*, Cambridge: Cambridge University Press, 1996; T. Söderqvist, *The Poetics of Biography in Science, Technology and Medicine*, Aldershot: Ashgate, 2007.

30 D. Sobel, *Longitude*, London; Fourth Estate, 1996.

31 M. Fichman, *An Elusive Victorian: The Evolution of Alfred Russel Wallace*, Chicago, IL: Chicago University Press, 2004; J. Morrell, *John Phillips and the Business of Victorian Science*, Aldershot: Ashgate, 2005.

32 A.F. Corcos and F.V. Monaghan, *Gregor Mendel's Experiments on Plant Hybrids: A Guided Study*, New Brunswick, NJ: Rutgers University Press, 1993.

33 A. Desmond, *Huxley*, London: Michael Joseph, 2 vols, 1994–7; P. White, *Thomas Huxley: Making the 'Man of Science'*, Cambridge: Cambridge University Press, 2003.

34 D.M. Knight, *Ideas in Chemistry: A History of the Science*, 2nd edn, London: Athlone, 1995.

35 J. Endersby, 'In the Bag', *TLS*, 13 July 2008, p. 12.

36 F. Burckhardt et al., eds, *The Correspondence of Charles Darwin*, Cambridge: Cambridge University Press, 1985–(in progress).

37 A.T. Gage and W.T. Stearn, *A Bicentenary History of the Linnean Society of London*, London: Academic Press, 1988; G.L. Herries-Davies, *Whatever is Under the Earth: The Geological Society of London, 1807 to 2007*, London: Geological Society, 2007.

38 L. Henson et al., eds, *Culture and Science in the Nineteenth-century Media*, Aldershot: Ashgate, 2004; G. Cantor and S. Shuttleworth, eds, *Science Serialised: Representations of the Sciences in Nineteenth-century Periodicals*, Cambridge, MA: MIT Press, 2004; G. Cantor et al., *Science in the Nineteenth-century Periodical*, Cambridge: Cambridge University Press, 2004; D.M. Knight, 'Snippets of Science', *Studies in History and Philosophy of Science* 36 (2005): 618–25.

39 J. Schummer, B. Bensaude-Vincent and B. Van Tiggelen, eds, *The Public Image of Chemistry*, London: World Scientific, 2007; D.M. Knight, *Public Understanding of Science: A History of Communicating Scientific Ideas*, London: Routledge, 2006.

40 S. Forgan and G. Gooday, 'Constructing South Kensington: The Buildings and Politics of T.H. Huxley's Working Environments', *BJHS*, 29 (1996): 435–68.

41 K. Hufbauer, *The Formation of the German Chemical Community, 1720–1795*, Berkeley, CA; California University Press, 1982; M.P. Crosland, *In the Shadow of Lavoisier: The* Annales de chimie *and the Establishment of a New Science*, Chalfont St Giles: BSHS, 1994; A. Bandinelli, '*Annales de Chimie* vs. *Observations sur la Physique/Journal de Physique, 1789–1803*: Scientific Communication during a Major Change in the Approach to Empirical Research', forthcoming in *Ambix*, 2009.

42 W.H. Brock and A.J. Meadows, *The Lamp of Learning: Taylor and Francis and the Development of Science Publishing*, London: Taylor and Francis, 1984.

43 D.M. Knight and H. Kragh, eds, *The Making of the Chemist: The Social History of Chemistry in Europe*, Cambridge: Cambridge University Press, 1998; D.M. Knight, 'Science and Culture in Mid-Victorian Britain: The Reviews, and William Crookes' *Quarterly Journal of Science*', *Nuncius* 11 (1996): 43–54; D.M. Knight, *Natural Science Books in English*, London: Batsford, 1972, pp. 190–232.

44 *The Magazine of Natural History* 1 (1828); and see R.B. Freeman, *British Natural History Books, 1495–1900: A Handlist*, Folkestone: Dawson, 1980.

45 W. St Clair, *The Reading Nation in the Romantic Period*, Cambridge: Cambridge University Press, 2004; A. Johns, *The Nature of the Book*, Chicago, IL: Chicago University Press, 1998; M. Frasca-Spada and N. Jardine, eds, *Books and the Sciences in History*, Cambridge: Cambridge University Press, 2000.

46 D.M. Knight, 'The Spiritual in the Material', in R.M. Brain, R.S. Cohen and O. Knudsen, eds, *Hans Christian Ørsted and the Romantic Legacy in Science*, Dordrecht: Springer, 2007, pp. 417–32.

47 H. Davy, *Consolations in Travel: Or the Last Days of a Philosopher*, London: John Murray, 1830, p. 20.

48 But see J.V. Pickstone, 'Working Knowledges Before and After circa 1800: Practices and Disciplines in the History of Science, Technology and Medicine, *Isis* 98 (2007): 489–516.

49 Good recent examples are P. Ball, *Elegant Solutions: Ten Beautiful Experiments in Chemistry*, London: Royal Society of Chemistry, 2005; W. Gratzer, *Terrors of the Table: The Curious History of Nutrition*, Oxford: Oxford University Press, 2005; R. Panek, *The Invisible Century: Einstein, Freud and the Search for Hidden Universes*, London: Fourth Estate, 2005; P. Pesic, *Sky in a Bottle*, Cambridge, MA: MIT Press, 2005.

50 A.D. Morrison-Low, *Making Scientific Instruments in the Industrial Revolution*, Aldershot: Ashgate, 2007; R.J. Richards, *The Romantic Conception of Life: Science and Philosophy in the Age of Goethe*, Chicago, IL: Chicago University Press, 2002; J. Secord, *Victorian Sensation: The Extraordinary Publication, Reception, and Secret Authorship of* Vestiges of the Natural History of Creation, Chicago, IL: Chicago University Press, 2000.

51 N. Jardine, J.A. Secord and E.C. Spary, eds, *Cultures of Natural History*, Cambridge: Cambridge University Press, 1996.

52 For example, C.A. Russell and G.K. Roberts, *Chemical History: Reviews of Recent Literature*, London: RSC, 2005.

53 J.L. Heilbron, ed., *The Oxford Companion to the History of Modern Science*, Oxford: Oxford University Press, 2003; D.M. Knight, *A Companion to the Physical Sciences*, London: Routledge, 1989; P. Clayton with Z.Simpson, eds, *The Oxford Handbook of Religion and Science*, Oxford: Oxford University Press, 2006.

54 E.J. Browne, W.F. Bynum and R. Porter, eds, *Dictionary of the History of Science*, London, 1981.

55 C.C. Gillispie et al., eds, *Dictionary of Scientific Biography*, New York: Scribner, 1970– (continuing in Supplements); B. Lightman et al., eds, *Dictionary of Nineteenth-century British Scientists*, Bristol: Thoemmes, 2004; a good one-volume one is T.I. Williams, ed., *A Biographical Dictionary of Scientists*, London: A. & C. Black, 1969.

56 F.A.J.L. James, ed., *The Correspondence of Michael Faraday*, London: Institution of Electrical Engineers, 1991–; F. Burckhardt et al., eds, *The Correspondence of Charles Darwin*, Cambridge: Cambridge University Press, 1985 –.

57 M. Beretta, C. Pogliano and P. Redondi, eds, *Journals and the History of Science*, Florence: Olschki, 1999.

Chapter 1   Science in and after 1789

1 A.J. Balfour, 'Presidential Address', *Report of the British Association*, Cambridge, 1904, pp. 3–14.

2 J.T. Merz, *A History of European Thought in the Nineteenth Century* [1904–12], 4 vols; reprint New York: Dover, 1965, vol. 1, pp. xi, 10–14.

3 R. Yeo, *Defining Science: William Whewell, Natural Knowledge and Public Debate in Early Victorian Britain*, Cambridge: Cambridge University Press, 1993.

4 W. Whewell, *History of the Inductive Sciences*, 3 vols, 3rd edn, London: Parker, 1857.

5 J.T. Merz, *A History of European Thought in the Nineteenth Century*, New York: Dover, 1965, vol. 1, p. 7.

6 J. Ravetz, *The No-nonsense Guide to Science*, Oxford: New Internationalist, n.d.; H. Collins and T. Pinch, *The Golem: What Everyone Should Know about Science*, Cambridge: Cambridge University Press, 1993.

7   A. Desmond and J. Moore, *Darwin*, London: Penguin, 1992; A. Desmond, J. Moore and J. Browne, *Charles Darwin*, Oxford: Oxford University Press, 2007.

8   R. Porter, *Flesh in the Age of Reason*, London: Penguin. 2003.

9   M. Shortland and R.Yeo, eds, *Telling Lives in Science: Essays on Scientific Biography*, Cambridge: Cambridge University Press, 1996; T. Söderqvist, ed., *The Poetics of Scientific Biography*, Aldershot: Ashgate, 2007.

10 C.C. Gillispie et al., eds, *Dictionary of Scientific Biography*, New York: Gale, 1970 – (continuing in supplements); B. Lightman, ed., *Dictionary of Nineteenth-century British Scientists*, Bristol: Thoemmes, 2004.

11 M.W. Rossiter, 'A Twisted Tale: Women in the Physical Sciences', and T. Shinn, 'The Industry, Research, and Education Nexus', in M.J. Nye, ed., *Cambridge History of Science*, Cambridge: Cambridge University Press, 2003, vol. 5, pp. 54–71, 133–53.

12 N.A.M. Rodger, *The Command of the Ocean: A Naval History of Britain, 1649–1815*, London: Penguin, 2004.

13 R. Yeo, 'Encyclopedias', in J.L. Heilbron, ed., *The Oxford Companion to the History of Modern Science*, Oxford: Oxford University Press, 2003, pp. 252–5.

14 C.C. Gillispie, ed., *A Diderot Pictorial Encyclopedia of Trades and Industry*, New York: Dover, 1959.

15 P. Thiry, Baron d'Holbach, *The System of Nature*, vol. 1, tr. H.H. Robinson, intr. M. Bush, Manchester: Cinamen, 1999.

16 J.R. Hofmann, *André-Marie Ampère: Enlightenment and Electrodynamics*, Cambridge: Cambridge University Press, 1995, pp. 356–65; M.P. Crosland, *Science under Control: The French Academy of Sciences, 1795–1914*, Cambridge: Cambridge University Press, 1995, pp. 192–202; L. de Broglie, in P. Duhem, *The Aim and Structure of Physical Theory*, tr. P.P. Wiener, New York: Athenaeum, 1962, p. xii.

17 J.L. Heilbron, ed., *The Oxford Companion to the History of Modern Science*, Oxford: Oxford University Press, 2003, pp. 1–5.

18 R. Fox and G. Weisz, ed., *The Organization of Science and Technology in Franmce, 1808–1914*, Cambridge: Cambridge University Press, 1980.

19 M.P. Crosland, *Science under Control: The French Academy of Sciences, 1795–1914*, Cambridge: Cambridge University Press, 1992; on Poisson, see p. 132.

20 See the entries on Instruments, Metrology and Standardization in J.L. Heilbron, ed., *The Oxford Companion to the History of Modern Science*, Oxford: Oxford University Press, 2003, pp. 406–17, 520–1, 774–6.

21 J.P.F. Deleuze, *History and Description of the Royal Museum of Natural History*, Paris: Royer, 1823.

22 W.E. Bynum et al., *The Western Medical Trdition, 1800–2000*, Cambridge: Cambridge University Press, 2006, pp. 37–53.

23 W.M. Jacob, *The Clerical Profession in the Long Eighteenth Century, 1680–1840*, Oxford: Oxford University Press, 2007, pp. 95–112.

24 M.P. Crosland, *The Society of Arcueil: A View of French Science at the Time of Napoleon I*, London: Heinemann, 1967.

25 J.B.J. Delambre, *Rapport Historique sur les Progrés des Sciences Mathematiques*, and G. Cuvier, *Rapport Historique sur les Progrés des Sciences Naturelles*, Paris: Imprimerie Imperiale, 1810.

26 H.G. Alexander, ed., *The Leibniz–Clarke Correspondence*, Manchester: University Press, 1956.

27 R. Hahn, 'Laplace', in J.L. Heilbron, ed., *The Oxford Companion to the History of Modern Science*, Oxford: Oxford University Press, 2003, pp. 446–7.

28 R. Holmes, *The Age of Wonder: How the Romantic Generation Discovered the Beauty and Terror of Science*, London: Harper, 2008, pp. 60–124, 163–210.

29 R.L. Numbers, *Creation by Natural Law: Laplace's Nebular Hypothesis in American Thought*, Seattle: Washington University Press, 1977; J.A. Secord, *Victorian Sensation: The Extraordinary Publication, Reception, and Secret Authorship of* Vestiges of the Natural History of Creation, Chicago, IL: Chicago University Press, 2000, pp. 57–61.

30 T.M. Porter, 'Statistics and Physical Theories', in M.J. Nye, ed., *Cambridge History of Science*, Cambridge: Cambridge University Press, 2003, vol. 5.

31 D.M. Knight, *Public Understanding of Science*, London: Routledge, 2006, pp. 106–18.

32 J. Bonnemains, E. Forsyth and B. Smith, eds, *Baudin in Australian Waters*, Oxford: Oxford University Press, 1988.

33 J.C. Beaglehole, ed., *The Journals of Captain James Cook: The Voyage of the* Endeavour, Cambridge: Cambridge University Press, 1968, pp. cxxxviii–cxl, 22–34.

34 T. Thomson, 'Tables of Weights and Measures', *Annals of Philosophy*, 1 (1813): 452–7.

35 J.F.W. Herschel, *Popular Lectures on Scientific Subjects*, London: Strahan, 1873, pp. 419–51.

36 U. Klein and W. Lefèvre, *Materials in Eighteenth-century Science: A Historical Ontology*, Cambridge, MA: MIT, 2007.

37 A. Donovan, *Antoine Lavoisier*, Cambridge: Cambridge University Press, 1993; B. Bensaude-Vincent and F. Abbri, eds, *Lavoisier in European Context*, Canton, MA: Science History, 1995.

38 V. Boanza, 'The Phlogistic Role of Heat in the Chemical Revolution, and the Origins of Kirwan's "Ingenious Modifications" . . . into the Theory of Phlogiston', and G. Taylor, 'Marking out a Disciplinary Common Ground: The Role of Chemical Pedagogy in establishing the Doctrine of Affinity at the Heart of British Chemistry', *Annals of Science* 65 (2008): 309–38 and 465–86.

39 H. Davy, *Collected Works*, London: Smith, Elder, vol. 2, pp. 311–26.

40 A.L. Lavoisier, *Elements of Chemistry*, tr. R. Kerr, Edinburgh: Creech, 1790, pp. xxii–vi, 175–6.

41 H. Cavendish, *Electrical Researches*, ed. J. Clerk Maxwell, Cambridge: Cambridge University Press, 1879, pp. 310–20, 433.

42 *Register of Arts and Sciences* 1 (1824): 3–5; D.M. Knight, 'Scientific Lectures: A History of Performance', *Interdisciplinary Science Reviews* 27 (2002): 217–24; P. Bertucci and G. Pancaldi, eds, *Electric Bodies: Episodes in the History of Medical Electricity*, Bologna: University Press, 2001.

43 M. Shelley, *Frankenstein* [1818], ed. D.L. Macdonald and K. Schere, 2nd edn, Peterborough, Ontario: Broadview, 1999, p. 84.

44 D.M. Knight, 'Davy and the Placing of Potassium among the Elements', *Historical Group Occasional Paper 4: Second Wheeler Lecture*, London: Royal Society of Chemistry, 2007.

45 W.F. Bynum, 'William Lawrence', in B. Lightman, ed., *Dictionary of Nineteenth-century British Scientists*, Bristol: Thoemmes, 2004, pp. 1196–8.

46 J.R. Bertomeu-Sánchez and A. Nieto-Galan, eds, *Chemistry, Medicine and Crime: Mateu J.B. Orfila (1787–1853) and His Times*, Sagamore Beach: Science History, 2006.

47 C.L.E. Lewis and S.J. Knell, eds, *The World's First Geological Society*, London: Geological Society, 2009.

48 G. Cuvier, *Discours sur les Révolutions du Globe*, ed. Dr Hoefer, Paris: Firmin-Didot, 1877.

49 M.J.S. Rudwick, *Scenes from Deep Time*, Chicago, IL: Chicago University Press, 1992.

50 T. Fulford, ed., *Science and Romanticism*, London: Routledge, 2002; R. Holmes, *The Age of Wonder*, London: Harper Press, 2008.

51 R.J. Richards, *The Romantic Conception of Life: Science and Philosophy in the Age of Goethe*, Chicago, IL: Chicago University Press, 2002.

52 S. Rushton, *Shelley and Vitality*, London: Palgrave Macmillan, 2005.

53 S.T. Coleridge, *Hints Towards the Formation of a More Comprehensive Theory of Life*, ed. S.B. Watson, London: Churchill, 1848; R. Holmes, *Coleridge: Darker Reflections*, London: HarperCollins, 1998, pp. 478–80.

54 W. Wordsworth and S.T. Coleridge, *Lyrical Ballads* [1798], London: Penguin, 1999, pp. 1–25; J.L. Lowes, *The Road to Xanadu: A Study in the Ways of the Imagination*, Boston, MA: Houghton Mifflin, 1927.

55 R. Joppien and B. Smith, *The Art of Captain Cook's Voyages*, New Haven, CT: Yale University Press, 2 vols, 1985; B. Smith and A. Wheeler, *The Art of the First Fleet and Other Early Australian Drawings*, New Haven, CT: Yale University Press, 1988.

56 F. Klingender, *Art and the Industrial Revolution*, ed. A. Elton, London: Evelyn, Adams & Mackay, 1968.

57 W. Vaughan, *Friedrich*, London: Phaidon, 2004.

58 J. Hamilton, *Turner and the Scientists*, London: Tate Gallery, 1998.

59 R.E.R. Banks et al., eds, *Sir Joseph Banks: A Global Perspective*, Kew: Royal Botanic Garden, 1994.

60 N. Chambers, ed., *The Letters of Sir Joseph Banks: A Selection, 1768–1820*, London: Imperial College, 2000; *The Scientific Correspondence of Sir Joseph Banks, 1765–1820*, London: Pickering and Chatto, 2006.

61 J. Gascoigne, *Joseph Banks and the English Enlightenment: Useful Knowledge and Polite Culture*, and *Science in the Service of Empire: Joseph Banks, the British State and the Uses of Science in the Age of Revolution*, Cambridge: Cambridge University Press, 1994 and 1998.

62 W. Paley, *Natural Theology* [1802], ed. M.D. Eddy and D.M. Knight, Oxford: Oxford University Press, 2006.

63 R. Holmes, *The Age of Wonder*, London: Harper, 2008, p. xvi.

64 H. Rossotti, ed., *Chemistry in the Schoolroom, 1806: Selections from Mrs Marcet's* Conversations on Chemistry, Bloomington, IN: Authorhouse, 2006.

65 E. Darwin, *The Botanic Garden* [1789–91], Menton: Scolar, 1973; C.U.M. Smith, ed., *The Genius of Erasmus Darwin*, Aldershot: Ashgate, 2005; D.M. Knight, *Public Understanding of Science*, London: Routledge, 2006, pp. 44–6.

Chapter 2 Science and its Languages

1 T. Sprat, *The History of the Royal-Society of London*, London: Martyn, 1667, p. 113; R. Burchfield, *The English Language*, 2nd edn, London: Folio Society, 2006, p. 36.

2 K. Pearson, *The Grammar of Science*, 2nd edn, London: A. & C. Black, 1900.

3 A. Lundgren and B. Bensaude-Vincent, eds, *Communicating Chemistry: Textbooks and their Audiences*, Canton, MA: Science History, 2000.

4 J.H. Mason, *The Value of Creativity: The Origins and Emergence of a Modern Belief*, Aldershot: Ashgate, 2003, p. 144; D. Brown, *God and Mystery in Words: Experience through Metaphor and Drama*, Oxford: Oxford University Press, 2008, pp. 24–33.

5 F.M. Barnard, ed., *J.G. Herder on Social and Political Culture*, Cambridge: Cambridge University Press, 1969, pp. 176, 156, 27.

6 J.G. Herder, *Outlines of a Philosophy of the History of Man*, tr. T. Churchill, 2nd edn, London: Johnson, 1803, vol. 1, pp. 367, 417–38.

7 G. Radick, *The Simian Tongue: The Long Debate about Animal Language*, Chicago, IL: Chicago University Press, 2007.

8 B. Hilton, *A Mad, Bad & Dangerous People? England 1783–1846*, Oxford: Oxford University Press, 2006, p. 145.

9 G. Catlin, *Letters and Notes on the Manners, Customs, and Condition of the North American Indians* [1841], Minneapolis: Ross & Haines, 1965, vol. 2, pp. 259–61; P. Stokes 'Journal' [1827], in H.K. Beals et al., eds, *Four Travel Journals*, London: Hakluyt Society, 2008, p. 221.

10 P.A. Lanyon-Orgill, *Captain Cook's South Sea Island Vocabularies*, London: for the author, 1979.

11 F.M. Barnard, ed., *J.G. Herder on Social and Political Culture*, Cambridge: Cambridge University Press, 1969, pp. 17–32, 117–77.

12 W. von Humboldt, *On Language: The Diversity of Human Language-structure and its Influence on the Mental Development of Mankind*, tr. P. Heath, intr. H. Aarsleff, Cambridge: Cambridge University Press, 1988; P.R. Sweet, *Wilhelm von Humboldt: A Biography*, Columbus, OH: Ohio State University Press, 1978–80.

13 F.M. Müller, *Lectures on the Science of Language*, London: Longman, 1861–4.

14 D. Kennedy, *The Highly Civilized Man: Richard Burton and the Victorian World*, Cambridge, MA: Harvard University Press, 2005, pp. 32–57, 116, 135.

15 H.F. Augstein, ed., *Race: The Origins of an Idea, 1760–1850*, Bristol: Thoemmes, 1996.

16 L. Koerner, 'Carl Linnaeus in His Time and Place', in N. Jardine, J.A. Secord and E.C. Spary, *Cultures of Natural History*, Cambridge, Cambridge University Press, 1996, pp. 145–62; J.L. Heller and J.M. Penhallurick, eds, *Index of Books and Authors Cited in the Zoological Works of Linnaeus*, London: Ray Society, 2008.

17 A.T. Gage and W.T. Stearn, *A Bicentenary History of the Linnean Society of London*, London: Academic Press, 1988.

18 I. Newton, *Opticks*, 4th edn [1730]; reprint New York: Dover, 1952, p. 369.

19 E. Darwin, *The Botanic Garden*, London: Johnston, 1791, 1789; reprint Menston: Scolar, 1973, vol. 2, p. 6. Curiously, *The Loves of the Plants* was the second part, but published first (in 1789).

20 C.U.M. Smith, ed., *The Genius of Erasmus Darwin*, Aldershot: Ashgate, 2005.

21 J. Ruskin, *Proserpina: Studies of Wayside Flowers*, Orpington: Allen, 1879, p. 6.

22 W. Blunt, *In for a Penny: A Prospect of Kew Gardens*, London: Hamish Hamilton, 1978; R. Desmond, *Kew; A History*, London: Harvill, 1995.

23 Anon, *Poetry of the Anti-Jacobin*, London: Wright, 1799, pp. 108–41.

24 J.H. Mason, *The Value of Creativity*, Aldershot: Ashgate, 2003, pp. 157–78.

25 B. Hilton, *A Mad, Bad & Dangerous People? England 1783–1846*, Oxford: Oxford University Press, 2006, pp. 439–68.

26 B. Bensaude-Vincent, 'Languages in Chemistry', in M.J. Nye, ed., *Cambridge History of Science*, Cambridge: Cambridge University Press, 2003, vol. 5, pp. 176–90; A.L. Lavoisier, *Elements of Chemistry*, tr. R. Kerr [1790]; reprint in D.M. Knight, ed., *The Development of Chemistry*, London: Routledge, 1998, vol. 2.

27 M.P. Crosland, *In the Shadow of Lavoisier: The* Annales de Chimie *and the Establishment of a New Science*, Chalfont St Giles: BSHS, 1994.

28 B. Bensaude-Vincent and F. Abbri, eds, *Lavoisier in European Context*, Canton, MA: Science History, 1995.

29 J. Simon, *Chemistry, Pharmacy and Revolution*, Aldershot: Ashgate, 2004.

30 D.M. Knight, *Humphry Davy: Science and Power*, 2nd edn, Cambridge: Cambridge University Press, 1998, pp. 28–41.

31 T.S. Kuhn, *The Structure of Scientific Revolutions*, 2nd edn, Chicago, IL: Chicago University Press, 1970; P. Hoyningen-Huene, 'Thomas Kuhn and the Chemical Revolution', *Foundations of Chemistry* 10 (2008): 101–16.

32 S. Parkes, *The Chemical Catechism*, 3rd edn, London: Lackington Allen, 1808; W. Pinnock, *Catechism of Geology*, 7 parts, London: Whittaker, Teacher, 1831; R.B. Kinraid, in B. Lightman, ed., *Dictionary of Nineteenth-century British Scientists*, Bristol: Thoemmes, 2004, pp. 1602–3.

33 T.S. Kuhn, 'The Function of Dogma in Scientific Research', in A.C. Crombie, ed., *Scientific Change*, Oxford: Oxford University Press, 1963.

34 H. Chang and C. Jackson, eds, *An Element of Controversy: The Life of Chlorine in Science, Medicine, Technology, and War*, British Society for the History of Science Monographs 13, 2007; H. Davy et al., 'The Elementary History of Chlorine' and 'The Early History of Chlorine', Alembic Club Reprints 9 and 13, Edinburgh: Alembic Club, 1894, 1897.

35 A.L. Smyth, *John Dalton, 1766–1844: A Bibliography of Works by and about Him, with an Annotated List of His Surviving Apparatus and Personal Effects*, 2nd edn, Aldershot: Ashgate, n.d. [1998].

36 W.H. Wollaston, 'A Synoptic Scale of Chemical Equivalents', *Philosophical Transactions* 104 (1814): 1–22, includes a plate of a chemical slide-rule; D.M. Knight, *Atoms and Elements*, 2nd edn, London: Hutchinson, 1970, and ed., *Classical Scientific Papers: Chemistry*, London: Mills & Boon, 1968.

37 H. Davy, *Collected Works*, London: Smith, Elder, 1840, vol. 7, p. 97.

38 J. Morrell and A. Thackray, *Gentlemen of Science*, Oxford: Oxford University Press, 1981, pp. 175–86, 485–91.

39 Alembic Club Reprints, 2: *Foundations of the Atomic Theory*, and 4: *Foundations of the Molecular Theory*, Edinburgh: Livingstone, 1961.

40 T. Thomson, 'On Oxalic Acid', and W.H. Wollaston, 'On Superacid and Sub-acid Salts', *Philosophical Transactions* 98 (1808): 63–95, 96–102.

41 J. Dalton, *A New System of Chemical Philosophy*, Part 1, Manchester: Bickerstaff, 1808, pp. 211–20.

42 T. Thomson, *An Attempt to Establish the First Principles of Chemistry by Experiment*, London: Baldwin, Craddock, 1825.

43 [W.Prout], 'On the Relation between the Specific Gravities of Gaseous Bodies and the Weight of their Atoms', *Annals of Philosophy* 6 (1815):

321–30; 7 (1816): 111–13. These and other papers by Thomson and Berzelius alluded to are reprinted in facsimile in D.M. Knight, ed., *Classical Scientific Papers: Chemistry*, 2nd series, London: Mills & Boon, 1970, pp. 1–70; quotation from p. 48. W.H. Brock, *From Protyle to Proton: William Prout and the Nature of Matter*, Bristol: Hilger, 1985.

44 C. Babbage, *Reflections on the Decline of Science in England, and on Some of its Causes*, London: Fellowes, 1830, pp. 167–83; quotation from p. 182.

45 S. Carnot, *Réflextions sur la Puissance Motrice du Feu* [1824], ed. R. Fox, Paris: Vrin, 1978.

46 C. Smith and N. Wise, *Energy and Empire: A Biographical Study of Lord Kelvin*, Cambridge: Cambridge University Press, 1989, pp. 167, 294.

47 A. Comte, *The Positive Philosophy*, ed. and tr. H. Martineau, London: Chapman, 1853; T.R. Wright, *The Religion of Humanity: The Impact of Comtean Positivism on Victorian Britain*, Cambridge, Cambridge University Press, 1986.

48 *Cambridge Problems: Being a Collection of the Printed Questions Proposed to the Candidates for the Degree of B.A. at the General Examinations from 1801 to 1820 Inclusive*, Cambridge: Deighton, 1821, p. 338.

49 J.F.W. Herschel, *A Treatise on Astronomy*, new edn, London: Longman, 1851, p. 5.

50 C. Brock, 'The Public Worth of Mary Somerville', *BJHS* 39 (2006): 255–72.

51 W. Whewell, *Astronomy and General Physics Considered with Reference to Natural Theology*, London: Pickering, 5th edn, 1836, pp. 326–42; quotation from p. 335.

52 W. Whewell, *Philosophy of the Inductive Sciences*, new edn, London: Parker, 1847; J.F.W. Herschel, *A Preliminary Discourse on the Study of Natural Philosophy* [1830], New York: Johnson, 1966.

53 G.G. Stokes, *Mathematical and Physical Papers*, vol. 2, Cambridge: Cambridge University Press, 1883, p. 97.

54 B.S. Gower, 'The Metaphysics of Science in the Romantic Era', in D.M. Knight and M.D. Eddy, eds, *Science and Beliefs*, Aldershot: Ashgate, 2005, pp. 17–29; D.M. Knight, 'The Spiritual in the Material', in R.M. Brain, R.S. Cohen and O. Knudsen, eds, *Hans Christian Ørsted and the Romantic Legacy in Science*, Dordrecht: Springer, 2007, pp. 417–32.

55 M. Faraday, *Experimental Researches in Electricity*, London: Taylor & Francis, 1839–55, vol. 2, pp. 284–93; 1855, vol. 3, pp. 447–52.

56 B. Mahon, *The Man Who Changed Everything*, Chichester: Wiley, 2003,

pp. 56–65; B.J. Hunt, 'Electrical Theory and Practice', in M.J. Nye, ed., *Cambridge History of Science*, Cambridge: Cambridge University Press, 2003, vol. 5, pp. 311–27.

57 W.H. Brock, ed., *The Atomic Debates*, Leicester: Leicester University Press, 1967.

58 They still exist, in the Science Museum in London.

59 A.W. Hofmann, 'On the Combining Power of Atoms', *Proceedings of the Royal Institution* 4 (1862–6): 401–30; quotation from pp. 419, 430. This and many other papers are reprinted in D.M. Knight, ed., *The Development of Chemistry*, London: Routledge, 1998, vol. 1.

60 M. van der S.Trienke, 'Selling a Theory: The Role of Molecular Models in J.H. van't Hoff's Stereochemistry Theory', *Annals of Science* 63 (2006): 157–77; J.H. van't Hoff, *The Arrangement of Atoms in Space*, tr. A. Eiloart [2nd edn, 1898], in D.M. Knight, ed., *The Development of Chemistry 1789–1914*, London: Routledge, 1998, vol. 10.

61 J.R. Cribb, 'Miller', in B. Lightman, ed., *Dictionary of Nineteenth-century British Scientists*, Bristol: Thoemmes, 2004.

62 M. Beretta, *Imaging a Career in Science: The Iconography of Antoine Laurent Lavoisier*, Canton, MA: Science History, 2002.

63 A. Greenberg, *From Alchemy to Chemistry in Picture and Story*, Hoboken, NJ: Wiley, 2008.

64 L. Jordanova, *Defining Features: Scientific and Medical Portraits, 1660–2000*, London: Reaktion, 2000.

65 W.H. Wollaston, 'On the Apparent Direction of Eyes in a Portrait', *Phil. Trans.* 114 (1824): 247–56.

66 M. Andrews, *Landscape and Western Art*, Oxford: Oxford University Press, 1999, pp. 197–9; D.M. Knight, '"Exalting Understanding without Depressing Imagination": Depicting Chemical Process', *Hyle* 9 (2003): 171–89.

67 E.R. Tufte, *The Visual Display of Quantitative Information*, Cheshire, CN: Graphics Press, 1983.

68 A. Coats, *The Book of Flowers: Four Centuries of Flower Illustration*, London: Phaidon, 1973; D.M. Knight, *Zoological Illustration*, Folkestone: Dawson, 1977.

69 F.E. Beddard, *Animal Coloration: An Account of the Principal Facts and Theories relating to the Colours and Markings of Animals*, New York: Macmillan, 1892.

70 A.M. Lysaght, *The Book of Birds: Five Centuries of Bird Illustration*, London: Phaidon, 1975.

71 W. Feaver, 'Pictures are Books', *Guardian, Review*, 27 May 2006, p. 14.

72 W.T. Stearn, ed., *The Australian Flower Paintings of Ferdinand Bauer*, intr. W. Blunt, London: Basilisk, 1976; S. Sherwood and M. Rix, *Treasures of Botanical Art: Icons from the Shirley Sherwood and Kew Collections*, London: Royal Botanic Garden, Kew, 2008.

73 J. Chalmers, *Audubon in Edinburgh, and His Scottish Associates*, Edinburgh: National Museums of Scotland, 2003.

74 A. Garrett, *A History of British Wood Engraving*, Tunbridge Wells: Midas, 1974; J. Uglow, *Nature's Engraver: A Life of Thomas Bewick*, London: Faber, 2006.

75 R. Richardson, *The Making of Mr Gray's Anatomy*, Oxford: Oxford University Press, 2008, pp. 168–228; D.M. Knight, *Zoological Illustration; An Essay towards a History of Printed Zoological Pictures*, Folkestone: Dawson, 1977, pp. 38–73.

76 P.H. Barrett, D.J. Weinshank, and T.T. Gottlieber, *A Concordance to Darwin's* Origin of Species, *First Edition*, Ithaca, NY: Cornell University Press, 1981.

77 W. Paley, *Natural Theology*, ed. M.D. Eddy and D.M. Knight, Oxford: Oxford University Press, 2006.

78 D.M. Knight, *Science and Spirituality: The Volatile Connection*, London: Routledge, 2004, pp. 53–73; another version of this chapter is in *Nuncius* 15 (2000): pp. 639–64.

79 A. Ford, *James Ussher*, Oxford: Oxford University Press, 2007.

80 B. Lightman, *Victorian Popularizers of Science*, Chicago, IL: Chicago University Press, 2007, pp. 219–94.

81 H. Miller, *The Testimony of the Rocks*, Edinburgh: Constable, 1857, p. 152; *Footprints of the Creator*, Edinburgh: A. & C. Black, 1861, pp.i–lxviii, 1–21.

82 R. Chambers, *Vestiges of the Natural History of Creation, and other Evolutionary Writings*, ed. J. Secord, Chicago, IL: Chicago University Press, 1994.

83 T. White, *Thomas Huxley: Making the 'Man of Science'*, Cambridge: Cambridge University Press, 2003; M.H. Cooke, *The Evolution of Nettie Huxley, 1825–1914*, Chichester: Phillimore, 2008; B. Lightman, *The Origins of Agnosticism: Victorian Unbelief and the Limits of Knowledge*, Baltimore, MD: Johns Hopkins University Press, 1987.

84 F. Temple et al., *Essays and Reviews: The 1860 Text and its Reading*, ed. V. Shea and W. Whitla, Charlottesville, VA: Virginia University Press, 2000.

85 D. Kennedy, *The Highly Civilized Man: Richard Burton and the Victorian World*, Cambridge, MA: Harvard University Press, 2006, pp. 131–63.

86 D. Pick, *Faces of Degeneration: A European Disorder, c.1848–c.1914*, Cambridge: Cambridge University Press, 1989.

87 W.M. Jacob, *The Clerical Profession in the Long Eighteenth Century, 1680–1840*, Oxford: Oxford University Press, 2008, pp. 31–63.

88 W. Clark, *Academic Charisma and the Origins of the Research University*, Chicago, IL: Chicago University Press, 2006; J. Retallack, ed., *Imperial Germany, 1871–1918*, Oxford: Oxford University Press, 2008, pp. 160–1, 235–7.

89 H. Hellman, *Great Feuds in Science, Great Feuds in Medicine*, and *Great Feuds in Technology*, Hoboken, NJ: Wiley, 1998, 2002, 2004.

90 F. James, 'An "Open Clash between Science and the Church?" – Wilberforce, Huxley and Hooker on Darwin at the BAAS, Oxford, 1860', in D.M. Knight and M.D. Eddy, eds, *Science and Beliefs*, Aldershot: Ashgate, 2005, pp. 171–93.

91 A. Laurent, *Chemical Method: Notation, Classification, & Nomenclature*, tr. W. Odling [1855], reprinted in D.M. Knight, ed., *The Development of Chemistry 1789–1914*, London: Routledge, 1998, vol. 7.

Chapter 3   Applied Science

1 H. Davy, *Collected Works*, vol. 2, London: Smith, Elder, 1839, p. 323.

2 A.H. Maehle, *Drugs on Trial: Experimental Pharmacology and Therapeutic Innovation in the 18th century*, Amsterdam: Rodopi, 1999.

3 R.L. Hills, *Power from Steam: A History of the Stationary Steam Engine*, Cambridge: Cambridge University Press, 1989, p. 172.

4 B. Marsden and C. Smith, *Engineering Empires: A Cultural History of Technology in Nineteenth-century Britain*, Basingstoke: Palgrave Macmillan, 2005.

5 T. Thomson, 'On Calico Printing', *Records of General Science* 1 (1835): 1–19, 161–73, 321–30.

6 D.M. Knight, 'Theory, Practice and Status: Humphry Davy and Thomas Thomson', in G. Emptoz, ed., *Between the Natural and the Artificial: Dyestuffs and Medicines*, Turnhout: Brepols, 2000, pp. 48–58.

7 A.W. Hofmann, 'On Mauve and Magenta, and the Colouring Matters Derived from Coal', *Proceedings of the Royal Institution* 3 (1858–62): 468–83.

8 R. Fox and A. Nieto-Galan, eds, *Natural Dyestuffs and Industrial Culture in Europe, 1750–1880*, Nantucket, MA: Watson, 1999; T. Travis,

'150th Anniversary of Mauve and Coal-Tar Dyes', *Gesellschaft Deutscher Chemiker, Fachgruppe Geschichte der Chemie, Mitteilungen* 19 (2007): 66–77; A. Simmons, 'Creating a New Rainbow: Dyes after Perkin', *Historical Group Newsletter*, Royal Society of Chemistry , August 2008, pp. 36–42.

9 M.R. Fox, *Dye-makers of Great Britain, 1856–1976: A History of Chemists, Companies, Products and Changes*, Manchester: ICI, 1987.

10 H. Davy, *Collected Works*, vol. 2, London: Smith, Elder, 1839, pp. 311–26.

11 F. James, 'How Big is a Hole? The Problems of the Practical Application of Science to the Invention of the Miners' Safety Lamp by Humphry Davy and George Stephenson in late Regency England', *Transactions of the Newcomen Society* 75 (2005): 175–227.

12 F.A.J.L. James, ed., *The Correspondence of Michael Faraday*, London: I.E.E., vol. 4, 1999, pp. 882–3.

13 J.J. McGann, ed., *The New Oxford Book of Romantic Period Verse*, Oxford: Oxford University Press, 1993, pp. 531–2.

14 C.A. Russell and G.K. Roberts, eds, *Chemical History: Reviews of Recent Literature*, London: Royal Society of Chemistry, 2005; esp. J. Hudson, 'Analytical chemistry', pp. 169–84.

15 W.H. Brock, *Justus von Liebig: The Chemical Gatekeeper*, Cambridge: Cambridge University Press, 1997; E. Homburg, 'Two Factions, One Profession: The Chemical Profession in German Society, 1789–1870', in D.M. Knight and H. Kragh, eds, *The Making of the Chemist*, Cambridge: Cambridge University Press, 1998.

16 J.R. Bertomeu-Sánchez and A. Nieto-Galan, eds, *Chemistry, Medicine, and Crime: Mateu J.B .Orfila and His Times*, Sagamore Beach, MA: Science History, 2006.

17 K.T. Hoppen, *The Mid-Victorian Generation: England 1846–1886*, Oxford: Oxford University Press, 1998, pp. 109–14.

18 H. Gay, 'Thorpe', in B. Lightman, ed., *Dictionary of Nineteenth-century British Scientists*, Bristol: Thoemmes, 2004; P.W. Hammond and H. Egan, *Weighed in the Balance: A History of the Laboratory of the Government Chemist*, London: HMSO, 1992.

19 H. Davy, *Collected Works*, London: Smith, Elder, 1840, vols 7 and 8.

20 J. Liebig, *Animal Chemistry: Or Organic Chemistry in its Application to Physiology and Pathology* [1842], tr. W. Gregory, intr. F.L. Holmes, New York: Johnson, 1964.

21 R. Poole, 'The March to Peterloo: Politics and Festivity in Late Georgian England', *Past and Present* 192 (2006): 109–54.

22 A. Ure, *The Philosophy of Manufactures*, London: Knight, 1835.

23 J. Tann, ed., *Selected Papers of Boulton and Watt. Vol. 1, the Engine Partnership, 1775–1825*, Cambridge, MA: MIT Press, 1981.

24 B. Bowers, *Sir Charles Wheatstone, 1802–1875*, London: HMSO, 1975; C. Wheatstone, *The Scientific Papers*, London: Physical Society, 1879.

25 R. Ashton, *142 Strand: A Radical Address in Victorian London*, London: Jonathan Cape, 2007, p. 35; J. Simmons, *The Victorian Railway*, London: Thames and Hudson, new edn, 1995, p. 346.

26 A. Nuvolari and B. Verspagen, 'Lean's *Engine Reporter* and the Development of the Cornish Engine: A Reappraisal', *Transactions of the Newcomen Society* 77 (2007): 167–89.

27 J. Simmons, *The Victorian Railway*, London: Thames and Hudson, 1991; L. James, *A Chronology of the Construction of Britain's Railways*, London: Ian Allan, 1983.

28 T. de Quincey, *The English Mail Coach and Other Writings*, Edinburgh: Black, 1863, pp. 287–352.

29 *Newcomen Society Links* 198 (2006) was devoted to Brunel.

30 J.C. Bourne, *Drawings of the London and Birmingham Railway* and *The History and Description of the Great Western Railway*, London: Bogue, 1839, 1846; reprint Newton Abbott: David and Charles, 1971).

31 T.H. Hoppen, *The Mid-Victorian Generation: England 1846–1886*, Oxford: Oxford University Press, 1998, pp. 289–93.

32 S. Brindle, *Brunel: The Man Who Built the World*, London: Weidenfeld and Nicolson, 2006; D.P. Miller, 'Principle, Practice and Persona in Isambard Kingdom Brunel's Patent Abolitionism', *BJHS* 41 (2008): 43–72.

33 H.K. Beals, R.J. Campbell, A. Savours and A. McConnell, eds, *Four Travel Journals, 1775–1874*, London: Hakluyt Society, 2008, p. 228n.

34 F.A.J.L. James, ed., *The Correspondence of Michael Faraday*, London: IEE, 1999, vol. 4, pp. 978–9.

35 M.C. Perry, *Narrative of the Expedition of an American Squadron to the China Seas and Japan*, New York: Appleton, 1856.

36 D.K. Brown, 'The Admirality; A Generalisation', *Newcomen Society Links* 204 (2007): 3.

37 K. Baynes and F. Pugh, *The Art of the Engineer*, London: Lutterworth, 1981.

38 M.R. Smith, *Harpers Ferry Armory and the New Technology: The Challenge of Change*, Ithaca, NY: Cornell University Press, 1977.

39 N. Rosenberg, ed., *The American System of Manufactures*, Edinburgh: Edinburgh University Press, 1969; D.J. Jeremy, *Transatlantic Industrial*

Revolution, Oxford: Blackwell, 1981; D.W. Howe, *What God Hath Wrought: The Transformation of America, 1815–1848*, Oxford: Oxford University Press, 2007, pp. 463–82, 525–69.

40 M.S. Seligman, ed., *Naval Intelligence from Germany, 1906–1914*, Aldershot: Ashgate for the Navy Records Society, 2007.

41 C. Babbage, *Passages from the Life of a Philosopher*, London: Longman, 1864, pp. 321–5.

42 F.A.J.L. James and M. Ray, 'Science in the Pits: Michael Faraday, Charles Lyell and the Home Office Enquiry into the Explosion at Haswell Colliery, County Durham, in 1844', *History and Technology* 5 (1999): 213–31.

43 C. McKean, *Battle for the North: The Tay and Forth Bridges and the 19th-century Railway Wars*, London: Granta, 2006.

44 J. Simmons, *The Victorian Railway*, London: Thames and Hudson, new edn, 1995, pp. 26–9.

45 There is a collection of such early nineteenth-century papers from Cambridge in Durham University Library.

46 D.M. Knight, 'The Poetry of Early Exam Papers in Science', *Paradigm* 2 (2000): 29.

47 N. Reingold et al., eds, *The Papers of Joseph Henry*, Washington, DC: Smithsonian, 1972–2008.

48 On Bohn, *Dictionary of Literary Biography*, Detroit: Gale, vol. 106, 1991, pp. 59–62; W. St Clair, *The Reading Nation in the Romantic Period*, Cambridge: Cambridge University Press, 2004, pp. 186–209.

49 J. Morrell and A. Thackray, *Gentlemen of Science: Early Years of the British Association for the Advancement of Science*, Oxford: Oxford University Press, 1981.

50 R. Anderson, 'What is Technology? Education through Museums in the 19th Century', *BJHS* 25 (1992): 169–84.

51 A. Simmons, 'Medicines, Monopolies and Mortars: The Chemical Laboratory and Pharmaceutical Trade at the Society of Apothecaries in the 18th Century', *Ambix* 53 (2006): 221–36; C.A. Russell, *Lancastrian Chemist: The Early Years of Sir Edward Frankland*, Milton Keynes: Open University Press, 1986.

52 W.H. Brock, *Fontana History of Chemistry*, London: HarperCollins, 1992, pp. 270–310; G. Emptoz and P.E. Aceves Pastrana, eds, *Between the Natural and the Artificial*, Turnhout: Brepols, 2000.

53 H. Hawkins, *Pioneer: A History of the Johns Hopkins University, 1874–1889*, Ithaca, NY: Cornell University Press, 1960.

54 *Royal Commission on Scientific Instruction and the Advancement of Science*, London: HMSO, 1872–5.

55 M. Sanderson, ed., *The Universities in the Nineteenth Century*, London: Routledge, 1975.

56 M.W. Travers, *A Life of Sir William Ramsay*, London: Arnold, 1956, pp. 73–80.

57 W.H. Brock, *Science for All*, Aldershot: Ashgate Variorum, 1996, article XI; and *Fontana History of Chemistry*, London: HarperCollins, 1992, pp. 396–435; D. Owen, *English Philanthropy, 1660–1960*, Cambridge, MA: Harvard University Press, 1965, pp. 269, 276–98.

58 D. Edgerton, *Science, Technology and the British Industrial 'Decline', 1870–1970*, Cambridge: Cambridge University Press, 1996.

59 D.P. Miller, *Discovering Water: James Watt, Henry Cavendish and the Nineteenth-century 'Water Controversy'*, Aldershot: Ashgate, 2005.

60 C.A. Russell, N.G. Coley and G.K. Roberts, *Chemists by Profession*, Milton Keynes: Open University Press, 1977; R.F. Bud and G.K. Roberts, *Science versus Practice: Chemistry in Victorian Britain*, Manchester: University Press, 1984.

61 C.W. Siemens, 'Measuring Temperature by Electricity', *Proceedings of the Royal Institution* 61 (1870–2): 438–48; H. Chang, *Inventing Temperature*, Oxford: Oxford University Press, 2004.

62 A.D. Morrison-Low, *Making Scientific Instruments in the Industrial Revolution*, Aldershot: Ashgate, 2007; A. McConnell, *Jesse Ramsden (1735–1800): London's Leading Scientific Instrument Maker*, Aldershot: Ashgate, 2007.

63 M. Faraday, *Chemical Manipulation*, 3rd edn, London: John Murray, 1842.

64 F. Accum, *System of Theoretical and Practical Chemistry*, 2nd edn, London: Kearsley, 1807.

65 A.Q. Morton, *Science in the 18th Century: The King George III Collection*, London: Science Museum, 1993; the museum also holds a portable laboratory of Faraday's.

66 P. Morris, ed., *From Classical to Modern Chemistry: the Instrumental Revolution*, London: Royal Society of Chemistry, 2002.

67 L. de Vries and J. Laver, *Victorian Advertisements*, London: John Murray, 1968; J. Lewis, *Printed Ephemera*, London: Faber, 1969.

68 T. Boon, *Films of Fact: A History of Science in Documentary Films*, London: Wallflower, 2008.

Chapter 4 Intellectual Excitement

1 O. Sacks, *Uncle Tungsten: Memoirs of a Chemical Boyhood*, London: Picador, 2001.

2 T.S. Kuhn, 'The Function of Dogma in Scientific Research', in A.C. Crombie, ed., *Scientific Change*, Oxford: Oxford University Press, 1963, pp. 347–69, and subsequent discussion.

3 P. Levi, *The Periodic Table*, tr. R. Rosenthal, London: Michael Joseph, 1985, p. 171.

4 S. Parkes, *The Chemical Catechism*, 4th edn, London: Lackington Allen, 1810.

5 C.A. Russell, *Lancastrian Chemist: The Early Years of Sir Edward Frankland*, Milton Keynes: Open University Press, 1986.

6 H. Hartley, *Humphry Davy*, London: Nelson, 1966, p. 148.

7 G.L. Herries Davies, *Whatever is Under the Earth: The Geological Society of London, 1807 to 2007*, London: Geological Society, 2007; C.L.E. Lewis and S.J. Knell, *The World's First Geological Society*, London: Geological Society, 2009, prints the papers from the conference celebrating the Society's bicentenary in November 2007.

8 K.A. Neeley, *Mary Somerville: Science, Illumination and the Female Mind*, Cambridge: Cambridge University Press, 2001.

9 C. Smith, 'Force, Energy, and Thermodynamics', in M.J. Nye, ed., *Cambridge History of Science*, Cambridge: Cambridge University Press, 2003, vol 5, pp. 289–310.

10 F.W.J. Schelling, *Ideas for a Philosophy of Nature*, tr. E.E. Harris and P. Heath, intr. R. Stern, Cambridge, Cambridge University Press, 1988; A. Cunningham and N. Jardine, eds, *Romanticism and the Sciences*, Cambridge: Cambridge University Press, 1990.

11 R.J. Richards, *The Romantic Conception of Life: Science and Philosophy in the Age of Goethe*, Chicago, IL: Chicago University Press, 2002.

12 J. van Wyhe, *Phrenology and the Origins of Victorian Scientific Naturalism*, Aldershot: Ashgate, 2004.

13 See the series of essays, 'Making Connections', in *Nature* 445 (2007), and subsequent issues: my 'Kinds of Minds' was published 10 May 2007, p. 149.

14 P. Bertucci, 'Revealing Sparks: John Wesley and the Religious Utility of Electrical Healing', *BJHS* 39 (2006): 341–62.

15 J. Priestley, *The History and Present State of Electricity* [1775], intr. R. Schofield, New York: Johnston, 1966, vol. 1, pp. xiv–xv.

16 M.P. Crosland, *Science Under Control: The French Academy of Sciences*,

1795–1914, Cambridge: Cambridge University Press, 1992, p. 382; D.M. Knight, 'Davy and the Placing of Potassium among the Elements', Royal Society of Chemistry, Historical Group Occasional Paper 4: the Second Wheeler Lecture.

17 M. Faraday, *Experimental Researches in Electricity*, London: Taylor, 1839–55, vol. 1, pp. 76–102 (paragraphs 265–360); p.136 (paragraph 482); vol. 2, pp. 211–17.

18 W. Whewell, *History of the Inductive Sciences*, London: Parker, 1837, vol. 3, p. 19.

19 See 'Photography', in J.L. Heilbron, ed., *The Oxford Companion to the History of Modern Science*, Oxford: Oxford University Press, 2003, pp. 636–8; H. Davy, *Collected Works*, London: Smith, Elder, 1839, vol. 2, pp. 240–5.

20 K. Jelved, A.D. Jackson and O. Knudsen, ed., *Selected Scientific Works of Hans Christian Ørsted*, Princeton, NJ: Princeton University Press, 1998, pp. 413–49.

21 J.R. Hofmann, *André-Marie Ampère: Enlightenment and Electrodynamics*, Cambridge: Cambridge University Press, 1995.

22 R. Boscovich, *A Theory of Natural Philosophy* [1763], tr. J.M. Child, Cambridge, MA: MIT Press, 1966.

23 M. Faraday, *Experimental Researches in Electricity*, London, Taylor, 1839–55, vol. 3, pp. 1–2. See also pp. 447–52; and vol. 2, pp. 284–93.

24 B. Mahon, *The Man Who Changed Everything: The Life of James Clerk Maxwell*, New York: Wiley, 2003; M.J. Nye, ed., *The Cambridge History of Science*, Cambridge: Cambridge University Press, 2003, vol. 5, pp. 271–324.

25 T.S. Kuhn, *The Essential Tension*, Chicago, IL: Chicago University Press, 1977.

26 J. Tyndall, *Fragments of Science*, 10th imp., London: Longman, 1899, vol. 1, pp. 422–38; 'Faraday as a Discoverer', *Proceedings of the Royal Institution* 5 (1866–9): 199–272.

27 J.P. Joule, *Scientific Papers*, London: Physical Society, 1884, vol. 1, pp. 298–328; 265–76.

28 D. Cahan, ed., *Hermann von Helmholtz*, Berkeley, CA: California University Press, 1993, pp. 291–460.

29 H. Helmholtz, *Popular Lectures on Scientific Subjects*, London: Longman, 1873, pp. xv–xvi, 153–96.

30 I.W. Morus, *When Physics Became King*, Chicago, IL: Chicago University Press, 2005.

31 S. Carnot, *Réflexions sur la Puissance Motrice du Feu* [1824], ed. R. Fox, Paris: Vrin, 1978; *Reflections on the Motive Power of Fire*, tr. R.H. Thurston and E. Mendoza, New York: Dover, 1960.

32 C. Smith and N. Wise, *Energy and Empire: A Biographical Study of Lord Kelvin*, Cambridge: Cambridge University Press, 1998, pp. 167, 294.

33 H.S. Kragh, *Entropic Creation: Religious Contexts of Thermodynamics and Cosmology*, Aldershot: Ashgate, 2008.

34 F. Burckhardt et al., eds, *The Correspondence of Charles Darwin*, vol. 7, Cambridge: Cambridge University Press, 1991, p. 423.

35 E. Darwin, *The Temple of Nature*, London: Johnson, 1803, p. 134.

36 *The Anti-Jacobin, or Weekly Examiner* [16 April 1798], 4th edn, London: Wright, 1799, vol. 2., pp. 170–3; E.L. de Montluzin, *The Anti-Jacobins, 1798–1800*, London: Macmillan, 1988.

37 D.M. Knight, 'Humphry Davy the Poet', *Interdisciplinary Science Reviews* 30/4 (2005): 356–72; the whole issue is devoted to science and poetry.

38 W. Paley, *Natural Theology* [1802], ed. M.D. Eddy and D.M. Knight, Oxford: Oxford University Press, 2006, p. 238.

39 D. Hume, *On Religion* [1779], ed. A.W. Colver and J.V. Price, Oxford: Oxford University Press, 1976, p. 241; T.R. Malthus, *An Essay on the Principles of Population*, London: Johnson, 1798, p. 48.

40 W. Wordsworth and S.T. Coleridge, *Lyrical Ballads* [1798], London: Penguin, 1999, p. 101; L. Daston and F. Vital, eds, *The Moral Authority of Nature*, Chicago, IL: Chicago University Press, 2004.

41 G. Cuvier, *Essay on the Theory of the Earth*, tr. R. Jameson, 5th edn, Edinburgh: Blackwood, 1827.

42 L. Oken, *Elements of Physiophilosophy*, tr. A. Tulk, London: Ray Society, 1847; K. von Reichenbach, *Researches on Magnetism, Electricity, Heat, Light, Crystallization, and Chemical Attraction, in their Relations to the Vital Force*, tr. W. Gregory, London: Taylor, Walton and Gregory, 1850; *Lectures on Od and Magnetism*, tr. F.D. O'Byrne, London: Hutchinson, 1926.

43 A. Desmond, *The Politics of Evolution: Morphology, Medicine and Reform in Radical London*, Chicago, IL: Chicago University Press, 1992.

44 H. Davy, *Collected Works*, London: Smith, Elder, 1839–40, vol. 7, pp. 35–44.

45 M.J.S. Rudwick, *Worlds Before Adam: the Reconstruction of Prehistory in the Age of Reform*, Chicago, IL: Chicago University Press, 2008.

46 J.A. Secord, *Victorian Sensation: The Extraordinary Publication, Reception, and Secret Authorship of* Vestiges of the Natural History of Creation,

Chicago, IL: Chicago University Press, 2000; [R.Chambers], *Vestiges of the Natural History of Creation* [1844], ed. J.Secord, Chicago, IL: Chicago University Press, 1994.

47 R.L. Numbers, *Creation by Natural Law: Laplace's Nebular Hypothesis in American Thought*, Seattle: Washington University Press, 1977.

48 L.A.J. Quetelet, *A Treatise on Man and the Development of his Faculties*, Edinburgh: Chambers, 1842.

49 A. Tennyson, *In Memoriam* [1850], ed. S. Shatto and M. Shaw, Oxford: Oxford University Press, 1982, p. 80 (section 56).

50 J.V. Thompson, *Zoological Researches and Illustrations, 1828–1834*, ed. A. Wheeler, London: Society for Bibliography of Natural History, 1968.

51 C. Darwin and A.R. Wallace, *Evolution by Natural Selection*, ed. G. de Beer, Cambridge: Cambridge University Press, 1958.

52 There are numerous modern editions, including facsimiles, of the *Origin of Species*, some based upon the first edition of 1859, others upon the sixth, the last revised by Darwin. For the words used, see P.H. Barrett, D.J. Weinshank and T.T. Gottleber, *A Concordance to Darwin's* Origin of Species, *first edition*, Ithaca, NY; Cornell University Press, 1981. See also R.B. Freeman, *The Works of Charles Darwin: An Annotated Bibliographical Handlist*, 2nd edn, Folkestone: Dawson, 1977, and *Charles Darwin: A Companion*, Folkestone: Dawson, 1978. Essential is F. Burkhardt et al. eds, *The Correspondence of Charles Darwin*, Cambridge: Cambridge University Press, 1985–(in progress).

53 A. Desmond and J. Moore, *Darwin*, London: Penguin, 1991; A. Desmond, J. Moore and J. Browne, *Charles Darwin*, Oxford: Oxford University Press, 2007.

54 A. Desmond, *Huxley*, London: Michael Joseph, 2 vols, 1994–7.

55 P. White, *Thomas Huxley: Making the 'Man of Science'*, Cambridge: Cambridge University Press, 2003; A.P. Barr, ed., *Thomas Henry Huxley's Place in Science and Letters*, Athens, GA: Georgia University Press, 1997.

56 F.A.J.L. James, 'An "Open Clash between Science and the Church"?', in D.M. Knight and M.D. Eddy, eds, *Science and Belief: From Natural Philosophy to Natural Science*, Aldershot: Ashgate, 2005, pp. 171–93.

57 C. Darwin, *The Descent of Man, and Selection in Relation to Sex*, ed. J. Moore and A. Desmond, London: Penguin, 2004.

58 H.W. Bates, *The Naturalist on the River Amazon*, intr. E. Clodd, London: John Murray, 1892; S.K. Naylor, 'Bates', in B. Lightman, ed., *Dictionary of Nineteenth-century British Scientists*, Bristol: Thoemmes, 2004.

59 W.S. Symonds, *Old Bones: Or Notes for Young Naturalists*, London: Hardwicke, 1861; N. Cooper, ed., *John Ray and his Successors: The Clergyman as Biologist*, Braintree: John Ray Trust, 2000.

60 V. Shea and W. Whitla, *Essays and Reviews: The 1860 Text and Its Reading*, Charlottesville, VA: Virginia University Press, 2000.

61 O. Chadwick, *The Victorian Church*, Oxford: Oxford University Press, 1966–70, vol. 2, pp. 1–150.

62 F. Temple, *The Relations between Religion and Science*, London: Palgrave Macmillan, 1885, pp. 97–123; and see P. Bowler, *Reconciling Science and Religion: The Debate in Early-twentieth-century Britain*, Chicago, IL: Chicago University Press, 2001.

63 C. Clark, 'Religion and Confessional Conflict', in J. Retallack, ed., *Imperial Germany, 1871–1918*, Oxford: Oxford University Press, 2008, pp. 83–105.

64 See John Lynch's entry on Mivart in B. Lightman, ed., *Dictionary of Nineteenth-century British Scientists*, Bristol: Thoemmes, 2004; D.M. Knight, *Science and Spirituality: The Volatile Connection*, London: Routledge, 2004, pp. 151–66.

65 A. Berlin, M.Z. Brettler and M. Fishbone, *The Jewish Study Bible*, Oxford: Oxford University Press, 2004.

66 T.H. Hoppen, *The Mid-Victorian Generation: England, 1846–1886*, Oxford: Oxford University Press, 1998, pp. 472–510.

67 T.H. Huxley, 'Romanes Lecture', in A. Barr, ed., *Major Prose of T.H. Huxley*, Athens, GA: Georgia University Press, 1997, p. 327.

68 H. Drummond, *The Ascent of Man*, London: Hodder, 1894, p. 435.

69 R.J. Richards, *The Romantic Conception of Life: Science and Philosophy in the Age of Goethe*, Chicago, IL: Chicago University Press, 2002.

70 E. Haeckel, *Kunstformen der Natur* [1904], ed. R.P. Hartmann, O. Breidbach and I. Eibl-Eibesfeldt, Munich: Prestel, 1998, p. 134.

71 D.L. Livingstone, *Darwin's Forgotten Defenders*, Edinburgh: Scottish Academic Press, 1987.

72 T.H. Huxley, *Man's Place in Nature* [1863], intr. D.M. Knight, London: Routledge, 2003; S. Coleman and L. Carlin, eds, *The Cultures of Creationism: Anti-evolutionism in English-speaking Countries*, Aldershot: Ashgate, 2004.

73 D.C. Lindberg and R. Numbers, eds, *Where Science and Christianity Meet*, Chicago, IL: Chicago University Press, 2003; J.H. Brooke and G. Cantor, *Reconstructing Nature: The Engagement between Science and Religion*, Edinburgh: T. & T. Clark, 1998; R.L. Numbers and

J. Stenhouse, eds, *Disseminating Darwinism*, Cambridge: Cambridge University Press, 1999.

74 J.W. Draper, *History of the Conflict Between Religion and Science*, London: King, 1875.

75 J.D. Burchfield, *Lord Kelvin and the Age of the Earth*, New York: Science History, 1975.

Chapter 5   Healthy Lives

1 P. Jenkins, *The New Faces of Christianity: Believing the Bible in the Global South*, Oxford: Oxford University Press, 2007, p. 184.

2 J. Morrell and A. Thackray, *Gentlemen of Science: Early Years of the British Association for the Advancement of Science*, Oxford, Oxford University Press, 1981, pp. 291–6.

3 A.P. Barr, ed., *The Major Prose of T.H. Huxley*, Athens, GA: Georgia University Press, 1997, pp. 154–73.

4 P. Bertucci, 'Revealing Sparks: John Wesley and the Religious Utility of Electrical Healing', *BJHS* 39 (2006): 341–62; W.H. Helfand, *Quack, Quack, Quack: The Sellers of Nostrums in Prints, Posters, Ephemera and Books*, New York: Grolier Club, 2002. For a short history of medicine, see R. Porter, *The Greatest Benefit to Mankind: A Medical History of Humanity from Antiquity to the Present*, London: HarperCollins, 1997.

5 N.A.M. Rodger, *The Command of the Ocean: A Naval History of Britain 1649–1815*, London: Penguin, 2004, pp. 51, 404–5, 527.

6 K. Bergdolt, *Wellbeing: A Cultural History of Healthy Living*, tr. J. Dewhurst, Cambridge: Polity, 2008, pp. 247–50; D. Gardner-Medwin, ed, *Medicine in Northumbria: Essays in the History of Medicine*, Newcastle: Pybus Society, 1993.

7 N. Black, *Walking London's Medical History*, London: Royal Society of Medicine, 2006.

8 F. Holmes, 'The Physical Sciences in the Life Sciences', in M.J. Nye, ed., *Cambridge History of Science*, Cambridge: Cambridge University Press, 2003, vol.5, pp. 219–36.

9 W. Babington, A. Marcet and W. Allen, *A Syllabus of a Course of Chemistry Lectures Read at Guy's Hospital*, London: Phillips, 1811; another edn, 1816, has a frontispiece of the laboratory there. There is such an annotated copy from 1811 in Durham University Library, SC 00826.

10 J.M. Bourgery and N.H. Jacob, *Atlas of Human Anatomy and Surgery* [1831–53], ed. J.M. le Minor and H. Sick, Köln: Taschen, 2005;

H. Gray, *Anatomy, Descriptive and Surgical*, illustrated. M.V. Carter, London: Parker, 1858; R. Richardson, *The Making of Mr Gray's Anatomy*, Oxford: Oxford University Press, 2008; M. Kemp, *The Human Animal in Western Art and Science*, Chicago, IL: Chicago University Press, 2007, pp. 17–52.

11 J. Hunter, *Essays and Observations in Natural History, Anatomy, Physiology, Psychology, and Geology*, ed. R. Owen, London: Van Voorst, 1841.

12 On chemical vitalism, see V. Klein and W. Lefèvre, *Materials in Eighteenth-century Science*, Cambridge, MA: MIT Press, 2007, pp. 251ff.

13 B.C. Brodie, *Psychological Inquiries; in a Series of Essays Intended to Illustrate the Mutual Relations of the Physical Organization and the Mental Faculties*, 2nd edn, London: Longman, 1855; *Autobiography*, 2nd edn, London: Longman, 1865.

14 M.F.X. Bichat, *A Treatise on the Membranes*, tr. J.G. Coffin, Boston, MA: Cummings and Hilliard, 1813, p. xv; *Physiological Researches on Life and Death*, tr. F. Gold, London: Longman, 1802, pp. 23–5, 81, 188–9, 228.

15 *Register of Arts and Sciences* 1 (1824): 3–5; D.M. Knight, *Public Understanding of Science: A History of Communicating Scientific Ideas*, London, Routledge, 2006, pp. 29–30.

16 H. Lonsdale, *A Sketch of the Life and Writings of Robert Knox, the Anatomist*, London: Macmillan, 1870, pp. 54–114.

17 R. Richardson, *The Making of Mr. Gray's Anatomy*, Oxford: Oxford University Press, 2008, pp. 117–39.

18 M.F.X. Bichat, *Recherches physiologiques sur la vie et la mort*, 4th edn, ed. F. Magendie, Paris: Gabon, 1822.

19 C. Bell, *Idea of a New Anatomy of the Brain* [1811], ed. E.A.O., London: Dawson, 1966.

20 *Synopsis of the Contents of the Museum of the Royal College of Surgeons of England*, London: RCS, 1850; this was available from the porter for sixpence.

21 W. Lawrence, *Lectures on Physiology, Zoology, and the Natural History of Man*, London: Benbow, 1822, pp. 9, 52–3.

22 W. St Clair, *The Reading Nation in the Romantic Period*, Cambridge: Cambridge University Press, 2004, pp. 337, 677, 679.

23 A. Desmond, *The Politics of Evolution: Morphology, Medicine, and Reform in Radical London*, Chicago, IL: Chicago University Press, 1989.

24 K. Bergdolt, *Wellbeing: A Cultural History of Healthy Living*, tr. J. Dewhurst, Cambridge: Polity, 2008, pp. 251–87.

25 J.A. Paris, *A Treatise on Diet*, 2nd edn, London: Underwood, 1827, p. 65.

26 J. Beresford, ed., *The Diary of a Country Parson, 1758–1802*, Oxford: Oxford University Press, 1949.

27 W. Gratzer, *Terrors of the Table: The Curious History of Nutrition*, Oxford: Oxford University Press, 2005.

28 B. Potter, *The Tale of the Flopsy Bunnies*, London: Warne, 1909.

29 W. Babington et al., *A Syllabus of a Course of Chemical Lectures at Guy's Hospital*, London: Phillips, 1816, p. 15.

30 J. Kidd, *On the Adaptation of External Nature to the Physical Condition of Man*, London: Pickering, 1833, pp. 226–8; A.H. Maehle, *Drugs on Trial: Experimental Pharmacology and Therapeutic Innovation in the Eighteenth Century*, Amsterdam: Rodopi, 1999, pp. 127–222.

31 N. Vickers, *Coleridge and the Doctors, 1795–1806*, Oxford: Oxford University Press, 2004.

32 R. Watson, *Chemical Essays*, 5th edn, London: Evans, 1796, vol. 4, pp. 253–5.

33 T. Trotter, *An Essay, Medical, Philosophical, and Chemical on Drunkenness and its Effects on the Human Body*, intr. R. Porter, London: Routledge, 1988.

34 C.T. Thackrah, *The Effects of Arts, Trades and Professions on Health and Longevity* [1832], ed. A. Meiklejohn, Edinburgh: Livingstone, 1957, p. 5.

35 J.P. Kay, *The Moral and Physical Condition of the Working Classes Employed in the Cotton Manufactures in Manchester*, London: Ridgeway, 1832.

36 P. Williamson, 'State Prayers, Fasts, and Thanksgivings: Public Worship in Britain, 1830–1847', *Past and Present* 200 (2008): 121–74.

37 J. Tyndall, 'Reflections on Prayer and Natural Law [1861]', *Fragments of Science*, 10th imp., London: Longman, 1899, vol. 2, pp. 1–7.

38 M. Harrison, *Disease and the Modern World, 1500 to the Present Day*, Cambridge: Polity, 2004, pp. 91–117.

39 J. Snow, *Snow on Cholera: Being a Reprint of Two Papers by John Snow together with a Biographical Memoir by B.W. Richardson*, New York: Hafner, 1965.

40 W. Griffith and A. Dronsfield, 'RSC Chemical Landmark Plaque to Dr John Snow', *Historical Group Newsletter, Royal Society of Chemistry*, August 2008, pp. 29–31.

41 P. Ackroyd, *Thames: Sacred River*, London: Chatto & Windus, 2007, pp. 270–5.

42 E. Chadwick, *Report on the Sanitary Condition of the Labouring Population of Great Britain* [1842], ed. M.W. Flinn, Edinburgh: Edinburgh University Press, 1965.

43 W.F. Bynum et al., *The Western Medical Tradition, 1800 to 2000*, Cambridge: Cambridge University Press, 2006, p. 92.

44 J. Simon, *English Sanitary Institutions*, London: Cassell, 1890.

45 F.A.J.L. James, ed., *The Correspondence of Michael Faraday*, London: I.E.E., vol. 4, 1999, p. 883.

46 Registrar General, *Annual Summary of Births, Deaths, and Causes of Death in London, 1874*, London: Eyre and Spottiswoode, 1875, pp. iii–viii.

47 M. Harrison, *Disease and the Modern World, 1500 to the Present*, Cambridge: Polity, 2004, p. 115; S.H. Preston and M.R. Haines, *Fatal Years: Child Mortality in Late-nineteenth-century America*, Princeton, NJ: Princeton University Press, 1991.

48 F. Nightingale, *Ever Yours, Florence Nightingale: Selected Letters*, ed. M. Vicinus and B. Nergaard, London: Virago, 1989, pp. 51–4.

49 W.H. Russell, *The War*, London: Routledge, 1855–6, vol. 1, p. 32; and the volumes are subsequently full of references to disease, dirt and calamity.

50 C. Kelly, ed., *Mrs Duberly's War: Journal and Letters from the Crimea*, Oxford: Oxford University Press, 2007.

51 F. Nightingale, *Notes on Nursing* [1859], London: Duckworth, 1970.

52 M.H. Frawley, *Invalidism and Identity*, Chicago, IL: Chicago University Press, 2004.

53 S. Jacyna, 'Medicine in Transformation, 1800–1849', and W.F. Bynum, 'The Rise of Science in Medicine, 1850–1913', in W.F. Bynum et al., *The Western Medical Tradition, 1800–2000*, Cambridge: Cambridge University Press, 2006, pp. 11–110, 111–245.

54 E. Darwin, *The Botanic Garden*, London: Johnson, 1791, Part 2, pp. 167–73.

55 W. Gratzer, *Terrors of the Table*, Oxford: Oxford University Press, 2005, pp. 118–34.

56 A.A. Russell and G.K. Roberts, *Chemical History: Reviews of the Recent Literature*, Cambridge: Royal Society of Chemistry, 2005.

57 T.S. Traill, *Outlines of a Course of Lectures on Medical Jurisprudence*, 2nd edn, Edinburgh: A. & C. Black, 1840.

58 J.R. Bertomeu-Sánchez and A. Nieto-Galan, eds, *Chemistry, Medicine*

and Crime: Mateu J.B. Orfila (1787–1853) and His Times, Sagamore Beach, MA: Science History Publications, 2006.

59 C.A. Russell, N.G. Coley and G.K. Roberts, *Chemists by Profession*, Milton Keynes: Open University Press, 1977.

60 F. Szabadváry, *History of Analytical Chemistry*, tr. G. Svehla, Oxford: Pergamon, 1966, pp. 349–74.

61 P.W. Hammond and H. Egan, *Weighed in the Balance: A History of the Laboratory of the Government Chemist*, London: HMSO, 1992.

62 W.F. Bynum et al., *The Western Medical Tradition, 1800–2000*, Cambridge: Cambridge University Press, 2006, has useful timelines; and Bynum's essay 'The Rise of Science in Medicine, 1850–1913' is very valuable indeed, being a fuller and expert discussion of topics central to this chapter; see also his *Science and the Practice of Medicine in the Nineteenth Century*, Cambridge: Cambridge University Press, 2007.

63 W.F. Bynum et al., *The Western Medical Tradition, 1800–2000*. Cambridge: Cambridge University Press, 2006, pp. 69–77.

64 J.W. Griffith and A. Henfrey, *The Micrographic Dictionary*, 2nd edn, London: Van Voorst, 1860; Journal of the Royal Microscopical Society, 1878 onwards.

65 G.L. Geison, *The Private Science of Louis Pasteur*, Princeton, NJ: Princeton University Press, 1995; B. Latour, *The Pasteurization of France*, Baltimore, MD: Johns Hopkins University Press, 1988.

66 M. Harrison, *Disease and the Modern World, 1500 to the Present Day*, Cambridge: Polity, 2004, pp. 124ff, 141.

67 C. Bernard, *An Introduction to the Study of Experimental Medicine*, tr. H.C. Greene, intr. I.B. Cohen, New York: Dover, 1957.

68 T.H. Huxley, *The Crayfish: An Introduction to the Study of Zoology*, London: Macmillan, 1879.

69 W. Osler, *A Way of Life*, intr. G.L. Keynes, Oxford: Oxford University Press, 1951.

70 W. Moorcroft and G. Trebeck, *Travels in the Himalayan Provinces of Hindustan and the Punjab from 1819 to 1825* [1841], ed. H.H. Wilson, intr. G.J. Alder, Oxford: Oxford University Press, 1979.

71 M. Foucault, *The Birth of the Clinic*, tr. A.M.S. Smith, London: Routledge, 1989; *Madness and Civilization: A History of Insanity in the Age of Reason*, tr. R. Howard, London: Tavistock, 1965.

72 W.F. Bynum et al., *The Western Medical Tradition, 1800–2000*, Cambridge: Cambridge University Press, 2006, pp. 197–203; C. Yanini, *The Architecture of Madness: Insane Asylums in the United States*,

Minneapolis: Minnesota University Press, 2007; J. Conolly, *An Inquiry Concerning the Indications of Insanity, with Suggestions for the Better Protection and Care of the Insane*, London: Taylor, 1830.

73 D. Pick, *Faces of Degeneration: A European Disorder, c.1848–c.1918*, Cambridge: Cambridge University Press, 1989.

74 R. Bivins, *Alternative Medicine? A History*, Oxford: Oxford University Press, 2007; K. Bergholt, *Wellbeing*, tr. J. Dewhurst, Cambridge: Polity, 2008, pp. 283–8.

75 W.H. Helfand, *Quack, Quack, Quack: The Sellers of Nostrums in Prints, Ephemera and Books*, New York: Grolier Club, 2002.

Chapter 6   Laboratories

1 L.M. Principe and L. Dewitt, *Transmutations: Alchemy in Art*, Philadelphia, PA: Chemical Heritage Foundation, 2002; L. Abraham, *A Dictionary of Alchemical Imagery*, Cambridge: Cambridge University Press, 1998.

2 M.P. Crosland, 'Early Laboratories, c.1600–1800, and the Location of Experimental Science' , *Annals of Science* 62 (2005): 233–54.

3 A.G. Debus, *The Chemical Promise: Experiment and Mysticism in the Chemical Philosophy, 1550–1800*, Sagamore Beach, MA: Science History, 2006; ed., *Alchemy and Early Modern Chemistry: Papers from Ambix*, London: Society for the History of Alchemy and Chemistry, 2004.

4 W.R. Newman, *Promethean Ambitions: Alchemy and the Quest to Perfect Nature*, Chicago, IL: Chicago University Press, 2004.

5 H. Ewing, *The Lost World of James Smithson: Science, Revolution and the Birth of the Smithsonian*, London: Bloomsbury, 2007, p. 274.

6 W.H. Brock, *William Crookes and the Commercialization of Science*, Aldershot: Ashgate, 2008.

7 J.A. Paris, *The Life of Sir Humphry Davy*, London: Colburn and Bentley, 1831, p. 179; the paper is in *Phil. Trans.* 97 (1807): 267–92.

8 M. Beretta, *Imaging a Career in Science: The Iconography of Antoine Laurent Lavoisier*, Canton, MA: Science History, 2002.

9 M.W. Rossiter, 'Scientific Marriages and Families', in M.J. Nye, *Cambridge History of Science*, Cambridge: Cambridge University Press, vol. 5, pp. 65–6.

10 H. Rosotti, ed., *Chemistry in the Schoolroom, 1806: Selections from Mrs Marcet's* Conversations on Chemistry, Bloomington, IN: Authorhouse, 2006, p. 112.

11 H. Davy, *Consolations in Travel, or the Last Days of a Philosopher*, London: John Murray, 1830, pp. 251–2.

12 J. Cottle, *Reminiscences of Samuel Taylor Coleridge and Robert Southey*, London: Houlston and Stoneman, 1847, p. 270.

13 W.T. Brande, *A Manual of Chemistry: Containing the Principal Facts of the Science, Arranged in the Order in Which they are Discussed and Illustrated in the Lectures at the Royal Institution*, London: John Murray, 1830, frontispiece to vol. 1; Faraday's drawing of a portable laboratory is the frontispiece to vol. 2.

14 F.A.J.L. James, ed., *The Common Purposes of Life: Science and Industry at the Royal Institution*, Aldershot: Ashgate, 2002; ed., *The Development of the Laboratory*, London: Macmillan, 1989.

15 P. and R. Unwin, '"A Devotion to the Experimental Sciences and Art": The Subscription to the Great Battery at the Royal Institution, 1808–9', *BJHS* 40 (2007): 181–203.

16 A.E. Jeffreys, *Michael Faraday: A List of his Lectures and Published Writings*, London: Royal Institution, 1960.

17 M. Faraday, *Chemical Manipulation* [1827], 3rd edn, London: John Murray, 1842.

18 B. Bowers and L. Symons, eds, *Curiosity Perfectly Satisfied: Faraday's Travels in Europe, 1813–15*, London: Peregrinus, 1991, pp. 23–30.

19 *The Quarterly Journal of Science, Literature, and the Arts* 10 (1821): plate 2 (facing p. 117).

20 B. Gee, 'Amusement Chests and Portable Laboratories', in F.A.J.L. James, ed., *The Development of the Laboratory*, London: Macmillan, 1989, pp. 37–59.

21 O. Sacks, *Uncle Tungsten: Memories of a Chemical Boyhood*, London: Picador, 2001.

22 T. Martin, ed., *Faraday's Diary: Being the Various Philosophical Notes of Experimental Investigation made by Michael Faraday*, 7 vols and index, London: Bell, 1932–6.

23 H. Chang, *Inventing Temperature: Measurement and Scientific Progress*, Oxford: Oxford University Press, 2004.

24 On Tennant, see M. Archer and C. Haley, eds, *The 1702 Chair of Chemistry at Cambridge: Transformation and Change*, Cambridge: Cambridge University Press, 2005; on Tennant and Wollaston, see M.C. Usselman, in B. Lightman, ed., *Dictionary of Nineteenth-century British Scientists*, Bristol: Thoemmes, 2004, vol. 4, pp. 1976–80, 2186–92.

25 W. Clark, *Academic Charisma and the Origins of the Research University*, Chicago, IL: Chicago University Press, 2006.

26 W.H. Brock, *Justus von Liebig: The Chemical Gatekeeper*, Cambridge: Cambridge University Press, 1997.

27 E. Homburg, 'Two Factions, One Profession: The Chemical Profession in German Society, 1780–1870', in D.M. Knight and H. Kragh, eds, *The Making of the Chemist: The Social History of Chemistry in Europe, 1789–1914*, Cambridge, Cambridge University Press, 1998; E. Homburg, A.S. Travis and H.G. Schrötter, eds, *The Chemical Industry in Europe, 1850–1914: Industrial Growth, Pollution, and Professionalization*, Dordrecht: Kluwer, 1998.

28 U. Klein, in J. Bertomeu-Sánchez and A. Nieto-Galan, eds, *Chemistry, Medicine and Crime: Mateu J.B. Orfila (1787–1853) and His Times*, Sagamore Beach, MA; Science History, 2006, pp. 79–100.

29 W.H. Brock, *The Fontana History of Chemistry*, London: HarperCollins, 1992, pp. 173–209.

30 T. Thomson, *An Attempt to Establish the First Principles of Chemistry by Experiment*, London: Baldwin, Cradock, 1825; D.M. Knight, ed., *Classical Scientific Papers: Chemistry*, 2nd series, London: Mills & Boon, 1970, pp. 15–70; J.B. Morrell, 'The Chemist Breeders: The Research Schools of Liebig and Thomas Thomson', *Ambix* 19 (1972): 1–46.

31 J.H. Brooke, *Thinking about Matter*, Aldershot: Ashgate Variorum, 1995, chapters 4 and 5.

32 D.M. Knight, *Ideas in Chemistry: A History of the Science*, London: Athlone, 2nd edn, 1995, pp. 112–27.

33 A.W. Williamson, *Papers on Etherification and the Constitution of Salts, 1850–1856*, Edinburgh: Alembic Club Reprints 16, 1949.

34 M. Faraday, *Chemical Manipulation* [1842], reprinted in D.M. Knight, ed., *The Development of Chemistry*, 10 vols, London: Routledge, 1998.

35 F. Szabadváry, *History of Analytical Chemistry*, tr. G. Svehla, Oxford: Pergamon, 1966, pp. 146–7, 161–92.

36 *João Jacinto de Magalhães (John Hyacinth de Magellan) Conference*, Coimbra: Museu de Fisica da Universidade de Coimbra, 1994.

37 W. Whewell, *History of the Inductive Sciences*, 3rd edn, London: Parker, 1857, vol. 3, pp. 141–52.

38 J.B. Daniell, *An Introduction to the Study of Chemical Philosophy*, London: Parker, 1839, p. vii.

39 H. Cavendish, *Electrical Researches*, ed. J.C. Maxwell, Cambridge: Cambridge University Press, 1879, p. lvii.

40　F.A.J.L. James and A. Peers, 'Constructing Space for Science at the Royal Institution of Great Britain', *Physics in Perspective* 9 (2007): 130–85, esp. 155.

41　N. Harte and J. North, eds, *The World of University College, London, 1828–1978*, London: UCL, 1978, p. 58; C.A.J. Chilvers, 'Thomas Thomson', in B. Lightman, ed., *Dictionary of Nineteenth-century British Scientists*, Bristol: Thoemmes, 2004, vol. 4, pp. 1999–2003.

42　B. Mahon, *The Man Who Changed Everything: The Life of James Clerk Maxwell*, Chichester: Wiley, 2003.

43　J. Morrell, *John Phillips and the Business of Victorian Science*, Aldershot: Ashgate, 2005, pp. 307–27.

44　*Sixth Report of the Royal Commission on Scientific Instruction and the Advancement of Science*, London: HMSO, 1875.

45　The Royal Society's *Notes and Records* 62 (2008): 1–148 was a special issue devoted to technicians; see also W.H. Brock, *The Fontana History of Chemistry*, London: HarperCollins, 1992, pp. 427–35.

46　C.A. Russell, *Lancastrian Chemist: The Early Years of Sir Edward Frankland*, Milton Keynes: Open University Press, 1986; and 'Frankland', in B. Lightman, ed., *Dictionary of Nineteenth-century British Scientists*, Bristol: Thoemmes, 2004, vol. 2, pp. 727–31.

47　H. Hartley, *Humphry Davy*, London: Nelson, 1966, p. 73.

48　M. Faraday, *Diary*, ed. T. Martin, London: Bell, 1932–6.

49　M.W. Travers, *A Life of Sir William Ramsay*, London: Arnold, 1956, p. 258.

50　See the volume by Berzelius reprinted in D.M. Knight, ed, *The Development of Chemistry*, 10 vols, London: Routledge, 1998.

51　P. Morris, *From Classical to Modern Chemistry: The Instrumental Revolution*, London: Royal Society of Chemistry, 2002.

52　F.A.J.L. James and A. Peers, 'Constructing Space for Science at the Royal Institution of Great Britain', *Physics in Perspective* 9 (2007): 161 (on teamwork, see pp. 168–9); and the special issue of *Centaurus* 39 (1997): 291–381 with international comparisons.

53　R. Hutchings, *British University Observatories, 1772–1939*, Aldershot: Ashgate, 2008.

54　J. Morrell, *John Phillips and the Business of Victorian Science*, Aldershot; Ashgate, 2005, pp. 307–27.

55　See F.A.J.L. James's essay in D.M. Knight and M.D. Eddy, *Science and Beliefs*, Ashgate: Aldershot, 2005.

56 W.H. Flower, *Essays on Museums and Other Subjects Connected with Natural History*, London: Macmillan, 1898, p. 43.

57 E.V. Brunton, *The* Challenger *Expedition, 1872–1876: A Visual Index*, 2nd edn, London: Natural History Museum, 2004, plates 60 and 821; R. Corfield, 'The Chemist Who Saved Biology', *Chemistry World* 5 (2008): 56–60.

58 H. de la Beche, 'Mineralogy', in J.F.W. Herschel, ed., *A Manual of Scientific Enquiry: Prepared for the Use of Officers in Her Majesty's Navy; and Travellers in General* [2nd edn, 1851], Folkestone: Dawson, 1974, p. 245.

59 N. Reingold, *Science in Nineteenth-century America: A Documentary History*, London: Macmillan, 1966.

60 His papers are reprinted in facsimile in D.M. Knight, ed., *Classical Scientific Papers; Chemistry*, 2nd series, London: Mills & Boon, 1970, pp. 71–125, 414–27.

61 J.J. Thomson, 'Cathode Rays', *Proceedings of the Royal Institution* 15 (1896–8): 419–32; another version, *Phil. Mag*, 5th series, 44 (1897): 293–316: reprinted in S. Wright, *Classical Scientific Papers: Physics*, London: Mills & Boon, 1964.

62 W. Thomson, Lord Kelvin, *Baltimore Lectures on Molecular Dynamics and the Wave Theory of Light*, Cambridge: Cambridge University Press, 1904.

63 J. Butt, ed., *The Poems of Alexander Pope*, London: Methuen, 1963, p. 516: 'Essay on Man', Epistle 2, line 2.

Chapter 7 Bodies, Minds and Spirits

1 C. Linnaeus, *Systema Naturae*, 10th edn, Holmiae: Salvii, 1758, vol. 1, pp. 20–4; P.B. Wood, 'The Science of Man', and M.T. Bravo, 'Ethnological Encounters', in N. Jardine, J. Secord and E. Spary, eds, *Cultures of Natural History*, Cambridge: Cambridge University Press, 1996, pp. 197–210, 338–57.

2 I.H.W. Engstrand, *Spanish Scientists in the New World: The Eighteenth Century*, Seattle: Washington University Press, 1981; J. Dunmore, ed., *The Journal of Jean-François de Galaup de la Pérouse, 1785–1788*, London: Hakluyt Society, 1994–5.

3 A. Day, *The Admiralty Hydrographic Service, 1795–1919*, London: HMSO, 1967; G.S. Ritchie, *The Admiralty Chart: British Naval Hydrography in the Nineteenth Century*, London: Hollis and Carter, 1967; N.A.M. Rodger, *The Command of the Ocean: A Naval History of Britain, 1649–1815*, London: Penguin, 2004.

4 J. Beaglehole, *The Life of Captain James Cook*, London: Hakluyt Society, 1974; ed., *The* Endeavour *Journal of Joseph Banks, 1768–1771*, Sydney: Angus and Robertson, 1962.

5 J.J. Rousseau, *Botany: A Study of Pure Curiosity*, tr. K. Ottevanger, illus. P.J. Redouté, London: Michael Joseph, 1979.

6 J. Beaglehole, ed., *The Journals of Captain James Cook: The Voyage of the* Resolution *and* Adventure, *1772–1775*, London, Hakluyt Society, 1969, pp. 749–52.

7 S.N. Mukherjee, *Sir William Jones: A Study in 18th-century British Attitudes to India*, Cambridge: Cambridge University Press, 1968; P.J. Marshall, ed., *The British Discovery of Hinduism in the 18th Century*, Cambridge: Cambridge University Press, 1970.

8 P.R. Sweet, *Wilhelm von Humboldt: A Biography*, Columbus, OH: Ohio State University Press, 1978–80; K.F. Schinkel, *The English Journey, 1826*, ed. D. Bindman and G. Riemann, tr. F.G. Walls, New Haven, CT: Yale University Press, 1993.

9 W. von Humboldt, *On Language: The Diversity of Human Language-structure and its Influence on the Mental Development of Mankind*, tr. P. Heath, intr. H.A Arsleff, Cambridge: Cambridge University Press, 1988.

10 Genesis 11: 1–9.

11 G. Radick, *The Simian Tongue: The Long Debate about Animal Language*, Chicago, IL: Chicago University Press, 2007.

12 I.B. Cohen, ed., *Isaac Newton's Papers and Letters on Natural Philosophy*, Cambridge: Cambridge University Press, 1958, pp. 298, 302.

13 J. Priestley, *Disquisitions Relating to Matter and Spirit*, 2nd edn, London: Johnson, 1782.

14 R.J. Boscovich, *A Theory of Natural Philosophy* [1763], tr. J.M. Child, Cambridge, MA: MIT Press, 1966.

15 R.E. Schofield, ed., *A Scientific Autobiography of Joseph Priestley (1733–1804)*, Cambridge, MA: MIT Press, 1966; P. Wood, *Science and Dissent in England, 1688–1945*, Aldershot: Ashgate, 2004.

16 J.C. Lavater, *Essays on Physiognomy: Designed to Promote the Knowledge and the Love of Mankind*, tr. T. Holcroft, 10th edn, London: Tegg, 1858.

17 M. Kemp, *The Human Animal in Western Art and Science*, Chicago, IL: Chicago University Press, 2007, pp. 212–42.

18 J. van Wyhe, *Phrenology and the Origins of Victorian Scientific Naturalism*, Aldershot: Ashgate, 2004.

19 J.G. Spurzheim, *The Physiognomical System of Drs Gall and Spurzheim; founded upon an Anatomical and Physiological Examination of the Nervous System in General and the Brain in Particular; and Indicating the Dispositions and Manifestations of the Mind*, 2nd edn, London: Baldwin, Cradock and Joy, 1815; S.R. Wells, *How to Read Character: A New Illustrated Handbook of Phrenology and Physiognomy for Students and Examiners, with a Descriptive Chart* [1871], Rutland, VT: Tuttle, 1971.

20 G. Combe, *The Constitution of Man Considered in Relation to External Objects* [1828], New York: Pearson, 1835: in this edition, it is the first part of a collection of similar works, in small type and double columns. More than 350,000 copies of the book had been sold around the world by 1899.

21 F.N. Egerton, *Hewett Cottrell Watson: Victorian Plant Ecologist and Evolutionist*, Aldershot: Ashgate, 2003; J.A. Secord, *Victorian Sensation: The Extraordinary Publication, Reception, and Secret Authorship of* Vestiges of the Natural History of Creation, Chicago, IL: Chicago University Press, 2000.

22 B. Smith, *European Vision and the South Pacific, 1768–1850: A Study in the History of Art and Ideas*, Oxford: Oxford University Press, 1960; A. Moyal, *A Bright and Savage Land: Scientists in Colonial Australia*, Sydney: Collins, 1986; T. Bonyhandy, *The Colonial Image: Australian Painting, 1800–1880*, Chippendale, Australia: Ellsyd Press, 1987; J. Hackforth-Jones, *The Convict Artists*, Melbourne: Macmillan, 1977; B. Berzins, *The Coming of the Strangers; Life in Australia, 1788–1822*, Sydney: Collins, 1988.

23 D.C. Lindberg and R.I. Numbers, eds, *Where Science and Christianity Meet*, Chicago, IL: Chicago University Press, 2003.

24 M. Harrison, *Disease and the Modern World: 1500 to the Present Day*, Cambridge: Polity, 2004, pp. 73–90.

25 T.R. Malthus, *First Essay on Population, 1798*, London: Macmillan, 1966, p. 48; G. Catlin, *O-Kee-Pa: A Religious Ceremony and Other Customs of the Mandans*, ed. J.C. Ewers, New Haven, CT: Yale University Press, 1967, p. 7, and *Letters and Notes of the Manners, Customs, and Conditions of the North American Indians* [1841], Minneapolis: Ross & Haines, 1965, vol. 1, p. 16; P.E. de Strzelecki, *Physical Description of New South Wales and Van Diemen's Land*, London: Longman, 1845, pp. 343–6.

26 D. Kennedy, *The Highly Civilized Man: Richard Burton and the Victorian World*, Cambridge, MA: Harvard University Press, 2005, pp. 131–63.

27 S. Schama, *Rough Crossings: Britain, the Slaves and the American Revolution*, London: BBC, 2005.

28 W. Wordsworth, *Poetry and Prose* [1807], ed. W.M. Marchant, London: Hart-Davis, 1955, p. 540.

29 M. Park, *Travels in the Interior Districts of Africa, . . . in the Years 1795, 1796 and 1797*, London: Bulmer, 1799.

30 T. Winterbottom, *An Account of the Native Africans in the Neighbourhood of Sierra Leone* [1803], London: Routledge, 1969.

31 T. de Quincey, *Confessions of an English Opium-eater, together with Selections from the Autobiography*, ed. E. Sackville-West, London: Cresset, 1950, pp. 74–6.

32 T.L. Peacock, *The Complete Novels*, ed. D. Garnett, London: Hart-Davis, 1963, vol. 1, pp. 91–343.

33 C. Bell, *The Hand: Its Mechanism and Vital Endowments as Evincing Design*, London: Pickering, 1837.

34 C. Bell, *Essays on the Anatomy of Expression in Painting*, London: Bell, 1806.

35 [G. Combe], *An Inquiry into Natural Religion: Its Foundation, Nature, and Applications*, Edinburgh: privately printed for confidential circulation, 1853.

36 J.M. Degérando, *The Observation of Savage Peoples* [1800], ed. E.E. Evans-Pritchard, London: Routledge, 1969; J. Bonnemains, E. Forsyth and B. Smith, eds, *Baudin in Australian Waters; The Artwork of the French Voyage of Discovery to the Southern Lands, 1788–1804*, Melbourne: Oxford University Press, 1988; N. Baudin, *Journal*, tr. C. Cornell, Adelaide: Libraries Board, 1974.

37 J. Waller, in B. Lightman, ed., *Dictionary of Nineteenth-century British Scientists*, Bristol: Thoemmes, 2004, pp. 1635–40.

38 J.C. Prichard, *Researches into the Physical History of Man*, ed. G.W, Stocking, Chicago, IL: Chicago University Press, 1973.

39 J.C. Prichard, *The Natural History of Man*, 3rd edn, London: Bailliere, 1848, pp. 191–3.

40 J. Herschel, ed., *Admiralty Manual of Scientific Enquiry* [1851], intr. D. Knight, Folkestone: Dawson, 1974, pp. 438–50, 166–204.

41 C. Darwin, *The Expression of the Emotions in Man and Animals*, London: John Murray, 1872.

42 T.H. Huxley, *Man's Place in Nature* [1863], in A.P. Barr, ed., *The Major Prose of T.H. Huxley*. Athens, GA: Georgia University Press, 1997, pp. 20–153.

43 C. Lyell, *The Geological Evidences for the Antiquity of Man, with Remarks on Theories of the Origin of Species by Variation*, London: John Murray, 1863, pp. 498–9, 506.

44 C. Darwin, *The Descent of Man and Selection in Relation to Sex*, London: John Murray, 1871; ed. J. Moore and A. Desmond, London: Penguin, 2004.

45 C. Darwin and A.R. Wallace, *Evolution by Natural Selection*, ed. G. de Beer, Cambridge: Cambridge University Press, 1958.

46 On Bates, see J. Dickenson, in *Oxford Dictionary of National Biography*, 2004; S.K. Naylor, in B. Lightman, ed., *Dictionary of Nineteenth-century British Scientists*, Bristol: Thoemmes, 2004, pp. ???.

47 A.R. Wallace, *My Life: A Record of Events and Opinions*, 2nd edn, London: Chapman & Hall, 1908, pp. 334–56.

48 F.A. Mesmer, *Le Magnétisme Animal*, ed. R. Amadou, F.A. Pattie and J. Vinchon, Paris: Payot, 1971.

49 N. Vickers, *Coleridge and the Doctors*, Oxford: Oxford University Press, 2004, pp. 63–78, 79–91; R. Holmes, *Coleridge: Darker Reflections*, London: HarperCollins, 1998, pp. 423–88.

50 R. Holmes, *Coleridge: Early Visions*, London: Hodder & Stoughton, 1989, and, ed., *Coleridge: Selected Poems*, London: HarperCollins, 1996, pp. 101, 81, 229. J.L. Lowes, *The Road to Xanadu: A Study in the Ways of the Imagination* [1927], Boston: Houghton Miflin, 1955.

51 S. Ruston, *Shelley and Vitality*, Basingstoke: Palgrave Macmillan, 2005; M. Shelley, *Frankenstein* [1818], ed. D.L. Macdonald and K. Scherf, 2nd edn, Peterborough, Ontario: Broadview, 1999.

52 S. Poggi and M. Bossi, *Romanticism in Science: Science in Europe, 1790–1840*, Dordrecht: Kluwer, 1994; R.J. Richards, *The Romantic Conception of Life: Science and Philosophy in the Age of Goethe*, Chicago, IL: Chicago University Press, 2002.

53 H.C. Ørsted, *The Soul in Nature, with Supplementary Contributions*, tr. L. and J.B. Horner, London: Bohn, 1852; see D.M. Knight, 'The Spiritual in the Material', in R.M. Brain, R.S. Cohen and O. Knudsen, eds, *Hans Christian Ørsted and the Romantic Legacy in Science*, Dordrecht: Springer, 2007, pp. 417–32.

54 K. von Reichenbach, *Researches on Magnetism, Electricity, Heat, Light, Crystallization, and Chemical Attraction, in their Relation to the Vital Force*, tr. W. Gregory, London: Taylor, Walton and Gregory, 1850.

55 W.R. Cross, *The Burned-over District: The Social and Intellectual History*

of Enthusiastic Religion in Western New York, 1800–1850, Ithaca, NY: Cornell University Press, new edn, 1982, pp. 345–52.

56 J. Oppenheim, *The Other World: Spiritualism and Psychical Research in England, 1850–1914*, Cambridge: Cambridge University Press, 1985.

57 R. Browning, 'Mr Sludge, "The Medium"', in *Dramatic Monologues*, ed. A.S. Byatt, London: Folio Society, 1991, pp. 206–52; on Home, see A. Gauld, in *Oxford Dictionary of National Biography*, 2004; P. Lamont, *The First Psychic: The Peculiar Mystery of a Notorious Victorian Wizard*, London: Little, Brown, 2005.

58 W. Crookes, 'Spiritualism Viewed by the Light of Modern Science', 'Experimental Investigation of a New Force', and 'Some Further Experiments on Psychic Force', *Quarterly Journal of Science* 7 (1870): 316–21; 8 (1871): 339–49, 471–93; W.H. Brock, *William Crookes and the Commercialization of Science*, Aldershot: Ashgate, 2008.

59 A. Gauld, *The Founders of Psychical Research*, London: Routledge, 1968.

60 E. Gurney, F.W.H. Myers and F. Podmore, *Phantasms of the Living*, London: SPR, 1886.

61 D.M. Knight, *Science in the Romantic Era*, Aldershot: Ashgate, 1998, pp. 317–24.

62 F.W.H. Myers, *Human Personality and its Survival of Bodily Death* [1903]; abridged editions, Norwich, Pelegrin, 1992; Charlottesville, VA: Hampton Roads, 2001.

63 J.M. Charcot, *Clinical Lectures on Senile and Chronic Diseases*, tr. W.S. Tuke, London: New Sydenham Society, 1881.

64 E. Haeckel, *Kunstformen der Natur* [1862, 1904], ed. R.P. Hartmann, O. Breidbach and I. Eibl-Eibesfeldt, München: Prestel, 1998.

65 A. Owen, *The Place of Enchantment: British Occultism and the Culture of the Modern*, Chicago, IL: Chicago University Press, 2004.

Chapter 8 The Time of Triumph

1 B. Hilton, *A Mad, Bad & Dangerous People?: England 1783–1846*, Oxford: Oxford University Press, 2006.

2 Registrar General, *Annual Summary of Births, Deaths, and Causes of Death in London, 1874*, London: Eyre and Spottiswoode, 1875, pp. iii–viii.

3 K.T. Hoppen, *The Mid-Victorian Generation, 1846–1886*, Oxford: Oxford University Press, 1998, pp. 275–315.

4 C. Babbage, *Reflections on the Decline of Science in England*, London: Fellowes, 1830, pp. 131–5.

5 J. Auerbach, *The Great Exhibition of 1851: A Nation on Display*, New Haven, CT: Yale University Press, 1999; H. Hobhouse, *The Crystal Palace and the Great Exhibition; Art, Science and Productive Industry: A History of the Royal Commission for the Exhibition of 1851*, London: Continuum, 2002.

6 J.R. Davis, *The Great Exhibition*, Sutton: Stroud, 1999.

7 *The Illustrated Exhibitor: A Tribute to the World's Industrial Jubilee*, London: Cassell, 1851, p. 346.

8 *The Art Journal Illustrated Catalogue of the Industry of All Nations* [1851], Newton Abbott: David & Charles, 1970.

9 I. Armstrong, *Victorian Glassworlds: Glass Culture and the Imagination, 1830–1880*, Oxford: Oxford University Press, 2008.

10 J. Buzard, J.W. Childers, and E. Gillooly, ed., *Victorian Prism: Refractions of the Crystal Palace*, Charlottesville, VA: University of Virginia Press, 2007.

11 K.T. Hoppen, *The Mid-Victorian Generation, 1846–1886*, Oxford: Oxford University Press, 1998, p. 2.

12 J. Scott Russell, 'On the Crystal Palace Fire', *Proceedings of the Royal Institution* 5 (1866–9): 18–24.

13 S. Forgan and G. Gooday, 'Constructing South Kensington: The Buildings and Politics of T.H. Huxley's Working Environments', *BJHS* 29 (1996): 435–68.

14 *The Illustrated Exhibitor*, London: Cassell, 1851, p. 3.

15 R. Hunt, *The Art Journal Illustrated Catalogue*, Newton Abbott: David & Charles, 1970, pp. i, iv, x, xiii, xvi.

16 The Society of Arts, *Lectures on the Results of the Great Exhibition of 1851*, London: Bogue, 1852, p. 33, later, Captain Washington, p. 547, and Playfair, p. 196.

17 L. Playfair, Society of Arts, *Lectures on the Results of the Great Exhibition of 1851*, London: Bogue, 1852, p.196.

18 D. Edgerton, *Science, Technology and the British Industrial 'Decline', 1870–1970*, Cambridge: Cambridge University Press, 1996; R. Bud, S. Niziol, T. Boon and A. Nahum, *Inventing the Modern World: Technology since 1750*, London: Science Museum, 2000.

19 C. Babbage, *The Exposition of 1851: Or, Views of the Industry, the Science, and the Government, of England*, London: John Murray, 1851, p. 189; see also his *Reflections on the Decline of Science in England, and on Some of its Causes*, London: Fellowes, 1830.

20 A.W. Hofmann, 'On the Combining Power of Atoms', *Proceedings of the*

Royal Institution 4 (1862–6): 401–30; reprinted in D.M. Knight, ed., *The Development of Chemistry*, London: Routledge, 1998, vol. 1; A.J. Rocke, 'The Theory of Chemical Structure and its Applications', in M.J. Nye, *Cambridge History of Science*, Cambridge: Cambridge University Press, 2003, vol. 5, pp. 266–71.

21 A.W. Hofmann, 'On Mauve and Magenta', *Proceedings of the Royal Institution* 3 (1858–62): 468–83.

22 W.H. Brock, *Fontana History of Chemistry*, London: HarperCollins, 1992, pp. 241–69; P.J. Ramberg, *Chemical Structure, Spatial Arrangement*, Aldershot: Ashgate, 2003.

23 W.H. Brock, ed., *The Atomic Debates*, Leicester: Leicester University Press, 1967.

24 S. Alvarez, J. Sales and M. Seco, 'On Books and Chemical Elements', *Foundations of Chemistry* 10 (2008): 79–100; H.W. Schütt, 'Chemical Atomism and Chemical Classification', in M.J. Nye, ed., *Cambridge History of Science*, Cambridge: Cambridge University Press, 2003, vol. 5, pp. 237–54; D.M. Knight, *Ideas in Chemistry: A History of the Science*, 2nd edn, London: Athlone, 1995, pp. 128–41, and ed., *Classical Scientific Papers: Chemistry*, 2nd series, *Papers on the Nature and Arrangement of the Chemical Elements*, London: Mills & Boon, 1970.

25 A.W. Hofmann, 'On Mauve and Magenta', *Proceedings of the Royal Institution* 3 (1858–62): 482–3; cf. 4 (1862–6): 430.

26 J. Golinski, *British Weather and the Climate of the Enlightenment*, Chicago, IL: Chicago University Press, 2007.

27 L. Jenyns, *Observations in Meteorology*, London: van Voorst, 1858; I. Wallace, ed., *Leonard Jenyns: Darwin's Life-long Friend*, Bath: Royal Literary and Scientific Institution, 2003; F. Arago, *Meteorological Essays*, intr. A. von Humboldt, trans. E. Sabine, London: Longman, 1855; R. FitzRoy, *The Weather Book*, London: Longman, 1863.

28 L.M. Antony, ed., *Philosophers without Gods: Reflections on Atheism and the Secular Life*, Oxford: Oxford University Press, 2007.

29 J. Sutherland, *Mrs Humphry Ward: Eminent Victorian, Pre-eminent Edwardian*, Oxford: Oxford University Press, 1990.

30 R. Ashton, *142 Strand: A Radical Address in Victorian London*, London: Vintage, 2008, pp. 26–32, 51–81.

31 F.M. Turner, *John Henry Newman: The Challenge to Evangelical Religion*, New Haven, CT: Yale University Press, 2002; contrast this with the more teleological account of Newman's life in S. Gilley, *Newman and His Age*, London: Darton, Longman and Todd, 1990.

32 J.A. Froude, *The Nemesis of Faith*, 2nd edn, London: Chapman, 1849; J.W. Goethe, *Novels and Tales*, London: Bohn, 1854.

33 F.W. Newman, *Phases of Faith* [1850], ed. U.G. Knoefflmacher, Leicester: Leicester University Press, 1970, p. 175.

34 K.T. Hoppen, *The Mid-Victorian Generation, 1846–1886*, Oxford: Oxford University Press, 1998, pp. 427–71; C. Clark, 'Religion and Confessional Conflict', in J. Retallack, ed., *Imperial Germany, 1871–1918*, Oxford: Oxford University Press, 2008, pp. 83–105.

35 A. Tennyson, *In Memoriam* [1850], ed. S. Shatto and M. Shaw, Oxford: Oxford University Press, 1982, pp. 114, 80, 136, 134, 148; on *Vestiges*, see J.A. Secord, *Victorian Sensation*, Chicago, IL: Chicago University Press, 2000.

36 H.L. Mansel, *The Limits of Religious Thought*, 4th edn, London: John Murray, 1859; B. Lightman, *The Origins of Agnosticism: Victorian Unvbelief and the Limits of Knowledge*, Baltimore: Johns Hopkins University Press, 1987, pp. 32–67. See also A. Pyle, ed, *Agnosticism: Contemporary Responses to Spencer and Huxley*, Bristol: Thoemmes, 1995.

37 B. Lightman, 'Scientists as Materialists: Tyndall's Belfast Address', and G. Dawson, 'Victorian Periodicals and the Making of William Kingdon Clifford's posthumous Reputation', in G. Cantor and S. Shuttleworth, ed., *Science Serialized*, Cambridge, MA: MIT Press, 2004, pp. 199–237, 259–84.

38 H. Maudsley, *Lessons of Materialism*, London: Sunday Lecture Society, 1879, p. 24.

39 Ruth Barton's book on the X-club will be published in 2010 by Ashgate.

40 A.W. Hofmann, 'On Mauve and Magenta', *Proceedings of the Royal Institution* 3 (1860–2): 468–83; swatches of fabric dyed with these colours are pasted in at the end of the article.

41 South Kensington Museum, *Free Evening Lectures, delivered in Connection with the Special Loan Collection of Scientific Apparatus*, and *Conferences held in Connection with the Special Loan Collection of Scientific Apparatus: Physics and Mathematics*, and *Chemistry, Biology, Physical Geography, Mineralogy, and Meteorology*, London: Chapman and Hall, 1876.

42 P. Morris, in J. Schummer, B. Bensaude-Vincent and B. van Tiggelen, *The Public Image of Chemistry*, London: World Scientific, 2007, pp. 297–303.

43 G.R. de Beer, *Sir Hans Sloane and the British Museum*, Oxford: Oxford University Press, 1953.

44 H.B. Carter, *Sir Joseph Banks,1743–1820*, London: British Museum (Natural History), 1988, has plans of the house on pp. 334–5.

45 M. Flinders, *A Voyage to Terra Australis*, London: Nicol, 1814, vol. 2, pp. 533–613.

46 N.A. Rupke, *Richard Owen: Victorian Naturalist*, New Haven, CT: Yale University Press, 1994.

47 *Synopsis of the Contents of the Museum*, London: Royal College of Surgeons, 1850; R. Owen, *Hunterian Lectures in Comparative Anatomy, May–June 1837*, ed. P.R. Sloan, London: Natural History Museum, 1992; J.W. Gruber and J.C. Thackray, *Richard Owen Commemoration*, London: Natural History Museum, 1992.

48 B. Lightman, *Victorian Popularizers of Science: Designing Nature for New Audiences*, Chicago, IL: Chicago University Press, 2007, pp. 167–218; G. Pancaldi, 'Museums', in J.L. Heilbron, ed, *The Oxford Companion to the History of Modern Science*, Oxford: Oxford University Press, 2003, pp. 550–1.

49 W.H. Flower, *Essays on Museums*, London: Macmillan, 1898.

50 H. Ewing, *The Lost World of James Smithson: Science, Revolution, and the Birth of the Smithsonian Institution*, London: Bloomsbury, 2007.

51 J. Chalmers, *Audubon in Edinburgh, and His Scottish Associates*, Edinburgh: National Museum of Scotland, 2003; J. Elphick, *Birds: The Art of Ornithology*, London: Scriptum, 2004; A.M. Coates, *The Book of Flowers: Four Centuries of Flower Illustration*, London: Phaidon, 1973.

52 C. Darwin, ed., *Zoology of the Voyage of* HMS Beagle *During the Years 1832–1836* [1838–43], Wellington, NZ: Nova Pacifica, 1980.

53 R. Desmond, *Great Natural History Books and Their Creators*, London: British Library, 2003.

54 M.J. Crowe, *The Extraterritorial Life Debate, 1750–1900: The Idea of a Plurality of Worlds from Kant to Lowell*, Cambridge: Cambridge University Press, 1986.

55 M.P. Crosland, *Science under Control: The French Academy of Sciences, 1795–1914*, Cambridge: Cambridge University Press, 1992, pp. 241, 404–5; A.J. Rocke, *Nationalizing Science: Adolphe Wurtz and the Battle for French Chemistry*, Cambridge, MA: MIT Press, 2001.

56 H. Hawkins, *Pioneer: A History of the Johns Hopkins University, 1874–1889*, Ithaca, NY: Cornell University Press, 1960.

Chapter 9   Science and National Identities

1 H. Trevor-Roper, *The Invention of Scotland: Myth and History*, ed. J.J. Cater, New Haven, CT: Yale University Press, 2008; W. St Clair, *The Reading Nation in the Romantic Period*, Cambridge: Cambridge University Press, 2004, pp. 632–44.

2 A. Walsham, 'Recording Superstition in Early Modern Britain', *Past and Present*, Supplement 2 (August 2008):178–206.

3 M. Girouard, *The Return to Camelot: Chivalry and the English Gentleman*, New Haven, CT, Yale University Press, 1981.

4 L. Colley, *Britons: Forging the Nation, 1707–1837*, New Haven, CT: Yale University Press, 1992; P. Langford, *A Polite and Commercial People: England, 1727–1783*, Oxford: Oxford University Press, 1989, pp. 389–459.

5 M. Beretta, *Imaging a Career in Science: The Iconography of Antoine Laurent Lavoisier*, Canton, MA: Science History, 2002.

6 A.J. Rocke, *Nationalizing Science: Adolphe Wurtz and the Battle for French Chemistry*, Cambridge, MA: MIT Press, 2001.

7 J. Retallack, ed., *Imperial Germany, 1871–1918*, Oxford: Oxford University Press, 2008.

8 M.D. Gordin, K. Hall and A. Kojevnikov, eds, 'Intelligentsia Science: The Russian Century, 1860–1960', *Osiris* 3, 2008; S. Dixon, 'Superstition in Imperial Russia', in S.A. Smith and A. Knight, eds, *The Religion of Fools? Superstition Past and Present*, *Past and Present*, Supplement 3 (2008): 207–28.

9 H.C. Ørsted, *Selected Scientific Works*, tr. and ed. K. Jelved, A.D. Jackson and O. Knudsen, Princeton, NJ: Princeton University Press, 1998; R.M. Brain, R.S. Cohen and O. Knudsen, eds, *Hans Christian Ørsted and the Romantic Legacy in Science*, Dordrecht: Springer, 2007.

10 D.M. Knight and H. Kragh, eds, *The Making of the Chemist: The Social History of Chemistry in Europe, 1789–1914*, Cambridge: Cambridge University Press, 1998.

11 A. Nieto-Galan, ed., *Science, Technology and the Public in the European Periphery*, Aldershot: Ashgate, 2009.

12 W. Roxburgh, *Plants of the Coast of Coromandel, Selected from Drawings and Descriptions Presented to the Court of Directors of the East India Company, and Published under the Direction of Sir Joseph Banks*, London: Nicol, 1795.

13 T.B. Macaulay, *Selected Letters*, ed. T. Pinney, Cambridge: Cambridge University Press, 1974, pp. 149–59.

14 C.A. Bayly, *Empire & Information: Intelligence Gathering and Social Communication in India, 1780–1870*, Cambridge: Cambridge University Press, 1996.

15 R. Desmond, *The India Museum, 1801–1879*, London: HMSO, 1982; M. Archer, *Natural History Drawings in the India Office Library*, London: HMSO, 1962; and *British Drawings in the India Office Library*, London: HMSO, 1969; and *Company Drawings in the India Office Library*, London: HMSO, 1972.

16 E. Kaempfer, *The History of Japan, together with a Description of the Kingdom of Siam, 1690–1*, tr. J.G. Scheuchzer, Glasgow, 1906; V.M. Golownin, *Memoirs of a Captivity in Japan, 1811–1813* [1824], intr. J. McMaster, Oxford: Oxford University Press, 1973.

17 S. Raffles, *Memoir of the Life and Public Services of Sir Thomas Stamford Raffles* [1830], intr. J. Bastin, Oxford: Oxford University Press, 1991; T.S. Raffles, *The History of Java* [1817], intr. J. Bastin, Oxford: Oxford University Press, 2nd edn, 1978; T. Horsfield, *Zoological Researches in Java, and the Neighbouring Islands* [1824], intr. J. Bastin, Oxford: Oxford University Press, 1990.

18 M. Flinders, *A Voyage to Terra Australis . . .*, London: Nicol, 1814; P.P. King, *Narrative of a Survey of the Intertropical and Western Coasts of Australia*, London: John Murray, 1827; J.L. Stokes, *Discoveries in Australia*, London: Boone, 1846; J. MacGillivray, *Narrative of the Voyage of HMS Rasttlesnake . . .*, London: Boone, 1852; T.H. Huxley, *Diary of the Voyage of HMS Rattlesnake*, ed. J. Huxley, New York: Doubleday, 1936.

19 E.H.J. Feeken, G.E.E. Feeken and O.H.K. Spate, *The Discovery and Exploration of Australia*, Melbourne: Nelson, 1970; G. Blainey, *The Tyranny of Distance: How Distance Shaped Australia's History*, Melbourne: Macmillan, 1968.

20 J. Molony, *The Native-Born: The First White Australians*, Melbourne: Melbourne University Press, 2000; R.W. Home, ed., *Australian Science in the Making*, Cambridge: Cambridge University Press, 1988; R. MacLeod, ed., *The Commonwealth of Science: ANZAAS and the Scientific Enterprise in Australasia*, Oxford: Oxford University Press, 1988; R. Home, *Science as a German Export to Nineteenth-century Australia*, London: Institute of Commonwealth Studies, 1995.

21 G.M. Caroe, *William Henry Bragg: Man and Scientist*, Cambridge: Cambridge University Press, 1978; J.M. Thomas and D. Phillips, eds, *Selections and Reflections: The Legacy of Sir Lawrence Bragg*, London: Royal Institution, 1990.

22 A.G. Hopkins, 'Rethinking Decolonization', *Past and Present* 200 (2008): 211–47.

23 J.H. Andrews, *A Paper Landscape: The Ordnance Survey in Nineteenth-Century Ireland*, Oxford: Oxford University Press, 1975; J.A. Secord, 'The Geological Survey of Great Britain as a Research School', *History of Science* 24 (1986): 223–75.

24 C. Molland, *It's Part of What We Are: Some Irish Contributors to the Development of the Chemical and Physical Sciences*, Dublin: Royal Irish Society, 2007.

25 W. Parsons, Lord Rosse, *Scientific Papers*, London: Lund Humphries, 1926.

26 J.P. Nichol, *The Architecture of the Heavens*, London: Parker, 1850.

27 These come from M. Magnussen, ed., *Chambers Biographical Dictionary*, Edinburgh: Chambers, 1996.

28 D.H. Frank and O. Leaman, ed., *History of Jewish Philosophy*, London: Routledge, 1997.

29 T. Shinn, 'The Industry, Research, and Education Nexus', in M.J. Nye, ed., *Cambridge History of Science*, Cambridge: Cambridge University Press, 2003, vol. 5, pp. 133–53.

30 M.P. Crosland, *The Society of Arcueil: A View of French Science at the Time of Napoleon I*, London: Heinemann, 1967; *Science under Control: The French Academy of Sciences, 1795–1914*, Cambridge: Cambridge University Press, 1992.

31 W. Clark, *Academic Charisma and the Origins of the Research University*, Chicago, IL: Chicago University Press, 2006.

32 R. Fox and A. Nieto-Galan, ed., *Natural Dyestuffs and Industrial Culture in Europe, 1750–1880*, Nantucket, MA: Watson, 1999.

33 W. Siemens, *Recollections*, tr. W.C. Coupland [1893], new edn *(Inventor and Entrepreneur)*, London: Lund Humphries, 1966.

34 S.E. Koss, *Sir John Brunner: Radical Plutocrat, 1842–1919*, Cambridge: Cambridge University Press, 1970.

35 P.J. Ramberg, *Chemical Structure, Spatial Arrangement*, Aldershot: Ashgate, 2003; O.T. Benfey, ed., *Classics in the Theory of Chemical Combination*, New York: Dover, 1963.

36 C. Smith and N. Wise, *Energy and Empire: A Biographical Study of Lord Kelvin*, Cambridge: Cambridge University Press, 1989, pp. 335–6, 632–3.

37 J.C. Adams, *Scientific Papers*, ed. W. Grylls, Cambridge: Cambridge University Press, 1896–1900; C. Waff, 'Adams', in B. Lightman, ed.,

Dictionary of Nineteenth-century British Scientists, Bristol: Thoemmes, 2004; L.P. Williams, *Album of Science: The Nineteenth Century*, New York: Scribners, 1978, p. 100..

38 E. Du Bois-Reymond, P. Diepgen and P.F. Cranefield, eds, *Two Great Scientists of the Nineteenth Century: Correspondence of Emil Du Bois-Reymond and Carl Ludwig*, tr. S. Lichtner-Ayèd, Baltimore, MD: Johns Hopkins University Press, 1982.

39 J. Retallack, ed., *Imperial Germany, 1871–1918*, Oxford: Oxford University Press, 2004, esp. the chapters by M. Hewitson and R. Chickering (pp. 40–60, 196–218); M.S. Seligmann, ed., *Naval Intelligence from Germany: The Reports of the British Naval Attachés in Berlin, 1906–1914*, Aldershot: Ashgate, for the Naval Records Society, 2007.

40 K.T. Hoppen, *The Mid-Victorian Generation, 1846–1886*, Oxford: Oxford University Press, 1998, pp. 174–5.

41 J. Egerton, *Turner: The Fighting Temeraire*, London: National Gallery, 1995.

42 G. Roberts, 'Sir John Anderson, 1814–86: The Unknown Engineer Who Made the British Empire Possible', *Transactions of the Newcomen Society* 78 (2008): 261–91.

43 M.R. Smith, *Harpers Ferry Armory and the New Technology*, Ithaca, NY: Cornell University Press, 1977.

44 C. Smith, 'Dreadnought Science', *Transactions of the Newcomen Society* 77 (2007): 191–215.

45 A.R. Hall, *The Abbey Scientists*, London: Roger & Roberts, Nicholson, 1966.

46 See the special issue, *Història Ciêncas, Saúde – Manquinhos* 8, supplement, (2001).

47 D.M. Knight, 'Travels and Science in Brazil', in that issue, pp. 809–22.

48 H.W. Bates, *The Naturalist on the River Amazon*, London: John Murray, 1863.

49 N. Jardine et al., *Cultures of Natural History*, Cambridge: Cambridge University Press, 1996.

50 F. Burckhardt et al., eds, *The Correspondence of Charles Darwin*, Cambridge: Cambridge University Press, 1999, vol. 11, pp. 358–61.

51 M.R.Sa, 'James William Helenus Trail: A British Naturalist in Nineteenth-century Amazonia', *Historia Naturalis* 1 (1998): 99–254, esp. 161–71.

52 W. Ellis, *Three Visits to Madagascar During the Years 1853–54–56*, London: John Murray, 1858.

53 J.C. Ross, *A Voyage of Discovery and Research in the Southern and Antarctic Regions, During the Years 1839–43* [1847], Newton Abbot: David and Charles, 1969.

54 T.H. Levere and R.A. Jarrell, eds, *A Curious Field-Book: Science and Society in Canadian History*, Toronto: Toronto University Press, 1974.

55 J. Franklin, *Narrative of a Journey to the Shores of the Polar Sea* [1823], reprint Rutland, VT: Tuttle, 1970; *Narrative of a Second Journey* [1828], reprint New York: Greenwood, 1969; R. Hood, *To the Arctic by Canoe, 1819–21*, ed. C.S. Houston, Montreal: McGill-Queen's University Press, 1974; J. Richardson, *Arctic Ordeal*, ed. C.S. Houston, Montreal: McGill-Queen's University Press, 1984.

56 W. Barr, ed., *Searching for Franklin: The Land Arctic Searching Expedition*, London: Hakluyt Society, 1999.

Chapter 10   Method and Heresy

1 R. Descartes, *A Discourse of a Method for the Well Guiding of Reason, and the Discovery of Truth in the Sciences*, London: Newcombe, 1649.

2 Aristotle, *De Partibus Animalium*, 639a, tr. W. Ogle, Oxford: Oxford University Press, 1911.

3 C. Babbage, *Reflections on the Decline of Science in England*, London: Fellowes, 1830, p. 206.

4 H.T. Buckle, 'On the Influence of Women in the Progress of Knowledge', *Proceedings of the Royal Institution* 2 (1854–8): 504–5.

5 P. Duhem, *The Aim and Structure of Physical Theory* [1914], tr. P.P. Wiener, New York: Athenaeum, 1962, pp. 55–104.

6 See the 'Connections' series in *Nature* 2007; my 'Kinds of Minds', 10 May issue, p. 149, was the last.

7 A. Comte, *The Positive Philosophy*, ed. and tr. H. Martineau, London: Chapman, 1853; R. Ashton, *142 Strand: A Radical Address in Victorian London*, London: Vintage, 2008.

8 T.R. Wright, *The Religion of Humanity: The Impact of Comtean Positivism on Victorian Britain*, Cambridge: Cambridge University Press, 1986.

9 B. Lightman, *Victorian Popularizers of Science*, Chicago, IL: Chicago University Press, 2007, p. 386; T.H. Huxley, 'The Scientific Aspects of Positivism' [1869], in *Lay Sermons, Addresses, and Reviews*, London: Macmillan, 1877, pp. 147–73; J. Tyndall, 'Scientific Use of the

Imagination' [1870], in *Fragments of Science*, 10th imp., London: Longman, 1899, vol. 2, pp. 101–34.

10 B. Bensaude-Vincent, 'Atomism and Positivism: A Legend about French Chemistry', *Annals of Science* 56 (1999): 81–94; G.G. Stokes, *Mathematical and Physical Papers*, Cambridge: Cambridge University Press, 1883, vol. 2, p. 97.

11 R. Holmes, *The Age of Wonder: How the Romantic Generation Discovered the Beauty and Terror of Science*, London: Harper Press, 2008.

12 S.E. Despaux, 'Mathematics Sent across the Channel and the Atlantic: British Mathematical Contributions to European and American Scientific Journals, 1835–1900', *Annals of Science* 65 (2008): 73–99.

13 J.F.W. Herschel, *A Preliminary Discourse on the Study of Natural Philosophy* [1830], intr. M. Partridge, New York: Johnson, 1966.

14 S. Ruskin, *John Herschel's Cape Voyage: Private Science, Public Imagination, and the Ambitions of Empire*, Aldershot: Ashgate, 2004; on scientific travel, see D.M. Knight, *Public Understanding of Science*, London: Routledge, 2006, pp. 106–18.

15 J.F.W. Herschel, *A Preliminary Discourse*, New York: Johnson, 1966, pp. 196–7.

16 W. Whewell, *Astronomy and General Physics*, 5th edn, London: Pickering, 1836, p. 326; M. Fisch, *William Whewell: Philosopher of Science*, Oxford: Oxford University Press, 1991; R. Yeo, *Defining Science: William Whewell, Natural Knowledge and Public Debate in Early Victorian Britain*, Cambridge: Cambridge University Press, 1993; M. Fisch and S. Schaffer, eds, *William Whewell: A Composite Portrait*, Oxford: Oxford University Press, 1991.

17 W. Whewell, *History of the Inductive Sciences: From the Earliest to the Present Time*, 3 vols, 3rd edn, London: Parker, 1857.

18 B. Powell, *The Connexion of Natural and Divine Truth; or, the Study of the Inductive Philosophy*, London: Parker, 1838; *Essays on the Spirit of the Inductive Philosophy, the Unity of Worlds, and the Philosophy of Creation*, London: Longman, 1855. P. Corsi, *Science and Religion: Baden Powell and the Anglican Debate, 1800–1860*, Cambridge: Cambridge University Press, 1988.

19 R.J. Richards, *The Romantic Conception of Life: Science and Philosophy in the Age of Goethe*, Chicago, IL: Chicago University Press, 2002.

20 J.W. Goethe, *Theory of Colours*, ed. and tr. C.L. Eastlake, London: John Murray, 1840; H. Helmholtz, *Popular Lectures on Scientific Subjects*, tr. E. Atkinson, J. Tyndall et al., London: Longman, 1873, pp. 33–59.

21 J.B. Stallo, *The Concepts and Theories of Modern Physics* [1881], ed. P.W. Bridgmen, Cambridge, MA: Harvard University Press, 1960.

22 R.M. Brain, R.S. Cohen and O. Knudsen, eds, *Hans Christian Ørsted and the Romantic Legacy in Science*, Dordrecht: Springer, 2007.

23 A. Schuster, *The Progress of Physics During 33 Years (1875–1908)*, Cambridge: Cambridge University Press, 1911, pp. 115–17.

24 W. Swainson, *Preliminary Discourse on the Study of Natural History*, London: Longman, 1834; D.M. Knight, *Science and Spirituality: The Volatile Connection*, London: Routledge, 2004, pp. 77–80.

25 D.M. Knight, *Ordering the World: A History of Classifying Man*, London: Burnett, 1981.

26 E. Scerri, *The Periodic Table: Its Story and its Significance*, Oxford: Oxford University Press, 2006; H.W. Schütt, 'Chemical Atomism and Chemical Classification', in M.J. Nye, ed., *Cambridge History of Science*, Cambridge: Cambridge University Press, 2003, vol. 5, pp. 237–54; various tables are reprinted in facsimile in D.M. Knight, ed., *Classical Scientific Papers: Chemistry*, series 2, London: Mills & Boon, 1970, pp. 200–349.

27 M.W. Travers, *The Discovery of the Rare Gases*, London: Edward Arnold, 1928.

28 W.H. Brock, 'Radiant Spectroscopy: The Rare Earths Crusade', Royal Society of Chemistry, *Historical Group Occasional Papers* 5, 2007.

29 J. Thackray, *To See the Fellows Fight: Eye Witness Accounts of Meetings of the Geological Society of London and its Club, 1822–1868*, British Society for the History of Science Monograph 12, 2003.

30 B. Lightman, *Victorian Popularizers of Science: Designing Nature for New Audiences*, Chicago, IL: Chicago University Press, 2007, pp. 318–51.

31 M.A. Salmon, 'Newlands', in B. Lightman, ed., *Dictionary of Nineteenth-century British Scientists*, Bristol: Thoemmes, 2004.

32 J. Morrell, *John Phillips and the Business of Victorian Science*, Aldershot: Ashgate, 2005, pp. 349–70.

33 T.H. Huxley, *Man's Place in Nature* [1863], intr. A. Montagu, Ann Arbor: Michigan University Press, 1959.

34 The papers are reprinted in D.M. Knight, ed., *Classical Scientific Papers, Chemistry*, 2nd series, London: Mills & Boon, 1970, pp. 15–70.

35 C. Babbage, *Reflections on the Decline of Science in England, and Some of its Causes*, London: Fellowes, 1830, pp. 174–83.

36 A. Quetelet, *A Treatise on Man*, Edinburgh: Chambers, 1842, plate 4.

37 H. Woolf, *The Transits of Venus: A Study of Eighteenth-century Science*, Princeton, NJ: Princeton University Press, 1959.

38 G. Mendel, *Experiments on Plant Hybridisation*, tr. and ed. R.A. Fisher, London: Oliver and Boyd, 1965; reprints Fisher's classic critique from *Annals of Science* 1 (1836): 115–37.

39 J. Hunter, *Essays and Observations on Natural History, Anatomy, Physiology, Psychology, and Geology*, ed. R. Owen, London: Van Voorst, 1861, vol. 2, pp. 493–502; N.G. Coley, 'Home', in *Oxford Dictionary of National Biography*, Oxford: Oxford University Press, 2004.

40 L.P. Williams, *Album of Science: The Nineteenth Century*, New York: Scribners, 1978, p. 100.

41 R. Chenevix, 'Enquiries Concerning the Nature of a Metallic Substance Lately Sold in London', *Philosophical Transactions* 93 (1803): 290–320; M.C. Usselman 'Richard Chenevix', in B. Lightman, ed., *Dictionary of Nineteenth-century British Scientists*, Bristol: Thoemmes, 2004, pp. 415–19.

42 D.S. Evans, T.J. Deeming, B.H. Evans and S. Goldfarb, eds, *Herschel at the Cape: Diaries and Correspondence, 1834–1838*, Austin: Texas University Press, 1969, pp. 236–7, 282, and plate 17.

43 [W. Whewell], *Of the Plurality of Worlds: An Essay*, London: Parker, 1853; D. Brewster, *More Worlds than One: The Creed of the Philosopher, and the Hope of the Christian*, London: John Murray, 1854; M.J. Crowe, *The Extraterritorial Life Debate, 1750–1900*, Cambridge: Cambridge University Press, 1986.

44 C. Lyell, *The Geological Evidences of the Antiquity of Man*, London: John Murray, 1863, p. 498.

45 J. Oppenheim, *The Other World: Spiritualism and Psychical Research in England, 1850–1914*, Cambridge: Cambridge University Press, 1985, pp. 26–7.

46 J. Van Wyhe, *Phrenology and the Origins of Victorian Scientific Naturalism*, Aldershot: Ashgate, 2004.

47 F.N. Egerton, *Hewett Cottrell Watson: Victorian Plant Ecologist and Evolutionist*, Aldershot: Ashgate, 2003.

48 I. Lakatos, *The Methodology of Scientific Research Programmes*, Cambridge: Cambridge University Press, 1978.

49 J. Huxley, *The Courtship Habits of the Great Crested Grebe* [1914], intr. D. Morris, London: Jonathan Cape, 1968.

50 H.E. Howard, *Territory in Bird Life*, London: John Murray, 1920; F.E. Beddard, *Animal Colouration*, London: Sonnenschein, 1892.

51 M. Harrison, *Disease and the Modern World, 1500 to the Present*, Cambridge: Polity, 2005, pp. 92–6.

52 A.R. Wallace, *My Life*, new edn, London: Chapman & Hall, 1908, pp. 332–3; M. Fichman, *An Elusive Victorian: The Evolution of Alfred Russel Wallace*, Chicago, IL: Chicago University Press, 2004.

53 C. Bernard, *An Introduction to the Study of Experimental Medicine*, tr. H.C. Green, intr. L.J. Henderson and I.B. Cohen, New York: Dover, 1957.

54 W.E. Bynum et al., *The Western Medical Tradition, 1800 to 2000*, Cambridge: Cambridge University Press, 2006, pp. 111–239.

55 See her entry by Barbara Caine in the *Oxford Dictionary of National Biography*, Oxford: Oxford University Press, 2004.

56 *Theodore Parker's Experience as a Minister, with Some Account of his Early Life, and Education for the Ministry*, London: Watts, 1859.

57 B. Lightman, *Victorian Popularizers of Science: Designing Nature for New Audiences*, Chicago, IL: Chicago University Press, 2007, pp. 95–165.

58 D.M. Knight, 'Why is Science so Macho?', *Philosophical Writings* 14 (2000): 59–65; H. Davy, *Consolations in Travel: Or, the Last Days of a Philosopher*, London: John Murray, 1830, p. 245.

59 A. Owen, *The Place of Enchantment: British Occultism and the Culture of the Modern*, Chicago, IL: Chicago University Press, 2004.

60 K. von Reichenbach, *Researches on Magnetism, Electricity, Heat, Light, Crystallization, and Chemical Attraction, in their Relations to the Vital Force*, tr. W. Gregory, London: Taylor, Walter & Maberly, 1850.

61 J. Oppenheim, *The Other World: Spiritualism and Psychical Research in England, 1850–1914*, Cambridge: Cambridge University Press, 1985.

62 W.H. Brock, *William Crookes (1832–1919) and the Commercialization of Science*, Aldershot: Ashgate, 2008.

63 F. Galton, *The Art of Travel: Or, Shifts and Contrivances Available in Wild Countries* [1872 edn], intr. D. Middleton, Newton Abbott: David & Charles, 1971.

64 W. Shakespeare, *The Tempest*, IV. I, lines 188–92.

65 F. Galton, *Hereditary Genius; An Inquiry into its Laws and Consequences* [1892 3d.], intr. C.D. Darlington, London: Fontana, 1962; J. Waller, 'Francis Galton', in B. Lightman, ed., *Dictionary of Nineteenth-century British Scientists*, Bristol: Thoemmes, 2004.

66 M. Hawkins, *Social Darwinism in European and American Thought, 1860–1945*, Cambridge: Cambridge University Press, 1997.

67 C. Lombroso, *The Female Offender*, London: Fisher Unwin, 1895;

D. Pick, *Faces of Degeneration: A European Disorder, c.1848–c.1918*, Cambridge: Cambridge University Press, 1989.

68 D.A. MacKenzie, *Statistics in Britain, 1865–1930: The Social Construction of Knowledge*, Edinburgh: Edinburgh University Press, 1981.

69 J. Henry, 'Historical and other Studies of Science, Technology and Medicine in the University of Edinburgh', *Notes and Records of the Royal Society* 62 (2008): 223–35, esp. 226–9.

Chapter 11   Cultural Leadership

1 F. Schleiermacher, *On Religion: Speeches to its Cultured Despisers*, ed. and tr. R. Crouter, Cambridge: Cambridge University Press, 1988.

2 C.E. McClelland, *State, Society, and University in Germany, 1700–1914*, Cambridge: Cambridge University Press, 1980, pp. 99–149; W. von Humboldt, *On Language*, tr. P. Heath, intr. H. Aarsleff, Cambridge: Cambridge University Press, 1988.

3 M.P. Crosland, *Science Under Control: The French Academy of Sciences, 1795–1914*, Cambridge: Cambridge University Press, 1992, pp. 11–18; on Academicians' religion, pp. 192–202.

4 J.R. Hofmann, *André–Marie Ampère: Enlightenment and Electrodynamics*, Cambridge: Cambridge University Press, 1995.

5 M. Pickering, *Auguste Comte: An Intellectual Biography*, Cambridge: Cambridge University Press, 1993.

6 R. Holmes, *The Age of Wonder: How the Romantic Generation Discovered the Beauty and Terror of Science*, London: Harper Press, 2008.

7 H. Davy, *Collected Works*, ed. J. Davy, London: Smith, Elder, 1839–40, vol. 9; D.M. Knight, *Humphry Davy: Science and Power*, 2nd edn, Cambridge: Cambridge University Press, 1998, pp. 154–83.

8 C. Babbage, *Reflections on the Decline of Science in England, and on Some of its Causes*, London: Fellowes, 1830.

9 J. Morrell, 'William Venables Vernon Harcourt', in *Oxford Dictionary of National Biography*, Oxford: Oxford University Press, 2004.

10 J. Morrell and A. Thackray, *Gentlemen of Science: Early Years of the British Association for the Advancement of Science*, Oxford: Oxford University Press, 1981; R. MacLeod and P. Collins, eds, *The Parliament of Science: The British Association for the Advancement of Science 1831–1981*, London: Science Reviews, 1981; J. Morrell, *John Phillips and the Business of Victorian Science*, Aldershot: Ashgate, 2005, pp. 39–128.

11 C. Withers, R. Higgitt and D. Finnegan, 'Historical Geographies of Provincial Science: Themes in the Setting and Reception of the British

Association for the Advancement of Science in Britain and Ireland, 1831–c.1939, *BJHS* 41 (2008): 385–416.

12 T.H. Huxley, 'The Reception of the "Origin of Species"', in F. Darwin, ed., *The Life and Letters of Charles Darwin*, London: John Murray, 1887, vol. 2, p. 186.

13 W.M. Jacob, *The Clerical Profession in the Long Eighteenth Century, 1680–1840*, Oxford: Oxford University Press, 2007; D.M. Knight, *Science and Spirituality: The Volatile Connection*, London: Routledge, 2004, pp. 151–66.

14 R. Ashton: *142 Strand: A Radical Address in Victorian London*, London; Vintage, 2008; Ruth Barton is writing a book about the X-club.

15 S. Butler, *Alps and Sanctuaries of Piedmont and the Canton Ticino* [1881], intr. R.A. Streatfield, Gloucester: Sutton, 1986, p. 66; B. Lightman, *Victorian Popularizers of Science*, Chicago, IL: Chicago University Press, 2007, pp. 289–94.

16 A. Hume, *The Learned Societies and Printing Clubs of the United Kingdom*, London: Willis, 1853.

17 B. Lightman, 'Scientists as Materialists in the Periodical Press: Tyndall's Belfast Address', in G. Cantor and S. Shuttleworth, eds, *Science Serialized*, Cambridge, MA: MIT Press, 2004, pp. 199–237.

18 K.T. Hoppen, *The Mid-Victorian Generation, 1846–1886*, Oxford: Oxford University Press, 1998, pp. 31–55, 375, 419.

19 J. Barwell-Carter, *Selections from the Correspondence of Dr George Johnston*, Edinburgh: Douglas, 1892, pp. 110, 122–3, and cf. 135, 491–2, 515.

20 D.E. Allen, *The Victorian Fern Craze*, London: Hutchinson, 1969; *The Naturalist in Britain: A Social History*, London: Penguin, 1978; *Naturalists and Society*, Aldershot: Ashgate Variorum, 2003.

21 *Principal Excursions of the Innerleithen Alpine Club During the Years 1889–94*, Galashiels: McQueen, 1895.

22 L. Oken, *Elements of Physiophilosophy*, tr. A.Tulk, London: Ray Society, 1847.

23 P.H. Gosse, *Omphalos: An Attempt to Untie the Geological Knot*, London: Van Voorst, 1857, has an advertisement at the back for such a holiday, and for books; A. Thwaite, *Glimpses of the Wonderful: The Life of Philip Henry Gosse*, London: Faber, 2002.

24 Walter White, *The Journals*, London: Chapman and Hall, 1898.

25 N. Reingold, ed., *Science in Nineteenth-century America: A Documentary History*, London: Macmillan, 1966.

26 J.C. Ross, *A Voyage of Discovery and Research in the Southern and Antarctic Regions*, London: John Murray, 1847.

27 H.K. Beals et al., eds, *Four Travel Journals*, London: Hakluyt Society, 2007, pp. 253–327.

28 E.V. Brunton, *The* Challenger *Expedition, 1872–1876: A Visual Index*, 2nd edn, London: Natural History Museum, 2004; H.N. Moseley, *Notes by a Naturalist on* HMS Challenger, London: John Murray, 1892.

29 J.A. Secord, ''The Geological Survey of Great Britain as a Research School, 1839–55', *History of Science* 24 (1886): 223–75; M.J.S. Rudwick, *The Great Devonian Controversy*, Chicago, IL: Chicago University Press, 1985; D.R. Oldroyd, *The Highlands Controversy: Constructing Geological Knowledge through Fieldwork in Nineteenth-century Britain*, Chicago, IL: Chicago University Press, 1990.

30 F.A.J.L. James, ed., *The Common Purposes of Life: Science and Society at the Royal Institution*, Aldershot: Ashgate, 2002

31 M.P. Crosland, *The Society of Arcueil: A View of French Science at the Time of Napoleon I*, London: Heinemann, 1967.

32 C.A. Russell, N.G. Coley and G.K. Roberts, *Chemists by Profession*, Milton Keynes: Open University Press, 1977.

33 L.T. Hobhouse, *Manchester Guardian*, 1 January 1901.

34 W.H. Brock and A.J. Meadows, *The Lamp of Learning: Taylor & Francis and the Development of Science Publishing*, London: Taylor & Francis, 2nd edn, 1988.

35 B. Lightman, *Victorian Popularizers of Science*, Chicago, IL: Chicago University Press, 2007; L. Henson et al., *Culture and Science in the Nineteenth-century Media*, Aldershot: Ashgate, 2004; G. Cantor et al., *Science in the Nineteenth-century Periodical: Reading the Magazine of Nature*, Cambridge: Cambridge University Press, 2004; G. Cantor and S. Shuttleworth, eds, *Science Serialized: Representations of Science in Nineteenth-century Periodicals*, Cambridge, MA: MIT Press, 2004.

36 D.M. Knight, 'Science and Culture in Mid-Victorian Britain: The Reviews, and William Crookes', *Quarterly Journal of Science', Nuncius* 11 (1996): 43–54; W.H. Brock, *William Crookes (1832–1919) and the Commercialization of Science*, Aldershot: Ashgate, 2008.

37 A. Fyfe, *Science and Salvation: Evangelical Popular Science Publishing in Victorian Britain*, Chicago, IL: Chicago University Press, 2004; B. Hilton, *A Mad, Bad & Dangerous People? England 1783–1846*, Oxford: Oxford University Press, 2006, pp. 332–42; K.T. Hoppen, *The Mid-*

Victorian Generation: England 1846–1886, Oxford: Oxford University Press, 1998, pp. 427–71.

38 R.M. Brain, R.S. Cohen and O. Knidsen, eds, *Hans Christian Ørsted and the Romantic Legacy in Science*, Dordrecht: Springer, 2007; my essay is on pp. 417–32.

39 [R.Chambers], *Vestiges of the Natural History of Creation*, ed. J.A. Secord, Chicago, IL: Chicago University Press, 1994; B. Lightman, *Victorian Popularizers of Science*, Chicago, IL: Chicago University Press, 2007, pp. 219–95.

40 M.J.S. Rudwick, *Scenes from Deep Time*, Chicago, IL: Chicago University Press, 1992.

41 K.T. Hoppen, *The Mid-Victorian Generation, England 1846–1886*, Oxford: Oxford University Press, 1998, pp. 472–510.

42 J. Sutherland, *Mrs Humphry Ward: Eminent Victorian, Pre-eminent Edwardian*, Oxford: Oxford University Press, 1990, pp. 83–131.

43 *Ernst-Haeckel-Haus der Universität Jena, Museum*, Braunschweig: Westermann, 1990.

44 J. Tyndall, *Fragments of Science*, 10th imp., London: Longman, 1899, vol. 2, pp. 101–34.

45 M.T. Brück, 'Smyth', in B. Lightman, ed., *Dictionary of Nineteenth-century British Scientists*, Bristol: Thoemmes, 2004.

46 J. Hamilton, *Turner and the Scientists*, London: Tate Gallery, 1998.

47 W.H. Brock, *William Crookes (1832–1919) and the Commercialization of Science*, Aldershot: Ashgate, 2008.

48 C. Kelly, ed., *Mrs Duberly's War: Journal and Letters from the Crimea*, Oxford: Oxford University Press, 2007, p. 312.

49 R. Liebreich, 'Turner and Mulready: On the Effect of Certain Faults of Vision on Painting, with especial Reference to their Works, *Proceedings of the Royal Institution* 6 (1870–2): 450–63.

50 H. Helmholtz, *Popular Lectures on Scientific Subjects*, tr. E. Atkinson et al., London: Longman, 1873, pp. 197–316.

51 J.W. Goethe, *Elective Affinities*, tr. R.J. Hollingdale, London: Penguin, 1978; S. Ruston, *Shelley and Vitality*, Basingstoke: Palgrave Macmillan, 2005.

52 M. Shelley, *Frankenstein: Or, the Modern Prometheus* [1818], ed. D.L. Macdonald and K.Scherf, 2nd edn, Peterborough, Ontario: Broadview, 1999.

53 W. Newman, *Promethean Ambitions: Alchemy and the Quest to Perfect Nature*, Chicago, IL: Chicago University Press, 2004; H. Collins and T.

Pinch, *The Golem: What Everyone Should Know about Science*, Cambridge: Cambridge University Press, 1993; J. Pelikan, *Faust the Theologian*, New Haven, CT: Yale University Press, 1995.

54 A. Ure, in *Register of Arts and Sciences* 1 (1824): 3–5; D.M. Knight, *Public Understanding of Science: A History of Communicating Scientific Ideas*, London: Routledge, 2006, pp. 29–30.

55 A. Tennyson, *In Memoriam*, ed. S. Shatto and M. Shaw, Oxford: Oxford University Press, 1982, sections 55, 120, 124.

56 L. Huxley, *The Life and Letters of Thomas Henry Huxley*, London: Macmillan, 1913, vol. 3, pp. 269–70; M.H. Cooke, *The Evolution of Nettie Huxley, 1825–1914*, Chichester: Phillimore, 2008, p. 114.

57 T.H. Huxley, *Major Prose*, ed. A.P. Barr, Athens, GA: Georgia University Press, 1997, pp. 283–344.

58 K.T. Hoppen, *The Mid-Victorian Generation: England 1846–1886*, Oxford: Oxford University Press, 1998, p. 394.

59 H. Helmholtz, *On the Sensations of Tone as a Physiological Basis for the Theory of Music* [1885], tr. A.J. Ellis, intr. H. Margenau, New York: Dover, 1954; quotation from introduction [p. 4] .

60 P. Fara, *Pandora's Breeches: Women, Science and Power in the Enlightenment*, London: Pimlico, 2004; M.W. Rossiter 'A Twisted Tale: Women in the Physical Sciences in the Nineteenth and Twentieth Centuries', in M.J. Nye, *Cambridge History of Science*, Cambridge: Cambridge University Press, 2003, vol. 5, pp. 54–71.

61 B. Lightman, *Victorian Popularizers of Science*, Chicago, IL: Chicago University Press, 2007, pp. 95–165.

62 L. de Vries, *Victorian Advertisements*, text by J. Laver, London: John Murray, 1968.

63 D.M. Knight, *Public Understanding of Science*, London: Routledge, 2006, pp. 135–52.

Chapter 12 Into the New Century

1 J. Krige and D. Pestre, *Science in the Twentieth Century*, Amsterdam: Harwood, 1997; A. Roland, 'Science, Technology and War', in M.J. Nye, *Cambridge History of Science*, Cambridge: Cambridge University Press, 2003, vol. 5, pp. 561–78.

2 A. Porter, ed., *The Oxford History of the British Empire: The Nineteenth Century*, Oxford: Oxford University Press, 1999; see esp. R.A. Stafford, 'Scientific Exploration and Empire', pp. 294–319.

3 H.B. Carter, *Sir Joseph Banks*, London: British Museum (Natural History), 1988, pp. 64–98, 101–3.

4 R.W. Home, ed., *Australian Science in the Making*, Cambridge: Cambridge University Press, 1988; *Science as a German Export to Australia*, London: Sir Robert Menzies Centre for Australian Studies, 1995.

5 F. Locher, 'The Observatory, the Land-based Ship and the Crusades: Earth Sciences in European Context, 1830–50, *BJHS* 40 (2007): 491–504.

6 A. von Humboldt, *Aspects of Nature, in Different Lands and Different Climates; with Scientific Elucidations*, tr. E. Sabine, London: Longman, 1850.

7 A. von Humboldt, *Cosmos: A Sketch of the Physical Description of the Universe*, tr. E.C. Otté, London: Bohn, 5 vols, 1849–58; K. Olensko in J.L. Heilbron, ed., *The Oxford Companion to the History of Modern Science*, Oxford: Oxford University Press, 2003, pp. 383–7.

8 J. Bonnemains, E. Forsyth and B. Smith, *Baudin in Australian Waters: The Artwork of the French Voyage of Discovery to the Southern Lands, 1800–1804*, Oxford: Oxford University Press, 1988; *The Journal of Post Captain Nicholas Baudin, Commander of the Corvettes* Géographe *and* Naturaliste, tr. C. Cornell, Adelaide: Libraries Board of S. Australia, 1974; D.B. Tyler, *The Wilkes Expedition: The First United States Exploring Expedition (1838–1841)*, Philadelphia, PA: American Philosophical Society, 1968.

9 R. Corfield, 'The Chemist Who Saved Biology', *Chemistry World* 5 (2008): 56–60; H.N. Moseley, *Notes by a Naturalist . . . Made During the Voyage of* HMS Challenger, London: John Murray, p. 445.

10 E.V. Brunton, *The* Challenger *Expedition, 1872–1876: A Visual Index*, 2nd edn, London: Natural History Museum, 2004.

11 *Scientific Memoirs: Selected from the Transactions of Foreign Academies of Science and Learned Societies, and from Foreign Journals*, 1 (1837)–7 (1853); reprint, New York: Johnson, 1966.

12 J.L.R. Agassiz, *Bibliographia Zoologiae et Geologiciae: A Catalogue of Books, Tracts and Memoirs on Zoology and Geology*, 4 vols, ed. and tr. H.E. Strickland, London, Ray Society, 1848–54.

13 Royal Society of London, *Catalogue of Scientific Papers, 1800–1900*, London: Royal Society, 1867–1925.

14 Lord Kelvin, 'Nineteenth-Century Clouds over the Dynamical Theory of Heat and Light', *Proceedings of the Royal Institution* 16 (1899–1901): 363–97; T. Kuhn, *Black-body Theory and the Quantum Discontinuity, 1894–1912*, Oxford: Oxford University Press, 1978.

15 See B.J. Hunt 'Electrical Theory and Practice in the Nineteenth Century', O. Darrigol 'Quantum Theory and Atomic Structure, 1900–1927', and J. Hughes 'Radioactivity and Nuclear Physics', in M.J. Nye, ed., *Cambridge History of Science*, Cambridge: Cambridge University Press, 2003, vol. 5, pp. 311–30, 331–49, 350–74.

16 T.H. Hoppen, *The Mid-Victorian Generation: England 1846–1886*, Oxford: Oxford University Press, 1998, pp. 492–3.

17 E. Rutherford and F. Soddy, 'The Cause and Nature of Radioactivity', *Philosophical Magazine*, 6th series, 4 (1902): 370–96.

18 F. Soddy, *Science and Life: Aberdeen Addresses*, London: John Murray, 1920.

19 L. Koenigsberger, *Hermann von Helmholtz*, tr. F.A. Welby, Oxford: Oxford University Press, 1906, pp. 330–1.

20 J.J. Thomson, *Recollections and Reflections*, London: Bell, 1936, p. 379.

21 W. Crookes, 'On Radiant Matter', *Nature* 20 (1879): 419–23, 436–40; 'Electricity in Transitu: From Plenum to Vacuum', *Chemical News* 63 (1891): 53–6, 68–70, 77–80, 89–93, 98–100, 112–14; reprinted in facsimile in D.M. Knight, *Classical Scientific Papers: Chemistry, 2nd series*, London: Mills & Boon, 1970, pp. 89–98, 102–123; W.H. Brock, *William Crookes*, Aldershot: Ashgate, 2008.

22 J.J. Thomson, *Recollections and Reflections*, London: Bell, 1936, pp. 325–71; 'Cathode Rays', *Philosophical Magazine*, 5th series, 44 (1897): 293–316. On technicians, see the special issue of *Notes and Records of the Royal Society* 62 (2008).

23 D.M. Knight, *Atoms and Elements*, London: Hutchinson, 1967.

24 These and other papers relevant to this story are reprinted in facsimile in S. Wright, ed., *Classical Scientific Papers: Physics*, London: Mills & Boon, 1964.

25 E. Rutherford, 'The Structure of the Atom', *Philosophical Magazine*, 6th series, 27 (1914): 488–98.

26 D.M. Knight, *Ideas in Chemistry: A History of the Science*, 2nd edn, London: Athlone, 1995, pp. 157–79.

27 P. Morris, ed., *From Classical to Modern Chemistry: The Instrumental Revolution*, London: Royal Society of Chemistry, 2002.

28 M. Bentley, *Lord Salisbury's World: Conservative Environments in Late-Victorian Britain*, Cambridge: Cambridge University Press, 2001.

29 A.J. Balfour, *A Defence of Philosophic Doubt*, London: Macmillan, 1879, p. 293.

30 A.J. Balfour, *The Foundations of Belief*, London: Macmillan, 2nd edn, 1895, p. 31.

31 B. Lightman, in A. Barr, ed., *Thomas Henry Huxley's Place in Science and Letters: Centenary Essays*, Athens, GA: Georgia University Press, 1997.

32 P. Metcalf, *James Knowles: Victorian Editor and Architect*, Oxford: Oxford University Press, 1980.

33 Synthetic Society, *Papers Read before the Synthetic Society, 1896–1908: and Written Comments thereon Circulated among the Members of the Society*, London: Synthetic Society, 1909.

34 Rayleigh, Lord, *Lord Balfour in Relation to Science*, Cambridge: Cambridge University Press, 1930.

35 F.W.H. Myers, *Human Personality and its Survival of Bodily Death* [1903 & 1919], Norwich: Pilgrim Books (Pelegrin Trust), 1992; J. Oppenheim, *The Other World: Spiritualism and Psychical Research in England, 1850–1914*, Cambridge: Cambridge University Press, 1985, pp. 249–66, esp. 255.

36 G. Makari, *Revolution in Mind: The Creation of Psychoanalysis*, London: Duckworth, 2008.

37 C. Chimisso, *Writing the History of the Mind: Philosophy and Science in France, 1900–1960s*, Aldershot: Ashgate, 2008.

38 P. Bowler, *Reconciling Science and Religion; The Debate in early Twentieth-century Britain*, Chicago, IL: Chicago University Press, 2001.

39 A. Gray, *Darwiniana: Essays and Reviews Pertaining to Darwinism* [1876], ed. A.H. Dupree, Cambridge, MA: Harvard University Press, 1963.

40 G. Cookson and C. Hempstead, *A Victorian Scientist and Engineer: Fleeming Jenkin and the Birth of Electrical Engineering*, Aldershot: Ashgate, 2000, pp. 166–8.

41 D.J. Kevkles, 'Heredity', in J.L. Heilbron, ed., *The Oxford Companion to the History of Science*, Oxford: Oxford University Press, 2003, pp. 361–3.

42 G. Mendel, *Experiments in Plant Hybridisation*, ed. J.H. Bennett, commentary by R.A. Fisher, biog. W. Bateson, Edinburgh: Oliver and Boyd, 1965; A.F. Corcos and F.V. Monaghan, *Gregor Mendel's Experiments on Plant Hybrids: A Guided Study*, tr. E.V. Sherwood, New Brunswick, NJ: Rutgers University Press, 1993; R. Olby, *Origins of Mendelism*, London: Constable, 1966.

43 G.R. Searle, *A New England? Peace and War, 1886–1918*, Oxford: Oxford University Press, 2004; esp. pp. 615–60.

44 K. Olesko, 'Kaiser Wilhelm/Max Planck Gesellschaft', in J.L. Heilbron,

ed., *The Oxford Companion to the History of Modern Science*, Oxford: Oxford University Press, 2003, pp. 433–4.

45 W.E.H. Lecky, *History of the Rise and Influence of the Spirit of Rationalism in Europe*, new imp., London: Longman, 1900, vol. 1, p. ix.

46 W.E.H. Lecky, *History of European Morals*, 13th imp., London: Longman, 1899, vol. 2, pp. 275–372.

47 *Census of England and Wales, 1911*, vol. 10, Part 1, HMSO, 1914, pp. v–x, xxiii.

48 D. Edgerton, *Science, Technology, and the British Industrial 'Decline', 1870–1970*, Cambridge: Cambridge University Press, 1996.

49 R. Holmes, ed., *The Oxford Companion to Military History*, Oxford: Oxford University Press, 2001.

50 T.H. Hoppen, *The Mid-Victorian Generation: England 1846–1886*, Oxford: Oxford University Press, 1998, p. 497.

51 R. Holmes, ed., *The Oxford Companion to Military History*, Oxford: Oxford University Press, 2001, pp. 199–201; H. Chang and C. Jackson, eds, *An Element of Controversy: The Life of Chlorine in Science, Medicine, Technology and War*, British Society for the History of Science Monographs 13, 2007.

52 D.M. Knight, *The Age of Science*, Oxford: Blackwell, 1992.

53 D.P. Miller, 'Seeing the Chemical Steam through the Historical Fog: Watt's Steam Engine as Chemistry', *Annals of Science* 65 (2008): 47–72.

54 B.T. Moran, *Distilling Knowledge*, Cambridge, MA: Harvard University Press, 2005, p. 166.

INDEX

Abel, Frederick 69
Abernethy, John 110–11
Académie des Sciences 45, 197, 251
  electricity 25
  elite salaried body 16
  laboratories 140
  under Napoleon 18–20
Accum, Frederick 77, 132
Adams, John Couch 208, 235, 265
Admiralty Manual of Scientific Enquiry (ed. J. Herschel) 163
African exploration 212–13
Agassiz, Louis 97
AGFA 59
agriculture
  Cuvier and 28
  Davy and 62
  fertilizers 62–3, 170
  political effects on 63–4
Albert, Prince 45, 206
  death of 186
  dyes and 59
  Great Exhibition *174*, 177
  modernizer 173–4
alchemy 2
alcohol 111
Alkali Act (1863) 62
Allen, William
  carbon and diamond 130
  Chemical Lectures (with Babington and Marcet) 108
Ambix journal 11
American Chemical Society 11
Ampère, André Marie 41
  Catholicism and 240
  electrodynamics 87
  works with visitors 205
anaesthetics 115, 119

Anatomy (Gray) 49, 107
Anatomy Act (1832) 110
Anderson, John 210
Andrews, Thomas 202
Animal Chemistry Club 26, 109
animal magnetism x, 14, 166–7
animal vivisection 232–3
Annales de Chimie journal 39
Annals of Philosophy 9, 42
Annals of Science 11
anthropology
  man's place in nature 163–6
  Piltdown hoax 228–9
  Prichard's ethnology 161–3
Anti-vaccination League 232
Antiquity of Man (Lyell) 164–5
apes
  evolution and 164–5
  man and 160–1
Apothecaries Act (1815) 107
Arago, Dominique François Jean 239
Archives of Natural History 11
Argand, Aimé 133
Aristotle 217
Arkwright, Richard 64
Armstrong, William, Baron 68–9, 210
art
  optics and vision 255
  see also illustration and painting; photography
Art Journal
  the Great Exhibition 175 (*see also* Imperial College)
astronomy 202, 208
  clusters of observation 42
  Laplace and 20–1
atoms and molecules
  attraction and repulsion 154–5

atoms and molecules (*cont.*)
  contradictions, and scepticism
    225
  Dalton's view of 40–1
  Einstein's breakthrough 104
  electrons 147
  Hofmann's models of 45–7, *46*
  Mendeleev's Periodic Tables 180
  models of 180, 207
  radioactive decay 149
  radioactivity and subatomic structures
    268–72
  weights of 226
Audubon, John James 191
  Birds of America 48
Australia 152
  aborigines 162
  colonial knowledge 201
  university in 188
Austria 198
Avogadro, Amadeo 41

Babbage, Charles 210, 240
  criticizes Great Exhibition 178–9
  inspects others 226, 227
  modernizing mathematics 43–4
  nearly collides with Brunel 70
  on precision 42
  visits Berlin 241
Babington, William 109
  Chemical Lectures (with Marcet and Allen)
    108
Bache, Alexander Dallas 248
Bacon, Francis xiv, 3
  application of science 56
  elements of modern world 5
  knowledge as power 195, 233
  method 220
Baer, Karl von 109
Baker, Benjamin 71
Balfour, Arthur J. 12, 13, 23
  career of 273–5
  A Defence of Philosophic Doubt 273–4
  Foundations of Belief 274
  science and faith 170
Banks, Joseph 30, 105, 153
  as administrator 248
  Botany Bay 201
  British Museum 189, 190
  colonial knowledge 200
  internationalism 266
  Kew Gardens 38
  learns Tahitian 36
  naked savages 152
  opposes Geological Society 83
Barrow, John 249
BASF 59, 281
Bates, Henry Walter 101, 165
  career of 211–12
Bateson, William
  Mendel's Principles of Heredity 278
Baudin, Nicholas 22, 161–2, 266
Bauer, Ferdinand 48
Bazalgette, Joseph
  London sewers 61, 116–17

HMS *Beagle* 266
  weather observations 181
Beche, Henry de la 146, 249
Becquerel, Antoine Henri 103, 268
Beddoes, Thomas 112
  Coleridge and 166
  method 220
  shocks frogs' legs 140
Belgium 199, 216, 280
Belgrand, Eugène 117
Bell, Charles 110, 111, 160–1
Bell, Joseph 121
Benbow, William 111
Bentham, Jeremy 115
benzene 137–8, 180
Bergman, Tobern 18
Bergson, Henri
  Creative Evolution 276
Berlin, University of 193
  linguistics 35
  neo-humanist philosophy 239
Bernard, Claude 169
  Experimental Medicine 232
  Study of Experimental Medicine 123–4
Berthelot, Marcellin 210, 240
  sceptical about atoms 45
  synthesis and analysis 138
Berthollet, Claude-Louis 18, 210
  beyond the Enlightenment 31
  bleach 57
  Society of Arcueil 18
Berwickshire Naturalists' Club 246
Berzelius, Jacob 41
  on Davy 83
  electricity 25
  on laboratories 130
  laboratory work 137
  mineral analysis 61
  organizing chemistry 86
  rows with Thomson 42, 53–4, 137, 226
  symbols adopted by BAAS 243
  works with visitors 205
Bessemer, Henry 67
Bewick, Thomas 48–9
Bichat, Xavier 109, 110
bicycles 80
biology
  classification 26–8
  Lamarck coins term 26, 84, 110
  life beyond Earth 228
  religion and 110
birds
  illustrations of 48–9
  preservation of 234
Birds of America (Audubon) 48
Birkbeck, George 73
Bismarck, Otto von 102, 106, 198
BJHS (journal) 10
Black, Joseph 18, 23
Blackwell, Elizabeth 125
Blake, William 28
Blumenbach, Johann Friedrich 25–6, 108, 158
Boerhaave, Hermann 205
Bohn, Henry 73
Bohr, Niels 271–2

Boole, George 45
Bosch, Carl 281
Boscovich, Roger J. 88, 154–5
The Botanic Garden (E. Darwin) 38
botanic gardens
    laboratories of 145
    see also Kew Gardens
Botany Bay 152–3, 201
Bourgery, Jean Marc
    Treatise 107
Boussingault, Jean-Baptiste 63
Boyle, Robert 203, 271
Bragg, Lawrence 201
Bragg, William Henry 19, 53, 201
Brande, William Thomas 53, 131, 133
Brazil 212
Brewster, David 192, 228
bridges 71
Bridgewater Treatises 253
    Bell 160
    Buckland 51, 97, *98*
Britain
    in 1789 195–6
    effect of French Revolution 30
    empire 200–2
    excitement about science 83
    grants for research 144
    hospitals 107, 124
    military technology 209–11
    naval supremacy 152
    non-metric measurement 22
    Paris exposition 187
    popularization 9
    refugee scientists 264
    Romantic movement 31
    state support for science 205
    Stuart dynasty 195
    thought 'backward' 178
    Whig history 2–4
British Association for the Advancement of
    Science (BAAS) 74, 262
    Balfour and 275
    'clerisy' of 244, 245
    formation and workings of 242–3
    membership 249
    phrenology and 230
    specializations within 250
British Meteorological Society 182
British Museum 189–91
British Society for the History of Science 10
Brodie, Benjamin I surgeon 26, 100
    Animal Chemistry Club 109
Brodie, Benjamin II chemist 45-6
Brooklyn Bridge 71
Brougham, Henry 218
Brown, John 162
Brown, Robert
    British Museum 189–90
    illustrations 48
Browning, Robert
    'Mr Sludge' 168
Brücke, Ernst 123
Brunel, Isambard Kingdom 65–7
    grand structures 80
    Great Exhibition and 177

    nearly collides with Babbage 70
    Saltash Bridge 71
Brunner, Sir John 75, 207
Buckland, Mary (née Morland) 52
Buckland, William
    biblical geology of *96*, 96–7
    Bridgewater Treatise 97, *98*
    God's design 51–2
Buckle, Henry 218
buildings and architecture
    Crystal Palace 175
    Eiffel's tower 188
    Laboratories 129-32, 139, 141-2, 144-5
    Museums 189–92
Bülow, Berhard von 209
Bunsen, Christian Karl Josias 162
Bunsen, Robert 78, 143, 206
Burton, Richard 159, 212
Butler, Samuel 244
Butterfield, Herbert 3

Cambridge University 53
    medicine 124
    see also Cavendish Laboratory
Canning, George
    parodies E. Darwin 32
Cannizzaro, Stanislao 41, 204, 208
Carlisle, Anthony 24
Carlyle, Thomas 7
Carnot, Lazare 197
    'organizer of victory' 248
    survives Revolution 239
Carnot, Sadi 89
    laws of thermodynamics 43, 91–3
Caro, Heinrich 59
Carpenter, William Benjamin 248
Carter, Henry Vandyke
    Gray's *Anatomy* 49, 107
Case Institute 148
Catalogue of Scientific Papers, 1800-1900 268
cathode rays 130, 147, *271*, *272*
Catlin, G. 158
Cavendish, Henry 16, 18
    electricity 23
    measures shocks 140
Cavendish, William (Duke of Devonshire) *see*
    Devonshire
Cavendish Laboratory 141
    electricity and atoms 269
    women and 259
Cavour, Camillo 204
Cellular Pathology (Virchow) 123
Chadwick, Edwin 115–16
Chagas, Carlos 212
HMS *Challenger* 249, 266
Chalmers, Thomas 245
Chambers, Ephraim
    Cyclopedia 15
Chambers, Robert
    deism 161
    phrenology and 157
    Vestiges 52, 97–100, 185, 192, 218, 230, 253,
        258
Champollion, François 36
Chapman, John 184, 219, 244

Charcot, Jean-Martin 126, 169, 275
Chemical Lectures (Babington, Marcet and
     Allen) 108
Chemical Manipulation (Faraday) 68, 132–3
Chemical News 252
Chemical Society of London 76, 142,
     250–1
chemistry
  analysis of materials 61–2
  apparatus 74, 76–9
  atoms and elements 23–4
  changing society 281–2
  colours and 57–9
  education and ix
  electricity and 85–6
  forensic 119–21
  gases and thermodynamics 92–3
  German language and 264–5
  hands-on science 22–3
  Hofmann's models and 46–7
  inorganic 224
  language of 38–9, 45
  Periodic Table 180, 224
  phlogiston 23
  physicists and chemists 269
  pollution 61
  predicted elements 224
  requires empirical knowledge 180–1
  social impact of xii
  structures and arrangements 137–8
  symbols of 41
  synthetic 180, 273
  see also atoms; laboratories
Chenevix, Richard 227–8
Chevallier, Temple 72
Chevreul, Michel 16
Chicago University 193
childlessness 105
China 200
chlorine 40, 281
cholera 113–15, 123
Christison, Robert 120
Church of Scotland 245
City and Guilds Institute 76
Clark, William 148
Clarkson, Thomas 159
Clausius, Rudolf 92
Clifford, William Kingdon 185, 280
Clift, William 227
Coast Survey and Topographical Engineers
     146
Cobbe, Frances Power 233–4
Coimbra, University of 139, 144
Cole, Henry 177
Coleridge, Samuel Taylor
  coins 'psychosomatic' 166
  Davy and 32
  feelings and nature 95
  Hints Towards...a Theory of Life 29
  influence on Mill 221
  interests 166–7
  spirit of Imagination 28–9
colonialism and imperialism
  East India Company 200
  materials of xi

resources 212
slavery 159
colours
  applied chemistry of 57–9
  Goethe's work on 255
Colt, Samuel 68
Combe, George 157, 161, 230
communications technology
  Marconi's radio message 215
  telegraphs 64–5
Comte, Auguste 5, 219
  mathematical knowledge 43
  positivism 236
  science over religion 240
Concepts and Theories of Modern Physics (Stallo)
     222
*Condition of the Working classes in England in
     1844* (Engels) 113
Congress of Vienna 197
Conolly, John 126
Conservatoire des Arts et Métiers 17
Consolations in Travel (Davy) 240
Constable, John 29, 48
  cloud studies 181, 255
consumer goods
  electricity and 80
  women and science 260–2
Conversations in Chemistry (Marcet) 130–1
Cook, James 22, 29
  learns Tahitian 36
  military intelligence 211
  sees effect of Europeans 158–9
  surveying skills 209
  voyages and discoveries 152–3
Copernicus, Nicholas 203
Copley Medal
  for Buckland 97
  for Chenevix 228
  for Davy 57, 59
  for Joule 90
  to Mayer 90
  no co-winners 130
Cornell University 193
Correns, Carl 278
Cottle, Joseph 131
Coulomb, Charles 22
Couper, Archibald Scott 265
Cradock, Baldwin *50*
Creative Evolution (Bergson) 276
Crimean War 280
  nursing and 117–18
  Russian defeat 199
Crookes, William 143
  electricity and gas 182, 269–71, *270*
  experiments of 146–7
  laboratory of 130
  photography and 255
  psychic forces 234
  publications of 251–2
  rare earth metals 224
  spiritualism 168, 234
Cruz, Oswaldo 212
Crystal Palace Exhibition *see* Great Exhibition
     (1851)
Cullen, William 162

culture
  science and arts 174-7, 262
  science as part of 238
  'two cultures' x, 53, 192
  values and x–xi; *see also* 'two cultures'
Curie, Marie 259, 268
Curie, Pierre 268
Cuvier, Georges 18, 22, 23
  beyond the Enlightenment 31
  contempt for evolution 95
  disagreements 84, 225, 276
  electricity 25
  God's design 51
  plant and animal physiology 26–8
  positivism 102
  royal menagerie and 16
  scepticism 218
  'Technologie' 28
Cyclopedia (Chambers) 15

D'Alembert, Jean le Rond 15
Dalton, John 19
  atomic theory 40–1, 137
  criticized by Thomson 42
  Laws of Chemical Combination 226
  Science Museum 189
  state pension 205
Daniell, J. F. 140
Darwin, Charles 179–80
  appalled by *Vestiges* 98
  biographical view of 6–7
  buried in Westminster Abbey 210
  collecting beetles 82
  Darwin Correspondence 7, 10
  debated ideas 54, 100–3, 226
  The Descent of Man 165–6
  development 99
  Expression of the Emotions 101, 163
  financial independence ix
  Lyell's *Geology and* 97
  medical treatments 127
  Origin of Species 50–1, 99, 101, 104
  overtaken by genetics 276–8
  religion and 52
  Sunday Lecture Society 186
  Wallace and 165–6, 225
  Wilberforce criticizes method 230
  Zoology of the Voyage of HMS Beagle 191
Darwin, Erasmus
  The Botanic Garden 38
  context of science 239
  eloquence 34
  grand vision 253
  influence on Mary Shelley 256
  The Loves of the Plants 32, 37, 94
  method 220
  mocked 93–4, 95
  poetry 84
  poisonous trees 119
  Shelley reacts to x
  The Temple of Nature 93
Darwin, Robert 112
Daubeny, Charles 62
Davy, Edmund 202
Davy, Humphry

agriculture and 62, 63
Animal Chemistry Club 109
antiphlogistic remedies 39
atoms and 41
'brilliant fragments' 83
chlorine and 40, 281
Coleridge and 29, 166
compared to Gay-Lussac 19
Consolations in Travel 240
Copley Medal for 25, 57
death of 86
disabled by explosion 131
electricity 23, 24, 85–6
Faraday and 140, 225
keen to discover 218
laboratory and apparatus 132, 142
laughing gas 166
method 227
miners' safety and 59–60, 81, 228
nitrous oxide 112
pantheistic verse 253
papers collected by Faraday 142–3
in Paris 208
popular lectures 31–2
Roman pigments 58
Salmonia 240
Science museum 189
on the seashore 82
Shelley reacts to x
shocks frogs' legs 140
standing and respect viii, 2, 6, 105, 210, 240
vision of science 56–7
women's education 233–4
works with visitors 205
De Quincey, Thomas 65, 160
A Defence of Philosophic Doubt (Balfour) 273–4
Degérando, Joseph Marié 161
Delambre, Jean-Baptiste 18, 22, 31
Denmark 199
dentistry 125
Desaga, C. 143
Descartes, René 3
  Discourse of a Method 217
  minds and bodies 154
  planetary vortices 2
  uses French language 35
The Descent of Man (Darwin) 165–6
Devonshire, Duke of (William Cavendish) 75,
      76, 191
  Cavendish Laboratory 141
  Devonshire Commission 76, 144
Dewar, James 225
Dialogue (Galileo) 35
Dickens, Charles 177
Dictionary of National Biography 6
Dictionary of Nineteenth-century British Scientists
      10
Diderot, Denis 30, 57
  Encyclopaedia 15
diet and nutrition 111
dinosaurs 227
  archaeopteryx 100
  Buckland and 97, 98
diphtheria 123
Discourse of a Method (Descartes) 217

Disraeli, Benjamin 106
Draper, John William 103
Drummond, Henry 103
Du Bois-Reymond, Emil 123
Duhem, Pierre 218–19
Dulong, Pierre 131
Dumas, Jean Baptiste 53–4, 239
Dürer, Albrecht 158
Durham, University of 72, 75
dyes
    education and 74–5
    Perkin's mauve 180

East India Companies 200
Eastlake, Charles 222
École Normale 17, 53
École Polytechnique 17, 53
    changing role of 192–3
    engineering 71
    first science education ix
    technical drawing 67–8
Edinburgh University 107
Edison, Thomas Alva 79, 187
education xi
    doctoral degrees 206
    for engineers 71
    evening classes 73
    external examiners 188
    increasing requirement of 244–5
    innovations at RI 173
    medical 123–6, 206
    national and international study
        205–6
    research universities ix, 136–7, 193
    rise of science in viii–x
    scientific culture and 241
    for women 233–4, 259–60
Ehrlich, Paul 124
Eiffel, Gustave 188
Einstein, Albert 149
    creative science 150
    Mach's influence 223
    matter and energy 104, 271, 272–3
Eldon, Lord 111
Elective Affinities (Goethe) 29, 184, 255
electricity
    in animals 85
    batteries 24–5
    in chemistry 85–6
    electromagnetism 86–9
    late eighteenth century 23
    from magic to physics 12
    social change and 79–80, 182, 262-3
electromagnetic spectrum 23, 147
    see also light
electromagnetism
    Faraday's experiments 182
Elements of Chemistry (Lavoisier)
    language of 38–9
    publication of 14
Elements of Physiophilosophy (Oken) 248
Encyclopaedia (Diderot) 15
energy
    Carnot's thermodynamics 91–3
    conservation of 85, 89

Einstein relates to matter 104, 271, 272–3
    heat and work 89–91
Engels, Friedrich
    Condition of the Working classes in England in
        1844 113
engineering
    education for 71–6
    precision work 146
    standardization 67–70
    technical drawing 67–8
    see also industry; transport
England
    institutions of 241
    medical education 124
    mental asylums 126
Enlightenment thought 219
    in France 15–16
    reason over empirical science 238
environmental pollution 61
ethics and morality 258
    see also religion
eugenics 236
European Morals (Lecky) 279
Everest, George 77
Evidence as to Man's Place in Nature (Huxley)
    163
evolution 101–2
    acceptance of 85
    anthropology and 164–6
    Darwin's development 93
    epic vision of 253–4
    European reception of 169
    expressions of emotion 101
    initial suspicion of 230–1
    as progress 102
    reception of ideas 100–3
    scepticism about 218
    untidy patterns 224
Evolution and Ethics (Huxley) 258
exhibitions
    early attitudes towards 173–4
    see also Great Exhibition; museums
Experimental Medicine (Bernard) 232
exploration
    Africa 212–13
    Brazil 211
    polar 213–14, 214
Expression of the Emotions in Man and Animals
    (Darwin) 163

Faraday, Michael
    accused of trespass 87
    benzene 137–8
    Chemical Manipulation 68, 77, 132–3
    childless 105
    consulted about Stink 116
    as Davy's assistant 131, 142–3
    education 72
    electromagnetism 86–9, 182
    falls out with Davy 225
    Faraday Correspondence 10
    gases and electricity 146
    Great Exhibition and 177
    laboratories and apparatus 132–3
    laboratory and apparatus 134

lamp fuels 64
method 227
money and viii
in Paris 208
'physical' chemistry 140
public lectures 53, 89
religious belief 130
safety and 71
Science museum 189
social class and 6
state pension 205
taxing electromagnetic induction 192
theories of electricity 86
Whewell helps with technical terms 44–5
Farr, William 114–15
Fenton, Roger 255
fertility *see* population and demographics
fertilizers 170
The Fighting Temeraire (Turner) 209
fingerprints
Galton and 104
Finsbury Park Technical College 75
First Principles (Thomson) 41–2
Fischer, Emil 180
FitzRoy, Captain Robert
Weather Book 181–2, *182*
Fleming, Alexander 106
Flexner, Abraham 125
flight
Wright brothers 215
zeppelins 210
Flinders, Matthew 48, 189, 266
Flower, William H. 145, 191
folklore 197–8
Food and Drugs Act (1860) 62
forensic science 119–21
forgery and plagiarism 227–8
Forster, George 266
Forster, Johann 266
Forth Bridge 71
Foster, George Carey 226
Foster, Henry 249
Foster, Michael 233
Cambridge physiology 124
Foucault, Michel 126
Foundations of Belief (Balfour) 274
Fourier, Jean Baptiste Joseph, Baron de 43, 259
flow of heat 92
under Napoleon 239
Fowler, John 71
Fox, Charles James
History of James the Second 3
France
African exploration 213
anti-clericalism 102
dread of revolution 170
Enlightenment thought 15–16
exhibitions 173
first research schools 12
Franco-Prussian War 188, 193
hospitals 107
laboratories 144
metric system 22
under Napoleon 18–19, 197–8

Napoleonic Wars 280
nursing Sisters of Charity 117–18
popularization 9
psychic research 169
Revolution 14–15, 30, 239–40, 280
role of polytechnics 192–3
specialization 83
state support for science 205
Third Republic 188
Franco-Prussian War 188, 193
Frankenstein (Shelley) x, 24, 29–30, 167, 255–6
Frankland, Edward 83, 142, 186, 206
Franklin, Benjamin 18, 85, 146, 149
electricity 23
the use of discoveries 192
Franklin, John
death in the ice 213–14, *214*
Franz Ferdinand, Archduke 199
Fraunhofer, Joseph 79, 272
Frere, Hookham 32
Fresnel, Augustin Jean 271–2
Freud, Sigmund 126, 170, 275–6
Friedrich, Caspar David 29
Froude, Hurrell 184
Froude, James Anthony
Nemesis of Faith 184

Galileo Galilei 3
biographical view of 6–7
Dialogue in Italian 35
literary style 34
Gall, Franz 155–6
Gall, J. G. 208
Galton, Francis 104, 278
Hereditary Genius 235–6
Galvani, Luigi 18
electricity 23
influence on Mary Shelley 256
shocks frogs' legs 140
Garnett, William 75
Gaskell, Elizabeth
North and South 113
Gauss, Karl Friedrich 18, 21, 42, 226
Gay-Lussac, Joseph-Louis 18, 41
compared to Davy 19
laboratory science 132
Liebig studies with 205
volumes of gases 42
Gehardt, Charles 54
Geological Society 83
Geological Survey 249
geology
Buckland's Flood *96*, 96–7
the Flood to evolution 96–9
George III 78
Germany
approaches to method 219, 222–3
Berlin chemical laboratory *145*
colonies 201, 213
education in 73–4, 75, 81
exhibitions 173
foreign students in 206–7
Franco-Prussian War 188, 193
Hitler and science 264
Hofmann returns to 187

hospitals 107
industrial chemists 136–7
Kulturkampf 102
laboratories 144
military technology 209–11
Naturphilosophie 169, 199
reception of evolution 102–3
research universities 193
Romantic movement 31
states of 198
support for science 205–6, 240–1
translated papers 268
unification 216
universities ix, 36, 53, 61-2
vitalism repudiated 123
before the wars 278–9
Gesellschaft Deutscher Naturforscher und
    Ärtze (GDNA) 240–1, 262
Gibbs, Josiah Willard 146
Gibbs, Willard 19
Giessen, University of ix
    chemical analysis 61–2, 74
Gilbert, Joseph Henry 62, 63
Gillman, James 166
Glaisher, James 181
Goethe, Johann Wolfgang von 219, 236
    Elective Affinities 29, 184, 255
    on rewriting history 13
    Theory of Colours 222, 255
Gosse, Philip 248
Gould, Elizabeth 48
Gould, John 48, 191
Graham, Thomas 189
Gray, Asa 99, 276
Gray, Henry
    Anatomy 107
Great Exhibition (1851) 68, *174*, 174–7, 193
    follow up 186–7
    lectures on 177–9
    message to public 186–7
Grimm, Jacob and Wilhelm 198
Guldberg, Cato Maximilian 138
Gurney, Edmund 168
Guy's Hospital 107, 121

Haber, Fritz 281
Haeckel, Ernst
    evolutionary monism 169
    leaves God out 254
    reception of evolution 103
Hakluyt Society 247
Harcourt, William Venables Vernon 138, 242,
    248
Harrison, Frederic 185
Harrison, John 7
Hautes Écoles 17
Haüy, Rene 26
Hawaii 153
health and disease
    applying science 106
    child mortality 105
heat *see* energy
Helmholtz, Hermann 89, 210
    atomic electricity 269
    electrons 147

Goethe's colours and 222
    influence on art 255
    investigations of a polymath 90–1
    repudiates vitalism 123
    Sensations of Tone 258–9
    thermometers 121
Henry, John 87
Henry, Joseph 72
    directs Smithsonian 146, 248
    weather observations 181
Herder, Johann Gottfried 3, 35
Hereditary Genius (Galton) 235–6
heredity
    after Darwin 276–8
    following evolution 104
    Galton and 235–6
    Mendel's peas 277–8
Herschel, Sir John 236
    defends inches, ounces and pounds 22
    dismisses natural selection 93
    edits *Admiralty Manual of Scientific Enquiry*
        163
    'Man the Interpreter of Nature' xiv
    modernizing mathematics 43–4
    Moon people hoax 228
    photography 86
    A Preliminary Discourse on method 84,
        220–2, 223
    refuses to specialize 83
Herschel, William 23
    identifies Uranus 21, 88
Hertz, Heinrich 182, 223
Hints Towards...a Theory of Life (Coleridge) 29
Hippocrates 111
History of England (Macaulay) 3
*History of European Thought in the Nineteenth
    Century* (Merz) 12–13
History of James the Second (Fox) 3
History of Rationalism (Lecky) 279
history of science
    biography and 6–8
    defining 9–11
    historians and science 1–2
    as progress 3
    Whig 2–4
History of Science Society 10
History of the Inductive Sciences (Whewell) 3–4
HMS *Challenger* 146
hoaxes 228
Hobhouse, L. T. 251
    specialization of science 192
Hoechst 59
Hoff, Jacobus van't 47, 54, 226
Hofmann, August Wilhelm von 78, 210
    Chemical Laboratories 145
    dyes 58
    laboratory 177
    molecular models 45–7, *46*, 180, 207
    pure and applied science 192
    returns to Germany 187
    and Royal College of Chemistry 173
    scientific foundations of chemistry 181
Holcroft, Thomas 155
Holloway, Thomas 127
Home, Daniel Dunglas 168

Home, Sir Everard 227
homeopathy 125
Hooke, Robert 34
Hooker, Joseph 99, 104, 190
    Kew and 190, 212
Hopkins University *see* Johns Hopkins
    University
Horsfield, Thomas 200
Howard, Luke 26, 181, 255
    meteorology 29
Hueffer, Francis 258
human beings
    anthropology of 161–3
    expression and body language 160–1
    in hierarchy of beings 151
    monogenesis *versus* polygenesis 159
    natural history view of 160–3
    place in chain of being 170
    races and ethnicities 158–60
    savages 152–3
Humboldt, Alexander von 18, 23, 24, 205–6
    collects skulls 158
    dynamic connections 219
    internationalism 266
    magnetic crusade 248–9, 266
    military intelligence 211
    in Paris 208
Humboldt, Wilhelm von 53, 193, 205–6
    On Language 153–4
    linguistics 35
    University of Berlin 239
Hume, David 95, 221, 273
Hunt, Robert 176–7
Hunter, John 25, 95
    museum of 190
    plagiarized 227
    surgical education 107–8
Hunter, William 108–9
Huskisson, William 70
Hutton, James 18
Huxley, Thomas Henry 7, 179–80
    agnosticism 185
    applies for colonial posts 188
    archaeopteryx 100
    Balfour and 274
    the 'church scientific' 243
    debates 54, 99–100, 226
    dismisses *Vestiges* 218
    Evidence as to Man's Place in Nature 163
    Evolution and Ethics 258
    German translations 268
    lectures to working men 106
    Man's Place in Nature 103
    medicine and 124, 127
    Owen and 190, 225, 226, 245
    positivism 219
    Privy Counsellor 210
    public lectures 53
    realism and 45
    religion and 52–3
    saved by *Archaeopteryx* 227
    Sunday Lecture Society 186
    vivisection and 233
    Wilberforce and 245

Idealism and science 13
illness and disease
    cholera 113–15
    Europeans spread 158–9
    mosquitoes and malaria 213
    recognition of germs 122–3
    sanitation and 118
    tropical 213
    vaccines and immunization 122, 123
Illustrated Exhibitor 175
illustration and painting 55
    nineteenth-century science and 47–50
    public access to natural history 191–2
    technical drawing 67–8
Imperial Chemical Industries 207
Imperial College 76
In Memoriam (Tennyson) 185, 257–8
India 200, 213
inductive method *see* method
industry
    British 'backwardness' 178
    early factories 64–5
    health and 112–13
Innerleithen Alpine Club 246
Institut de France 31
Institute of Chemistry 251
Institute of Midwives 125
instruments and apparatus
    development of laboratories and 131–4
Interdisciplinary Science Reviews journal 11
International Congress, Karlruhe (1860) 137
*Introduction to the Study of Experimental
    Medicine* (Bernard) 123–4
Ireland
    education and science in 201–3
    Great Famine 170, 202
Isis 10
Italy
    science in states of 203–4

Jacob, Nicholas Henri 107
James, William
    psychical research 234
    Varieties of Religious Experience 170
James VII and II 3
Janet, Pierre 234
Japan 215
Jardin des Plantes 197
Jenkin, Fleeming 276
Jenyns, Leonard 181
Jewish culture 102, 203
Jex-Blake, Sophia 125
Johns Hopkins University 75, 125, 148, 193
Johnston, George 246–7, 248
Johnston, J. F. W. 241
Jones, William 36, 153, 154
Joule, James 208
    heat and work 89, 90
    memorial in Westminster Abbey 210
Journey to the Centre of the Earth (Verne) 257
Jussieu, Antoine Laurent de 15, 16, 27, 224

Kane, Sir Robert 202, 203
Kant, Immanuel 157
Karlsruhe Conference (1860) 180

Karlsruhe Congress (1860) 208
Kay-Shuttleworth, James 113
Keats, John 28
  disenchantment of 252–3
  medicine and 29, 107
  Newton's rainbow 30
Keble, John 184
Kekulé, August 138, 210
  dispute with Couper 265
  foreign studies 207
  ring structure 207
  synthetic chemistry 180
Kelvin, Lord (William Thomson) 79, 90, 103
  dissipation of energy 92
  Faraday and 44–5
  honoured 210
  Ireland and 202
  at Johns Hopkins University 148
  matter and energy 268
  thermodynamics and 208
Kew Gardens xi, 190, 216
  Banks and 38
  colonial knowledge 201
  illustration and 48
  open to public 191
  world sources 212
Kidd, John 111–12
King, Philip Parker 201
Kingsley, Charles 101, 248
Kipp, Petrus Jacobus 139
Kirchhoff, Gustav R. 78, 143
Knight, Richard 132
Knowles, James
  The Nineteenth Century 275
Knox, Robert 110
Koch, Robert 123
Kolbe, Hermann 54, 226
Koyré, Alexandre 3
Kuhn, Thomas S. 4, 39–40, 82

laboratories 149–50
  animals and 232
  apparatus and equipment 131–4, 138–40, 189
  botanic gardens 145
  conditions in 139, 142–4
  Crookes' experiments 146–7
  dangers of 131, 143
  development of 120–31
  English educational institution 141–2
  museums 145
  physics and 140–2
  shipboard 146
  skills of 133–7
  various activities in 144–6
Laënnec, René-Théophile 121
Lagrange, Joseph Louis
  Mécanique analytique 20
Lamarck, Jean-Baptiste
  coins 'biology' 109
  disagreements 84, 225, 276
  evolution and 84, 95
  grand vision 253
  Lyell and 97
  royal menagerie and 16

The Lancet
  Food and Drugs Act 62
  Wakely and 111
languages 36, 54–5
  Austrian empire 198–9
  Biblical myths of 154
  families of 153–4
  Faraday and Whewell's neologisms 44–5
  of God's design 51–3
  human beings and 153–5, 160
  international English 264–5
  Lavoisier and chemistry 38–9
  local and regional 197
  of mathematics 42–5
  new scientific paradigms and 39–40
  translated papers 268
  visual 45–7
Laplace, Pierre-Simon 210
  beyond the Enlightenment 31
  Mechanique celeste 20–1
  Society of Arcueil 18
  under Napoleon 18, 20
  works with Lavoisier 23
Latin language
  classification with 37
  use of 35–6
Laurent, Auguste 54, 138
Lavater, Johann Kaspar
  Physiognomy 155
Lavoisier, Antoine Laurent de
  atoms and elements 23
  beheaded in Revolution 16, 197
  beyond the Enlightenment 31
  conservation of matter 84
  David's portrait of 47
  Elements of Chemistry 14, 38–9
  heat as a fluid 92
  laboratory and apparatus 129, 133, 149
  launches journal 9
  names oxygen 40
  phlogiston and 54
  positivism 102
  supported Revolution 239
  the vital flame 25
Lavoisier, Marie Anne (née Paulze) 130
Lawes, John Bennet 62, 63
Lawrence, Thomas 47
Lawrence, William 166, 208
  biology text without religion 110–11
  denounced for blasphemy 26
Le Verrier, Urbain 208
Lear, Edward 48, *49*, 191
Lecky, William
  European Morals 279
  History of Rationalism 279
Leibniz, Gottfried
  foundation for mathematics 20
Leverrier, Urbain
  Neptune and 265
Lewis, Meriwether 148
libraries 249
Liebig, Justus von ix, 19, 210
  in Britain 207
  chemical analysis 62
  fertilizers 62–3, 170

invents research students 136–7
  method 236
  against Naturphilosophie 84, 223
  occupations for chemists 74
  state support 205–6
  teaching laboratories 149
Liebig's Annalen 136
light
  Fraunhofer lines 272
  gas 64
  measuring speed of 148
  quantum theory 271–2
  wave theory 218, 220, 271–2
  see also optics
Limits of Religious Thought (Mansel) 185
Linnaeus, Carl 15, 27, 36
  baseline for naming 179
  displaced by Jussieu 224
  entertaining names 132
  language of classification 37–8, 39
  Systema Naturae 151
Linnean Society 99
Lister, Joseph 122, 210
Literary and Philosophical Societies 72, 249
Livingstone, David 212
Lockyer, Norman 75
  disagrees with Proctor 225–6
  Nature 251
Lodge, Oliver 235
Logic (Mill) 221
Lombroso, Cesare 236
London
  Great Stink 116
  sanitation 115–17
  scientific institutions 16, 74, 83, 140, 173,
    176, 189
London Missionary Society 213
London School of Medicine for Women 125
London University
  Birkbeck and 73
  external examiners 188
Longitude 7, 21-2
L'Ouverture, Toussaint 159
The Loves of the Plants (E. Darwin) 32, 37, 94
Lowell Lectures
  Drummond on evolution 103
Ludwig, Karl 123
Luttrell, Henry 61
Lyell, Charles
  Antiquity of Man 164–5
  on hominid fossils 229
  neutral about evolution 99
  Principles of Geology 97, 98
  safety 71

Macaulay, Thomas Babington
  History of England 3
McGonagall, William 71
Mach, Ernst
  opposes metaphysics 223
  positivism 236
Macmillan, Harold 278
Macmillan publishers 251
Macpherson, James
  Ossian poems 195

Madagascar 213
Magellan, John Hyacinth 139–40
Magendie, François 27–8, 110, 111
magnetism 22
  electro- 86–9
  North Pole 213
Malthus, Thomas R. 13, 95, 158
'Man the Interpreter of Nature' (Herschel) xiv
Man's Place in Nature (Huxley) 103
Mansel, Henry
  Limits of Religious Thought 185
Marcet, A.
  Chemical Lectures (with Babington and Allen)
    108
Marcet, Jane 133
  Conversations in Chemistry 130–1
Marconi, Guglielmo 182, 202, 215
Marsh, James 120
Martineau, Harriet 219
Martineau, James 233
Marxism 5
Mary II 3
Maskelyne, John Nevil 229
mass production 177
Massachusetts Institute of Technology 75
materialism 169, 170
mathematics
  barrier for most people 32
  engineering and 72
  language of 42–5
  models for the unseen 45
  precision and accuracy 42
  in 1790s 19–22
  statistics 21
  university education and viii
Maury, Matthew 66
Maxwell, James Clerk
  dynamical theory of gases 92–3, 106
  electrodynamics 269
  equations over theory 223
  Faraday and 44–5, 88
  lectures at Cavendish Laboratory 141
  memorial in Westminster Abbey 210
Mayer, Julius Robert 89–90
Maynooth Seminary 202
McClintock, J.
  The Fate of Sir John Franklin 213–14, *214*
Mécanique analytique (Lagrange) 20
Mechanics' Institutes 64
  curiosity and 83
  engineering education 72–3
  laboratory technicians 142
  phrenology and 230
Mechanique celeste (Laplace) 20–1
medicines and pharmacology
  anaesthetics 115, 119
  apparatus and 74
  education for 17, 107–11, 123–6
  grave robbing 109–10
  instruments 121–2
  international studies 208
  medicinal plants 111–12
  nursing 117–19
  occupations and 112–13
  poisons and 119–21

tropical disease hospitals 213
university education and viii
vaccines and immunization 122, 123
  see also public health
Meister, Joseph 122
Meldola, Raphael 277
Melincourt (Peacock) 160
Mendel, Gregor 7, 19, 104, 231, 277–8
Mendeleev, Dmitri 19
  Periodic Table 180, 224
Mendel's Principles of Heredity: a Defence
  (Bateson) 278
mental health
  asylums 126
  legal definitions of insanity 126
  Prichard's humane view 163
mercury poisoning 112
Merz, John Theodore 80
  History of European Thought in the Nineteenth
    Century 12–13
Mesmer, Franz-Anton 166–7
  mesmerism x
Metaphysical Society 275
method ix, 236–7
  Berzelius's precision and 42
  conservatism 218
  Darwin's 230–1
  defining theory 220–1
  experiment and observation 221
  Herschel's *Discourse* 220–2, 223
  inductive 3–4, 220–2
  misdemeanours 226–9
  Naturphilosophie 222–3
  positivism 217, 223
  precision and accuracy 42
  rows and rivalries 225–6
  synthetic and analytic minds 218–19
Methods of Ethics (Sidgwick) 273
metric system 16
  British Association and 70
  Delambre 22
Michelson, Albert 148–9
microscopes 121–2
Middlesex Asylum, Hanwell 126
midwifery 125–6
Mill, John Stuart 220-2
  Logic 221
  scepticism 273
Miller, Hugh 52
Miller, J. S. 94
Miller, William Hallowes 47
Millikan, Robert 271
mining safety 59–60, 81
Mivart, George Jackson 102
models
  believing in the unobserved 45
  of molecules 180, 207
  ring structures 207
  of the unseen 55
  visual 45–7
Molland, Charles 202
Monboddo, Lord (James Burnett)
  The Origin and Progress of Language 160
Mond, Ludwig 75, 207
Monge, Gaspard 67

Moorcroft, William 125
Morley, Edward 148–9
Morgan, Thomas Hunt 278
Morton, William 119
Moses, Revd Stainton 168
motors, electromagnetism and 87–8
Müller, Johannes 123
Müller, Max 35
Murchison, Roderick 201, 225
Murdock, William 64
Murray, John 211, 267
Museum of Natural History, Paris 16, *17*
Museum of Practical Geology 176
museums
  Crystal Palace and 176
  laboratories in 145
  London 176, 189
  open to public 190–2
  popularization of science 8–9
  scientific culture and 241
  see also individual museums
music 258–9
Myers, Frederic 168–9, 170

Napoleon Bonaparte 280
  abolishes Holy Roman Empire 198
  English mind 219
  hundred day comeback 60
  Laplace and God 20
  science flourishes under 17–18, 239
  repression in Haiti 159
  training for engineers 192–3
Napoleon III 69, 188
national parks 232
natural history 22–8
  God's design 50–3
  illustration and 47–50
  love of nature 234
  museums 145–6
  as a profession 246–7, *247*
Natural History Museum 145, 176, 189,
  190–1
Natural Theology (Paley) 30, 51
naturalism 157
Nature 251, *252*, 265
Naturforscher *see* Gesellschaft Deutscher
  Naturforscher und Ärzte
Naturphilosophie 84, 87
naval matters 213-4, 266
  arms race 265–6
  battleships 209–11
  HMS *Dreadnought* 210
  submarines 210
  surveys 21-2, 66-7, 152,
navigation 16
  chronometers 21–2
Nemesis of Faith (Froude) 184
Neptune 208, 265
Netherlands 199, 200
New Sydenham Society 128
New York Industrial Exhibition (1853)
  68
New Zealand 152, 153
Newcomen Society 11
Newlands, John 226

Newman, Francis
    Phases of Faith 184–5
Newman, John Henry 184–5
Newton, Isaac 3
    brute matter 154
    description 37
    foundation for mathematics 20
    fundamental particles 271
    influence of 15
    language of works 35
    poets' views of 28, 30
    prism investigations 129, 132
    reasoning of 218, 219, 220
    shifts paradigm 40
Newtonian science
    electromagnetism and 88
    overthrow of 149
Nichol, John Pringle 203
Nicholson, William 24
Nicholson's Journal 9
Nightingale, Florence 117–18
    Notes on Nursing 118
nitrous oxide (laughing gas) 112, 119, 166
Nobel, Alfred 69, 215, 216
Nobel Prize
    Becquerel and Curie 268
    for Einstein 272
    Rutherford 268
North and South (Gaskell) 113
Norway 200
Notes on Nursing (Nightingale) 118
nuclear energy *see* atoms and molecules
nursing 117–19
    midwives 125–6

occultism and magic x, 229, 234, 254
oceanography 146
Odling, William 207
Oken, Lorenz 73, 95–6, 219, 223, 236
    as administrator 248
    Elements of Physiophilosophy 248
    formation of GDNA 240–1
Old Bones (Symonds) 101
On Language (Humboldt) 153–4
opium 111
Opticks (Newton) 35
optics 22
    instruments 79
    microscopes 121–2
    Newton's prism 129
Orfila, Mateu 120, 121, 127, 208
The Origin and Progress of Language and *Ancient Metaphysics* (Monboddo) 160
Origin of Species (Darwin) 104
    fifty years on 276
    illustration 50–1
    publication of 99–100
    reception of 101
Ørsted, Hans Christian 87, 223, 254
    The Soul in Nature 167, 199, 253
Osiris 10
Osler, William 125
Ostwald, Wilhelm 45, 223, 272
Owen, Richard 54, 99, 190, 225, 226, 245
Oxford University 53, 96–7

    laboratories 139, 145
    medicine 124

Paget, Rose (later Thomson) 259
Palaeontographical Society 247
Paley, William 2, 95
    benevolent design 94–5
    Darwin and 52
    long argument 99
    Natural Theology 30, 51, 253
Palmerston, Lord 186
Panizzi, Anthony 190
Paracelcus (Theophrastus B. von Hohenheim) 119, 129
Paris
    Exhibition (1867) 187
    Exposition (1900) 80
    mental asylums 126
    Revolutionary and Napoleonic 15–26, 31, 45, 107, 110, 120-1,125, 161-2
    sanitation 117
Paris, John Ayrton 111
Paris Zoo 17
Park, Mungo 159–60, 212
Parker, Theodore 233
Parkinson, James 26
Parsons, Charles 202
Parsons, William (Lord Rosse) 202
Pasteur, Louis 106, 122, 123
Pasteur Institute 122
Pauli, Wolfgang 146
Paxton, Joseph 174
Peacock, George 43
Peacock, Thomas Love
    Melincourt 160
Pearson, Karl 34
Peel, Robert 63, 192
Pepys, William Haseldine 130, 132
Periodic Table 180, 224
Perkin, William Henry 58–9, 180
Pérouse, Comte de la 152–3
Perry, Matthew 67
Phases of Faith (F. Newman) 184–5
Phillips, John 145, 226
philosophy 4–5
    see also ethics and morality; method; religion
phlogiston 23, 39, 54
photography 255
    light and chemistry 86
    meaning for science 47–8
phrenology x, 14, 170
    development of 155–8
    failed science 229–30
    respectability of 84–5
physics
    apparatus 273
    Einstein's matter and energy 271, 272–3
    laboratories 140–2, 149
    physicists and chemists 269
    quantum theory 271–2
    sub-atomic structures 268–72
Physiognomy (Lavater) 155
Piltdown man 228–9
Pinel, Philipppe 126
Pitt, William 32

Planck, Max 271
Planck Institute 278–9
platinum 135–6
Playfair, Lyon 178
Podmore, Frank 168
poisons 119–21
Poisson, Denis 16, 18, 146
polar exploration 213–14, *214*
Pombal, Marquis of 139
Poor Laws 115
Pope, Alexander 150
Popper, Karl 4–5
popularization of science xi
population and demographics 105
portraits' eyes 37
Portugal 199
positivism 217, 236
  in France 102, 124
  reaction to Enlightenment 219
  Vienna Circle 223
Poulton, Edward 277
Powell, Baden 222
A Preliminary Discourse (Herschel) 84–5, 220–2
*A Preliminary Discourse on the Study of Natural
  History* (Swainson) 223–4
Prestwich, Joseph 229
Prichard, James Cowles
  ages of time 164
  ethnology 161–3
  Researches into... Man 162
Priestley, Joseph 18, 59
  biography of 6
  carbon dioxide 25
  context of science 239
  effect of science 31
  electricity 23
  the French Revolution 30
  key to new world 85
  laboratory of 149
  Lavoisier and 39
  method 220
  natural theology 253
  phlogiston 23
  photosynthesis 86
  Unitarian religion 154–5
  Wedgwood apparatus 133
  writings of 2
Principia Mathematica (Newton) 35
Principles of Geology (Lyell) 97, 98
Proctor, Richard 225–6
Prout, William 42, 224
psychology
  Freud's psychoanalysis 275–6
  Wundt 276
public health
  cholera 113–15
  workers' lives 113
Public Health Act (1848) 115–16
public interest xi
  Great Exhibition and 186–7
  museums and 190–2
publications xi
  early popularization 8–9
  history of science and 10–11
  international efforts 266–7

specialization and 251–2
  technology of 73
Punch magazine *256, 260*
  Faraday holds his nose 116–7
Pussin, Jean-Baptiste 126
Pythagoras 259

quantum theory 271–2
Quarterly Journal of Science 133, 251
Quetelet, Lambert 97, 106, 199, 226
quinine 111, 213

radio
  Marconi's message 215
radioactivity
  Becquerel 103, 268
  Curie 268
radium 268
Rae, John 214
Raffles, Stamford 200
railway transport
  bridges 71
  early development of 65–6
  first fatality 70
  standard technology for 69–70
  underground 80
Ramsay, William 54, 225
Rankine, William 92
Ray Society 247, 268
Rayleigh, Lord (John Strutt) 75, 235
Reform Bill (1832) 3, 113, 170
Reichenbach, Karl von 96, 167
religion
  African mission 213
  agnosticism 185–6
  argument of design 50–3, 94–5, 102
  biology and 110
  the 'church scientific' 243, 244–5
  doubt and Christianity 183–7
  doubt within science 183–6
  effect of French Revolution 239
  evolutionary monism 169
  Napoleon and Laplace 20
  nineteenth century x
  punishment by disease 114
  reception of evolution 101–2
  scepticism applied to 273–4
  science and faith 170
  spirituality and science 252–4, 262
Religious Tract Society 253
Researches into the Physical History of Man
  (Prichard) 162
Reynolds, Joshua 57
Richards, Theodore 149
Ritter, Johann 23, 25
Robert Elsmere (Ward) 183, 254
Robespierre, Maximilien 197
Rodriguez, Barbosa 212
Roebling, John 71
Roman Catholic Church
  evolution and 102
  in France 188
  revolutionary France and 15
  undermined by French Revolution 239
Romanticism 84, 166–7

in Britain 195
in Germany 103
Italian nationalism 204
Röntgen, Wilhelm 147
Rosetta Stone 36
Ross, Captain James Clark 213, 266
Ross, Ronald 213
Rosse, Lord (William Parsons) 202, *202*
Round the World in Eighty Days (Verne) 257
Rousseau, Jean-Jacques 15, 30, 152
Rowland, Henry Augustus 79, 146
Royal Academy 57
Royal College of Chemistry 173, 176
  see also Imperial College
Royal College of Music 176
Royal Colleges of Physicians and Surgeons
  243, 245
Royal Colleges Physicians and Surgeons
  collection of anatomy and fossils 190
  founding of 107
Royal Commission on Scientific Instruction
  141
Royal Cork Institution 202
Royal Geographical Society 248
Royal Holloway College 127
Royal Horticultural Society 191
Royal Institute of Chemistry 76, 142, 245
Royal Institution 74
  apparatus of 131–2
  compared to Paris 19
  Davy's lectures 31, 240
  Faraday's lectures 88
  lecture theatre *241*
  lectures *250*
  membership of 249
  military matters 69
  Prince Albert and 45
  social class and 6
  stinks of 139
Royal Society 83, 244, 248
  Animal Chemistry Club 26
  Copley Medal 25
  Davy and 240
  gentlemen's club 16
  members of 243
  membership 251
  Notes and Records 11
Royal Society for the Prevention of Cruelty to
  Animals (RSPCA) 232
Royal Society for the Protection of Birds
  234
Royal Society of Chemistry 11
rubber 212
Rumford, Count (Benjamin Thompson) 173
Ruskin, John 38
Russell, Bertrand 123
Russell, William 117–18
Russia 199, 206–7
Rutherford, Adelaide 19
Rutherford, Ernest 19, 54, 149
  alpha and beta rays 268–9
  sub-atomic structure 271

Sabine, Edward 249, 266
Sabine, Elizabeth (née Leeves) 266

Sacks, Oliver 133
safety, miners and 59–60
St Bartholomew's Hospital 107
St-Hilaire, Étienne Geoffroy 95
Saint-Hilaire, Geoffroy 276
St-Hilaire, Isidore 95
St Thomas's Hospital 107
Salisbury, 3rd Marquess (Robert Cecil) 273
Salmonia (Davy) 240
sanitation 106
  cholera 113–15
  pollution 61
Sanskrit language 153
Scheele, Carl Wilhelm 18
Schelling, Friedrich 84, 236
Schinkel, Karl Friedrich 153
Schleiden, Matthias 109
Schleiermacher, Friedrich 239
Schrödinger, Erwin 202
Schuster, Arthur 223, 225
Schwann, Theodor 109
science
  applied 56–7, 170, 186–7, 192
  within artistic culture 254–9
  'church scientific' 243, 244–5, 263, 282
  defining 14
  defining history of 9–11
  doubt as feature of 183–6
  education for viii–x
  excitement of 82–4
  failed ideas 229–31
  historians and 1–2
  internationalism 265–7
  language gaps 264–5
  likened to a church 263
  misdemeanours 237
  national traditions and 197–208
  nations and 215–16
  philosophy of knowledge 4–5
  popularization of 8–9
  scientific language 33–5, 35–6
  specialization 192
  spirituality and 252–4, 262
  twentieth century 264–5
  values and x–xi
  in war 280–1
  see also biology; chemistry; method; physics;
    scientists
science fiction 255–7
Science Gossip 260–2, *261*
Science Museum, London 176
Scientific Memoirs 267
scientists 262–3
  an occupation 279–80
  backgrounds of 244–5, 246
  bad behaviour 226–9
  biographical histories of 6–8
  as clerisy 263
  concept of 34
  conservatism 218
  disagreement 225–6
  formation of label viii
  as professionals 179, 245–8
  quarrels among 53–4
  specialization 83–4, 251

scientists (*cont.*)
    term coined at BAAS 242–3
    white coats 130–1
Scotland
    in 1789 195–6
    education in 72
    medical education 124
    phrenology 157
    strong institutions 241
Scott, Walter 195
scurvy 112
Seacole, Mary 118
Sedgwick, Adam 97, 225
Sensations of Tone (Helmholtz) 258–9
Shakespeare, William
    The Tempest 235
Sharp, Granville 159
Shelley, Mary
    Frankenstein x, 24, 29–30, 167, 255–6
Shelley, Percy Bysshe 29, 167, 255
shipping
    shrinking the world 69
    steam engines 66–7
    see also naval matters
Sidgwick, Henry 168, 234, 275
    Methods of Ethics 273
Sidgwick, Nora 168, 170, 234, 259,
    275
Siemens, Werner 207
Siemens, William 207
Simon, John 116
Simpson, James Young 119
slavery
    Africa 212–13
    African 159
    Darwin and 163
    Huxley and 163
Sloane, Sir Hans 189–90
Smiles, Samuel 7
Smith, Francis Pettit 67
Smith, Robert Angus 62
Smith, William 189
Smithson, James 130, 133
Smithsonian Institution 72, 146
    establishment of 191
    Henry's administration of 248
Smyth, Charles Piazzi 254
Snow, John
    anaesthesia 115, 119
    identifies cholera source 114–15, 123
Sobel, Dava
    Longitude 7
social Darwinism 13
Social History of Medicine journal 11
social sciences
    philosophy of knowledge 5
    scientists uneasy with 106
societies and institutions xi, 11
    formation and membership of 248–52
    group efforts 248–9
    scientific culture and 241
    'the church scientific' 244
    see also individual organizations
Society for Psychical Research 168, 234–5,
    275

Society for the History of Alchemy and
    Chemistry 11
Society for the History of Natural History 11
Society for the History of Technology 11
Society for the Social History of Medicine 11
Society of Apothecaries 74
Society of Arts 248
Soddy, Frederick 149, 268–9
Solander, Daniel 266
Somerville, Mary 83, 251
    language of mathematics 43–4
The Soul in Nature (Ørsted) 167, 199, 253
Spain 199
spectroscopy 78–9, 273
Spencer, Herbert 5, 13, 186
spiritualism
    communicating with dead 167
    investigations into 168–9
    Mesmer and 167
    psychical research 170
    reaction to 19th c. science x
Spurzheim, J. G. *156*, 156–7, 230
Stallo, Johann Bernhard
    Concepts and Theories of Modern Physics
    222
Stanley, Henry Morton 212
Stanley, Owen 201
statistics
    Disraeli on 106
    public health 114
steam power 79
Stenhouse, John 207
Stephenson, George
    miners' lamp 60, 227–8
    railway transport 65
Stephenson, Robert 65, 72
    Britannia Bridge 71
stethoscopes 121
Stevenson, Robert Louis
    The Strange Case of Dr Jekyll and Mr Hyde
    169
Stewart, Balfour 235
Stokes, George Gabriel 202
Stokes, J. Lort 201
Stoney, George Johnstone 147, 269
Storia della Scienza 10
The Strange Case of Dr Jekyll and Mr Hyde
    (Stevenson) 169
Strickland, Hugh Edwin 179, 268
Strzelecki, P. E. de 158
Suhm, Rudolph von Willemoes 267
Sunday Lecture Society 185–6
surveying equipment 77
Swainson, William
    Preliminary Discourse 223–4
    Quinary System 179–80
Swan, Joseph 79
Sweden 200
Swedenborg, Emanuel 167
Switzerland 144
Symonds, William Samuel
    Old Bones 101
Synthetic Society 275
syphilis 124
Systema Naturae (Linnaeus) 151

Tahiti 152, 153, 158–9
Tait, Peter Guthrie 80, 208
Tanganyika 213
taxonomy
    effect of evolution 231
    Jussieu 224
    methods 223–4
    nationalism and 214–15
    Swainson and 179–80
    see also Linnaeus, Carl
Tay Bridge disaster 71
Taylor, Richard 267
Taylor & Francis publishers 251
technology
    cultural change and xi
    industrialized world 262–3
    mass production 68
    as new concept 74
    pattern of applying science 81
    standardization of materials 67–70
    transferring 195
    transforming society 282
Technology and Culture journal 11
telegraphic communication
    shrinking the world 69
    Wheatstone devises 64–5
temperature
    Celsius and Fahrenheit 77
    Wedgwood's thermometers 134, 135
The Tempest (Shakespeare) 235
Temple, Frederick, Archbishop of Canterbury
    religion and science 101
The Temple of Nature (E. Darwin) 93
Tennant, Smithson 135
Tennyson, Alfred, Lord
    In Memoriam 98, 185, 257–8
Terra del Fuego 152
textiles
    bleaches and dyes 57–9
    first factories 64
Thackrah, Charles Turner 112
thallium 130
Thénard, Louis-Jacques 120, 121
thermodynamics
    controversies 208
    laws of 43, 91–3
    Second Law of 43
Thompson, Benjamin (Count Rumford)
    173
Thomson, Charles Wyville 266–7
Thomson, J. V. (naval surgeon) 99
Thomson, Joseph John 147
    electricity 182, 269, 271
    physicists and chemists 269
    psychical research 234, 275
    women in lab 259
Thomson, Thomas 137
    on dyes and colouring 58
    First Principles 41–2
    hydrogen as base 224
    laboratory 140
    mineral analysis 61–2
    quarrels with Berzelius 42, 53–4, 137,
        226
Thomson, William see Kelvin, Lord

time
    Einstein and 272–3
    railway time 65
The Time Machine (Wells) 257, 274
Tirpitz, Alfred 209
Trail, James 212
transport 64–7
    see also railway transport; shipping
Trew, Gottfried 84
Trotter, Thomas 112
Tschermak, Erich von 278
tuberculosis 123
Tuke, William 126
Turner, Frank 184
Turner, J. M. W. 29, 222
    The Fighting Temeraire 209
Twenty Thousand Leagues Under the Sea (Verne)
    257
Tyndall, John 202
    applies for colonial posts 188
    childlessness 105
    German translations 268
    in Germany 206, 208
    imagination in science 254
    public lectures 53
    reason and faith 185, 186, 273
    scientific naturalism 245
typhoid 123

Unitarianism 154-5, 233
United States
    applied science 187
    Civil War 209, 280
    experiments 148–9
    grants for research 144
    Ivy League sciences 53
    medical education 125
    nationalist taxonomy 214–15
    natural history museums 146
    privately endowed universities 193
    refugee scientists 264
universities see education
University College London 140–1, 141
uranium 149, 268–9
Urban VIII, Pope 5
Ure, Andrew 109, 256
Ussher, James 51

Vaccination a Delusion (Wallace) 232
vaccination and immunization
    objections to 232
    principle of 231–2
values
    cultural changes x–xi
    expectation of neutrality 106
    'Victorian' 13
Varieties of Religious Experience (James) 170
Venus, Transit of (1769) 226–7
Verne, Jules 257
Vestiges of the Natural History of Creation
    (Chambers) 52, 97–100, 185, 192, 218,
        230
    deism of 253
    life elsewhere and 228
    materialism of 258

Victoria, Queen 111
  Great Exhibition *174*
  in mourning 59, 187
Victoria and Albert Museum 176, 189
Vienna Circle 223
Virchow, Rudolf 102
  Cellular Pathology 123
vitalism 108
  changing world view 138, 149
  repudiated 123
vivisection 232–3
Volta, Alessandro 18, 85, 203
  electricity 24–5
  influence on Mary Shelley 256
  shocks frogs' legs 140
Voltaire 15, 30
Voorst, John van 247
Vries, Hugo de 278

Waage, Peter 138
Wagner, Richard 198, 258
Wakley, Thomas 111
Walker, Adam *58*
Wallace, Alfred Russel 99, 101, 179–80
  biographical view of 7
  Darwinism 104, 165–6
  shock to Darwin 225
  specimen collection 211–12
  spiritualism x, 166, 168
  Vaccination a Delusion 232
Ward, Mrs Humphrey (Mary)
  Robert Elsmere 183, 254
Watson, Hewett Cottrell 157, 230
Watson, Richard 112
Watt, James 73, 92, 210
weapons and warfare 280
  application of science 209–11
  battleships 69
  mass production 69
  military technology 210
  scientists and 280–1
  see also naval matters
weather 181–2
Weather Book (FitzRoy) 181–2, *182*
Wedgwood, Josiah 76–7, 133, 134, 136
Wedgwood, Thomas 86
Weismann, August 104, 276–7
Wells, H. G.
  The Time Machine 257, 274
Wells, Horace 119
Welsbach, Carl Auers von 79
Wesley, John 107
Wheatstone, Charles 64
  Science museum 189
Whewell, William 13, 105
  Babbage and 179
  coins 'scientist' 243
  controversies 192
  against Darwin's conclusions 231
  discipline of science 86
  on Faraday 140

God's design 51
  Great Exhibition lectures 178
  helps Faraday with language 44–5
  History of the Inductive Sciences 3–4, 221
  modernizing mathematics 43–4
  plurality of worlds argument 228
White, Gilbert 34
White, Walter 248
Whitworth, Joseph 68–9, 177
Wickham, Henry 212
Wilberforce, Bishop Samuel 5
  against Darwin's method 231
  debates evolution 52–3, 54, 99, 245
Wilberforce, William 159
Wilhelm II 209
Wilkes, Charles 146, 266
Wilkins, John 34
William of Orange 3
Williamson, Alexander 138, 207
Wilson, George 74
Winterbottom, Thomas 160
Wittgenstein, Ludwig 223
Wöhler, Friedrich 138
Wolf, Joseph 191
Wollaston, William Hyde 23
  apparatus 133, 135–6
  avoids error 218
  cultural respect 240
  electricity 25
  Faraday and 87
  laboratories 137
  palladium and Chenevix 228
  on portrait eyes 47
  secretiveness 173
  slide-rules 40
  Thomson and 41–2
Wolsey, Thomas 263
women
  Cobbe and feminism 233
  education for 233–4
  Lecky on 279
  Marie Curie 268
  in medicine 125–6
  in science 259–60, 262
  Somerville and mathematics 43–4
Wordsworth, William 30
  on L'Ouverture 159
  moral good of nature 95
  romantic Newton and 28
World War I 14, 15, 195
Wright, Orville and Wilbur 215
Wundt, Wilhelm 169, 170, 276

X-rays 147
  crystallography 201, 273

Young, Thomas 36, 218, 271

zoology 223–4
Zoology of the Voyage of HMS Beagle (Darwin)
  191